Lecture Notes in Physics

Springer-Verlag
Berlin Heidelberg GmbH

The Editorial Policy for Proceedings

The series Lecture Notes in Physics reports new developments in physical research and teaching – quickly, informally, and at a high level. The proceedings to be considered for publication in this series should be limited to only a few areas of research, and these should be closely related to each other. The contributions should be of a high standard and should avoid lengthy redraftings of papers already published or about to be published elsewhere. As a whole, the proceedings should aim for a balanced presentation of the theme of the conference including a description of the techniques used and enough motivation for a broad readership. It should not be assumed that the published proceedings must reflect the conference in its entirety. (A listing or abstracts of papers presented at the meeting but not included in the proceedings could be added as an appendix.)

When applying for publication in the series Lecture Notes in Physics the volume's editor(s) should submit sufficient material to enable the series editors and their referees to make a fairly accurate evaluation (e.g. a complete list of speakers and titles of papers to be presented and abstracts). If, based on this information, the proceedings are (tentatively) accepted, the volume's editor(s), whose name(s) will appear on the title pages, should select the papers suitable for publication and have them refereed (as for a journal) when appropriate. As a rule discussions will not be accepted. The series editors and Springer-Verlag will normally not interfere with the detailed editing except in fairly obvious cases or on technical matters.

Final acceptance is expressed by the series editor in charge, in consultation with Springer-Verlag only after receiving the complete manuscript. It might help to send a copy of the authors' manuscripts in advance to the editor in charge to discuss possible revisions with him. As a general rule, the series editor will confirm his tentative acceptance if the final manuscript corresponds to the original concept discussed, if the quality of the contribution meets the requirements of the series, and if the final size of the manuscript does not greatly exceed the number of pages originally agreed upon. The manuscript should be forwarded to Springer-Verlag shortly after the meeting. In cases of extreme delay (more than six months after the conference) the series editors will check once more the timeliness of the papers. Therefore, the volume's editor(s) should establish strict deadlines, or collect the articles during the conference and have them revised on the spot. If a delay is unavoidable, one should encourage the authors to update their contributions if appropriate. The editors of proceedings are strongly advised to inform contributors about these points at an early stage.

The final manuscript should contain a table of contents and an informative introduction accessible also to readers not particularly familiar with the topic of the conference. The contributions should be in English. The volume's editor(s) should check the contributions for the correct use of language. At Springer-Verlag only the prefaces will be checked by a copy-editor for language and style. Grave linguistic or technical shortcomings may lead to the rejection of contributions by the series editors. A conference report should not exceed a total of 500 pages. Keeping the size within this bound should be achieved by a stricter selection of articles and not by imposing an upper limit to the length of the individual papers. Editors receive jointly 30 complimentary copies of their book. They are entitled to purchase further copies of their book at a reduced rate. As a rule no reprints of individual contributions can be supplied. No royalty is paid on Lecture Notes in Physics volumes. Commitment to publish is made by letter of interest rather than by signing a formal contract. Springer-Verlag secures the copyright for each volume.

The Production Process

The books are hardbound, and the publisher will select quality paper appropriate to the needs of the author(s). Publication time is about ten weeks. More than twenty years of experience guarantee authors the best possible service. To reach the goal of rapid publication at a low price the technique of photographic reproduction from a camera-ready manuscript was chosen. This process shifts the main responsibility for the technical quality considerably from the publisher to the authors. We therefore urge all authors and editors of proceedings to observe very carefully the essentials for the preparation of camera-ready manuscripts, which we will supply on request. This applies especially to the quality of figures and halftones submitted for publication. In addition, it might be useful to look at some of the volumes already published. As a special service, we offer free of charge LaTeX and TeX macro packages to format the text according to Springer-Verlag's quality requirements. We strongly recommend that you make use of this offer, since the result will be a book of considerably improved technical quality. To avoid mistakes and time-consuming correspondence during the production period the conference editors should request special instructions from the publisher well before the beginning of the conference. Manuscripts not meeting the technical standard of the series will have to be returned for improvement.

For further information please contact Springer-Verlag, Physics Editorial Department II, Tiergartenstrasse 17, D-69121 Heidelberg, Germany

J. Klamut   B. W. Veal   B. M. Dabrowski
P. W. Klamut   M. Kazimierski (Eds.)

Recent Developments in High Temperature Superconductivity

Proceedings of the 1st Polish–US Conference
Held at Wrocław and Duszniki Zdrój,
Poland, 11–15 September 1995

Springer

Editors

Jan Klamut
Institute of Low Temperature and Structure Research, PAS
International Laboratory of High Magnetic Fields and Low Temperatures
95 Gajowicka Str., PL-53-529 Wrocław, Poland

Boyd W. Veal
Argonne National Laboratory
9700 S. Cass Ave., Argonne, IL 60439, USA

Bogdan M. Dabrowski
Department of Physics, Northern Illinois University
De Kalb, IL 60115, USA

Piotr W. Klamut
Maciej Kazimierski
Institute of Low Temperature and Structure Research, PAS
P.O. Box 937, PL-50-950 Wrocław, Poland

Cataloging-in-Publication Data applied for.

Recent developments in high temperature superconductivity :
proceedings of the 1st Polish US conference, held at Wrocław
and Duszniki Zdrój, Poland, 11 - 15 September 1995 / J. Klamut
... (ed.).
  (Lecture notes in physics ; Vol. 475)
  ISBN 978-3-662-14096-3   ISBN 978-3-540-70695-3 (eBook)
  DOI 10.1007/978-3-540-70695-3
NE: Klamut, Jan [Hrsg.]; GT

ISSN 0075-8450
ISBN 978-3-662-14096-3

Typesetting: Camera-ready by authors/editors
Cover design: *design & production* GmbH, Heidelberg
SPIN: 10520117   55/3142-543210 - Printed on acid-free paper

Preface

This volume contains the proceedings of the First Polish–US Conference on High Temperature Superconductivity which was held September 11–15, 1995 in Wrocław and Duszniki Zdrój (Sudety mountains), in the southwest of Poland.

After almost ten years of extensive research that started with the pioneering discovery of the first superconducting copper oxide by J.G. Bednorz and K.A. Müller in 1986, scientists in many countries continue to devote substantial effort investigating properties in the still-growing family of copper oxides, and in seeking a microscopic description of high temperature superconductivity (HTSC). Since the very early days of HTSC, research has been conducted on HTS materials in Polish laboratories and universities. More than one hundred Polish scientists now work in the field. This group recognized and expressed the need for a conference on HTSC that would bring together, in Poland, leading experts from the international community. The conference format was further developed by cooperating scientists from the Institute of Low Temperature and Structure Research PAS and the Argonne National Laboratory. As perceived by the organisers, the conference would provide a forum for US and Polish scientists to discuss topical research issues, to exchange ideas, and to develop personal scientific contacts and collaborations. The educational content of the conference was also seen to be especially important for many young polish researches.

We were gratified to receive an enthusiastically favorable response from an outstanding group of US scientists who were invited to participate. This provided an opportunity to organize a unique meeting to present the current status of HTSC research and to address most of the current challenges in the physics of high temperature superconductivity. At the conference, a friendly atmosphere and long discussions among the participants contributed much to its success.

The conference featured 31 invited speakers (19 from US, 11 from Poland) plus more than sixty contributed poster presentations, involving more than one hundred participants. In addition to US and Polish attendees, we were honored by the presence of Nobel laureate K.A. Müller (IBM Zürich) who presented an invited lecture to the conference. The full list of lectures and poster presentations is included at the end of this volume.

The articles included in this book represent the majority of the invited lecturers. The articles cover a wide range of subjects in HTSC physics, from the most fundamental theoretical problems, through materials properties and synthesis issues, to an overview of current trends in industrial applications of HTSC materials. The editors anticipate that the book will become a valuable resource addressed to a large readership including university students and researchers who seek reviews of the most current problems in high temperature superconductivity physics. We thank all the authors for their contributions.

The conference could not have been successful without the outstanding cooperative effort provided by the staff from the Institute of Low Temperature and Structure Research, Polish Ac. Sci. (Wrocław, Poland). We were honored to work as a team with S. Gołąb, H. Misiorek, J. Olejniczak, T. Plackowski, A. Sikora, Cz. Sułkowski, D. Włosewicz and A.J. Zaleski. We are particularly grateful to the US Department of Energy and to the Polish State Committee for Scientific Research, which cooperatively financed the meeting within the framework of the Maria Skłodowska–Curie Joint Fund II (grant PAN/DOE-94-206). The conference was organized under the patronage of the Polish Physical Society and the Physics Committee of the Polish Academy of Sciences. We are also grateful for financial support from these institutions. We are indebted to Springer-Verlag for publication of the Proceedings.

Wrocław and Argonne   *The Editors*
February 1996

Table of Contents

IX

Conference Participants

Seen in the photograph:

1. D.C.Johnston
2. V.J.Emery
3. M.R.Norman
4. D.K.Finnemore
5. C.W.Chu
5. M.Thomson
7. D.Larbalestier
8. K.Wysokiński
9. J.Karpiński
10. J.Stankowski
11. M.Surma
12. K.A.Müller
13. D.Pines
14. Z.Galasiewicz
15. J.D.Jorgensen
16. J.Klamut
17. J.T.Markert
18. H.Szymczak
19. R.Micnas
20. M.Suenaga
21. P.M.Grant
22. R.Horyń
23. R.Szymczak
24. H.Drulis
25. M.Cieplak
26. A.Golnik
27. J.Czerwonko
28. L.Kowalewski
29. M.Kazimierski
30. M.Wołcyrz
31. A.J.Zaleski
32. A.Paszewin
33. T.Cichorek
34. M.Bałanda
35. T.Skośkiewicz
36. J.Olejniczak
37. P.Wróbel
38. T.K.Kopeć
39. A.Sikora
40. R.J.Radwański
41. Z.Bukowski
42. R.Jabłoński
43. A.Morawski
44. Z.Karpińska
45. A.Pajączkowska
46. P.W.Klamut
47. M.Wesołowska
48. M.Thomas
49. Z.Malinowski
50. P.Tekiel
51. A.Wojakowski
52. J.Bała
53. J.J.Wnuk
54. T.Krzysztoń
55. Z.Tomkowicz
56. A.Jeżowski
57. W.Pachla
58. A.Wiśniewski
59. S.Koleśnik
60. A.Shengelaya
61. Cz.Sułkowski
62. H.Misiorek
63. B.Dabrowski
64. D.Włosewicz
65. K.Rogacki
66. B.Kusz
67. A.Nabiałek
68. M.Gazda
69. B.Susła
70. I.Jacyna-Onyszkiewicz
71. M.Drozdowski
72. G.Crabtree
73. W.Sadowski
74. S.Robaszkiewicz
75. B.Bułka
76. Z.Henkie
77. A.Kołodziejczyk
78. W.Typek
79. A.B.Szytuła
80. M.Ciszek
81. B.Wójcicki
82. Z.Tarnawski
83. T.Łada

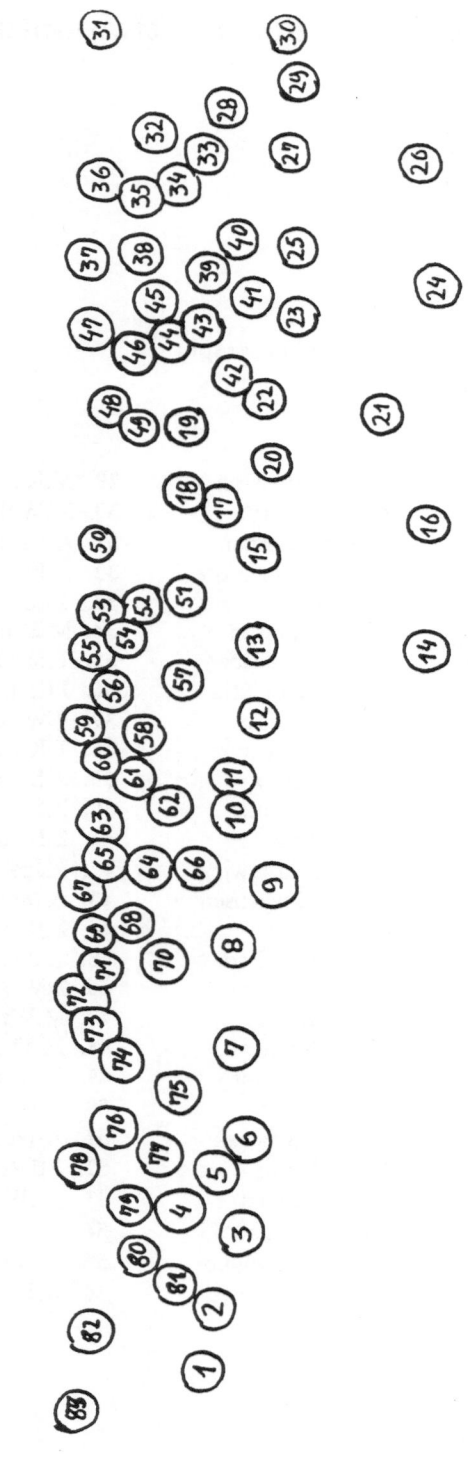

Structural Features that Optimize
High Temperature Superconductivity

J. D. Jorgensen[1], D. G. Hinks[1], O. Chmaissem[1], D. N. Argyriou[1], J. F. Mitchell[1], and B. Dabrowski[2]

[1] Materials Science Division and Science and Technology Center for Superconductivity, Argonne National Laboratory, Argonne, IL 60439, USA
[2] Physics Department, Northern Illinois University, DeKalb, IL 60115, USA

Abstract. Studies of a large number of compounds have provided a consistent picture of what structural features give rise to the highest T_c's in copper-oxide superconductors. For example, various defects can be introduced into the blocking layer to provide the optimum carrier concentration, but defects that form in or adjacent to the CuO_2 layers will lower T_c and eventually destroy superconductivity. After these requirements are satisfied, the highest T_c's are observed for compounds (such as the $HgBa_2Ca_{n-1}Cu_nO_{2n+2+x}$ family) that have flat and square CuO_2 planes and long apical Cu-O bonds. This conclusion is confirmed by the study of materials in which the flatness of the CuO_2 plane can be varied in a systematic way. In more recent work, attention has focused on how the structure can be modified, for example, by chemical substitution, to improve flux pinning properties. Two strategies are being investigated: (1) Increasing the coupling of pancake vortices to form vortex lines by shortening or "metallizing" the blocking layer; and (2) the formation of defects that pin flux.

1 Introduction

The structural complexity of the copper-oxide high-T_c superconductors provides the opportunity to optimize the superconducting properties. Over fifty distinct compounds have been discovered in the last ten years. For each of these, chemical substitution or variations in oxygen composition at a crystallographic site that supports variable occupancy can be used to "tune" the properties. For a number of years, the primary goal of such studies has been to achieve a high superconducting transition temperature, T_c. Steady progress has been made, although one might argue that new record-high T_c's were simply the result of exploring the variety of chemical compositions that could form rather than the result of a directed effort to achieve a crystal structure that had been predicted to yield a higher T_c. In the course of this work, however, it has become clear how T_c is related to various structural features and there is consensus concerning what structures yield the best high-T_c superconductors. A summary of what has been learned is presented in this paper.

More recently attention has turned to how one might adjust the structural properties to improve the critical current behavior (J_c) of these materials.[1] This work has been driven by the need to achieve better J_c's in order to use high-T_c materials in current carrying applications. The bulk J_c of a conductor depends on both extrinsic properties, such as grain alignment and the nature of grain boundaries, and on intrinsic properties, such as the flux pinning behavior of the material. Dramatic differences in the intrinsic flux pinning of the various compounds suggests that crystal structure, including both the "ideal" features of the structure and the presence of defects, can play an important role. For example, $YBa_2Cu_3O_{6+x}$ (hereafter abbreviated as Y-123) displays much better intrinsic flux pinning than Bi-Sr-Ca-Cu-O materials (BSCCO). However, the latter materials are preferred for wire applications because the grains align and form favorable grain boundaries for conduction during extrusion and firing processes. Clearly, if one could improve the intrinsic flux pinning behavior of BSCCO, the results would be of great importance. Recent work in several laboratories has focused on chemical substitution as a means of improving flux pinning behavior by modifying the crystal structure. This paper will review some of this work and outline a strategy for improving flux pinning through chemical substitution.

2 Optimizing T_c

It was realized early in the study of high-T_c materials that the superconductivity is a property of the CuO_2 layers that are present in every high-T_c structure.[2] These CuO_2 layers can occur singly or in groups separated by metal atoms such as Y or Ca. The intervening space between these groups of layers, which we will call the blocking layer, can be filled with a wide variety of structural elements (e.g. Cu-O chains, Tl-O layers, Bi-O layers, Hg-O layers, CO_3 ions, etc.). This flexibility is what has given rise to the large number of high-T_c compounds now known. For any of these compounds, certain conditions must be satisfied to achieve superconductivity.

First, the carrier concentration in the CuO_2 planes must be adjusted to the optimal level.[3, 4, 5, 6, 7] This is typically done by chemical substitution on a metal site (other than a Cu site in the CuO_2 layers) or by varying the oxygen content in a crystallographic site that supports variable occupancy.[2, 8, 9, 10] For example, the T_c of $La_{2-x}Sr_xCuO_4$ (LSCO) varies with the amount of Sr^{2+} substituted on the La^{3+} site, with the maximum T_c achieved for $x \approx 0.15$.[5, 11] For low values of x, T_c decreases and the material enters an insulating state; while for values of x higher than the optimum (near 0.15) T_c decreases and the material goes to a nonsuperconducting metallic state. This behavior appears to be true for all copper-oxide high-T_c materials[3, 4, 7, 12, 13], but in some compounds limitations in what compositions will form prevents the whole range of behavior from being accessed. For example, the $YBa_2Cu_3O_{6+x}$ compound exhibits its maximum T_c near the maximum oxygen concentration of 7 oxygen atoms per formula unit and is under doped for lower oxygen contents[14, 15, 16], while the $Tl_2Ba_2CuO_{6+x}$ compound exhibits its maximum T_c for x near zero and is over doped as the oxygen content increases.[17, 18]

The second requirement is that there be no defects in or near the CuO_2 layers. Defects such as oxygen atoms next to the Y or Ca atoms that separate adjacent CuO_2 layers have been shown to lower T_c and destroy superconductivity.[2, 8, 9, 19] Substitution of other metal atoms (e.g., Ni, Fe, or Zn) on the Cu sites in the CuO_2 layers in very low concentrations also destroys superconductivity.[20, 21, 22] Thus, chemical variables in the blocking layer are typically used to control T_c. (It is possible to modify carrier concentration with appropriate chemical substitution at the Y/Ca site.[12, 23, 24])

After these requirements are satisfied, there is a maximum T_c possible for each compound. These maximum T_c's vary widely (e.g., about 40 K for LSCO[11] vs. 90 K for Y-123[15] and 135 K for Hg-1223[25, 26]), suggesting that structural properties control the maximum T_c that can be achieved. Comparison of the structures of a large number of copper-oxide superconductors has led to the conclusion that the highest T_c is achieved in structures with flat and square CuO_2 layers (which implies a tetragonal crystal structure) and long apical copper-oxygen bonds connecting to these layers.[26, 27, 28, 29, 30, 31] The $HgBa_2Ca_{n-1}Cu_nO_{2n+2+x}$ (HBCCO) compounds, which have the highest T_c's yet observed for n=1, 2, and 3 layer compounds, also have the flattest CuO_2 layers and the longest Cu-O apical bond lengths.[26, 28] The relevant structural parameters are listed in Table 1.

Table 1. Copper-oxygen apical bond length (Cu-O) and buckling angle of the CuO_2 plane (Cu-O-Cu) for $HgBa_2Ca_{n-1}Cu_nO_{2n+2+x}$ compounds.[26, 27, 28]

No. of layers, n	T_c (K)	Cu-O (Å)	Cu-O-Cu (°)
1	95	2.78	180
2	126	2.78	179
3	135	2.74	178

The systematic correlation between buckling angle (the Cu-O-Cu angle in the CuO_2 plane), Cu-O apical bond length, and T_c for compounds with two CuO_2 layers (for which the most data are available) is shown in Fig. 1. For consistency, the figure is based on data for compounds in which Ca separates the two CuO_2 layers and the composition (carrier concentration) has been optimized to achieve the maximum T_c for each compound.[15, 19, 28, 32, 33, 34] For these compounds, the correlation between T_c and the buckling angle or apical bond distance is clear. However, Y-123[15], $YBa_2Cu_4O_8$ (Y-124)[35, 36], and $Pb_2Sr_2YCu_3O_8$ (PSYCO)[37] (and HBCCO compounds at high pressure, which will be discussed later) are exceptions. We believe that the most important difference for these compounds is the presence of metallic copper in the blocking layer as well as in the CuO_2 planes. The importance of this structural feature for obtaining high T_c will be discussed later.

The apical bond distance and buckling angle are closely related variables. This is illustrated in Fig. 2, where the apical distance is plotted vs. the buckling angle for the same compounds. Short apical distances lead to buckled CuO_2 planes because of oxygen-oxygen repulsion, while long apical distances allow the planes to approach a nearly flat condition.

An instructive confirmation of the relationship between buckling angle and T_c comes from the investigation of a single compound in which the buckling angle can be varied at constant carrier concentration. Such experiments can be done in the $La_{2-x}M_xCuO_4$ (M=Sr,Ba,Ca,Nd, etc.) system. In this structure, buckling of the CuO_2

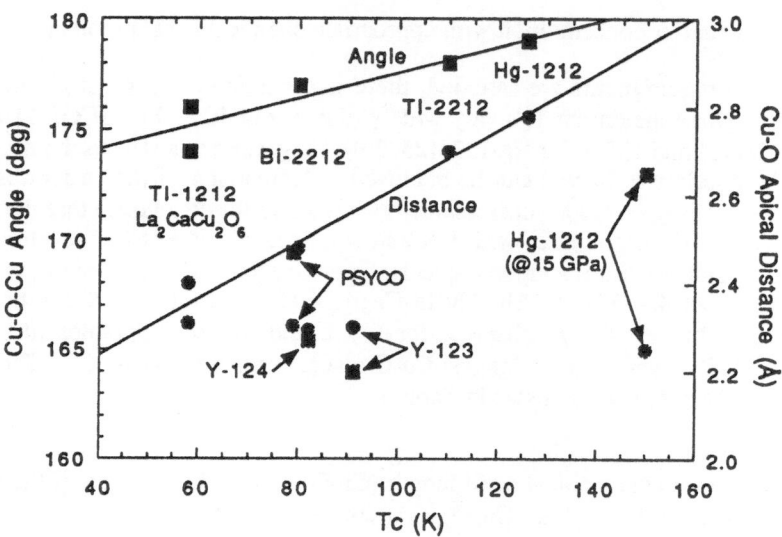

Fig. 1. Relationship between T_c and the Cu-O-Cu buckling angle for the CuO_2 planes and the Cu-O apical bond distance for compounds with two CuO_2 layers. Square symbols are angles; round symbols are distances.[data from Refs. 15, 19, 28, 32, 33, 34]

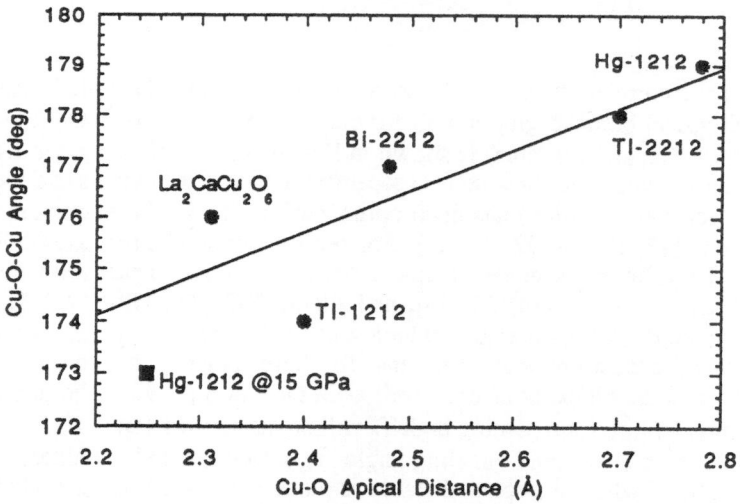

Fig. 2. Cu-O apical bond distance vs. Cu-O-Cu buckling angle for compounds with two CuO_2 layers where Ca separates the two layers.

plane results from a nearly rigid coordinated tilting of CuO_6 octahedra. The degree of tilting (and buckling) can be varied by the application of hydrostatic pressure or by substitutional chemistry on the La site. The application of pressure reduces the buckling angle and raises T_c.[38] For a composition with the maximum T_c, ($T_c \approx 40K$ for x=0.15 Sr substitution), T_c increases systematically as the buckling angle is reduced by the application of pressure.[39] Eventually, the buckling angle becomes 180° (flat CuO_2 planes), resulting in a transformation from orthorhombic to tetragonal symmetry. Further increases in applied pressure do not raise T_c.

Comparison of the Ca substituted system with the Sr substituted system leads to the same conclusion. At the same carrier concentration (x=0.15) the Ca substituted system has a larger amount of plane buckling and a lower T_c.[40] Through the use of the combined substitution of Nd, Ca, and Sr on the La site, a wide range of buckling angles can be achieved at constant carrier concentration.[41] Such experiments have yielded the quantitative relationship between buckling angle and T_c. T_c decreases 20 K for an increase in the buckling angle of 8°.

The highest T_c's yet achieved were observed by applying pressure to the HBCCO compounds.[42, 43, 44, 45] By the application of pressures on the order of 15-30 GPa, an increase of nearly 25 K is achieved for the T_c's of the n=1, 2, and 3 compounds; giving a record-high T_c for Hg-1223 of 160 K at ~20 GPa.[43] The structures of these compounds at high pressure are particularly interesting. Fig. 3 compares the structure of Hg-1212 at 15 GPa with the structure at ambient pressure.[46]

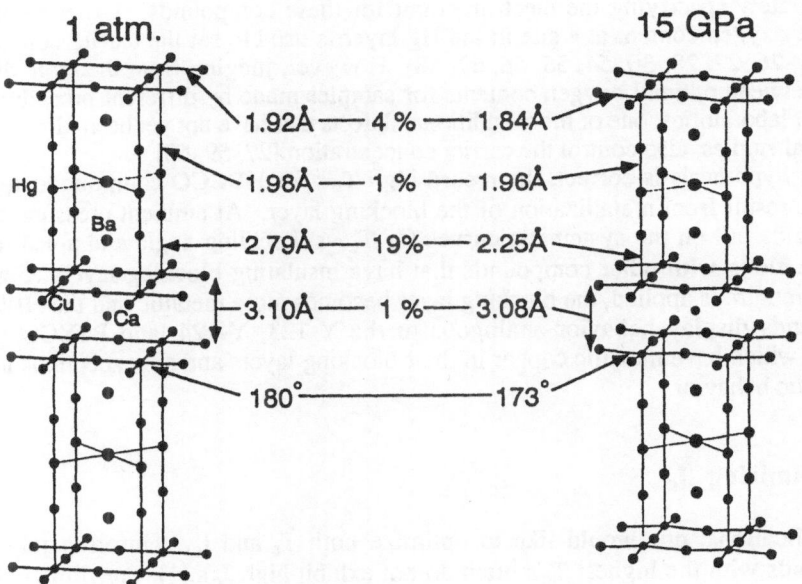

Fig. 3. Comparison of the structures of Hg-1212 at ambient pressure and 15 GPa. Values for selected bond lengths and angles, and the percentage change over this pressure range, are shown. The 15 GPa structure is extrapolated from accurate measurements at 0.6 GPa.

(The most accurate structural data at high pressure are obtained for the one- and two-layer compounds because impurity phases in samples of the three-layer compound reduce the precision available from structural refinements.) The largest pressure-induced structural change is a dramatic shortening of the apical Cu-O bond, from 2.79 to 2.25 Å. As would be expected, this is accompanied by a buckling of the CuO_2 planes. The relationship between apical bond length and plane buckling angle nicely follows the behavior observed for other compounds, as shown in Fig. 2.

Clearly, the large pressure-induced increase in T_c cannot be explained in terms of the systematic relationship between buckling angle, apical bond length, and T_c presented in Fig. 1. One important difference between the Hg-1212 structure at high pressure and the other two-layer structures is the length of the in-plane Cu-O bond. Pressure shortens this bond to 1.84 Å for Hg-1212 at 15 GPa, whereas it is around 1.92Å for all of the two-layer compounds at ambient pressure. This shortened Cu-O in-plane bond has important implications for the electronic structure, which remains very two dimensional, with a well-defined van Hove singularity in the density of states, even though the apical bond length is dramatically shortened by pressure.[47, 48] A peak in the density of states resulting from the van Hove singularity is thought to correspond to the optimum doping level.[49, 50, 51, 52] We speculate, however, that another difference in the electronic structure may be more important. Band structure calculations for the high-pressure structures show that the application of pressure either produces or significantly increases the overlap of a band involving Hg with the Fermi energy, resulting in an electronic structure where there are more carriers associated with the Hg-O layer.[47, 48, 53] These calculations are not accurate enough to determine exactly when the Hg-O layer first becomes metallic because of the difficulty of accurately specifying the electron count for these compounds. Experimentally, variable oxygen content at a site in the Hg layer is used to set the carrier concentration.[25, 26, 27, 28, 30, 54, 55, 56, 57, 58] However, judging from the large differences between optimal oxygen contents for samples made by different procedures in different laboratories, one or more additional defects that have not yet been identified in structural studies, also control the carrier concentration.[27, 59, 60]

If our hypothesis is correct, the record high T_c's for HBCCO compounds at high pressure result from metallization of the blocking layer. At ambient pressure, these compounds fall on the systematic curves for T_c vs. buckling angle and apical bond distance along with other compounds that have insulating blocking layers (Fig. 1). When pressure is applied, the blocking layer becomes more metallic and the HBCCO compounds display behavior analogous to the Y-123, Y-124, and PSYCO compounds, which have metallic copper in their blocking layers and are exceptions to the systematic behavior.

2 Optimizing J_c

For applications, one would like to optimize both T_c and J_c. Unfortunately, the compounds with the highest T_c's often do not exhibit high J_c's.[1] For film applications, Y-123 is the most widely used compound. High J_c's can be achieved through proper processing, such as Ion Beam Assisted Deposition (IBAD). BSCCO is the preferred material for bulk applications. Because of the micaceous nature of the material, the crystallites cleave and orient in a favorable way when wires, consisting of BSCCO powder in metal sheaths, are drawn. Unfortunately, the J_c is limited by the relatively poor flux pinning of BSCCO. Relatively poor flux pinning of the HBCCO compounds limits their usefulness for applications.[61, 62, 63, 64]

Work to improve J_c has concentrated for some time on issues such as achieving favorable orientation of crystalline grains, for example, by the cleaving process that occurs naturally in BSCCO or the recently developed IBAD process for Y-123 films.[1] Considerable effort has been spent investigating the effects of grain boundaries, which act as weak links in the conduction process. Work to understand and improve the intrinsic flux pinning behavior of copper-oxide superconductors began more recently. Some of the first work focused on ways to improve flux pinning in HBCCO compounds. Shimoyama et al. reported significant improvements in the flux pinning of HBCCO compounds where Hg has been partially replaced by Re or Cr and Ba replaced by Sr.[65, 66, 67] These compounds also exhibit improved chemical stability.

These early studies of chemically substituted HBCCO compounds have provided a framework for understanding how flux pinning might be improved through chemical substitution. Shimoyama et al., have argued that at least two features of the modified structures may enhance flux pinning: (1) The substitution of Sr for Ba significantly shortens the blocking layer distance[65, 67], and (2) the substitution (e.g., Re or Cr) at the Hg site may make the blocking layer more metallic.[68] Both of these are thought to increase the coupling of pancake-like vortices to form vortex lines along the c axis. This increases flux pinning because fewer pinning centers are required to pin extended vortex lines than to pin an equivalent amount of flux in the form of individual pancake vortices.[69, 70] In a recent paper, Chmaissem et al. also explored the possibility that chemical substitution (Cr for Hg) in Hg-1201 can form extended defects that could act as pinning centers.[71] We have recently performed some of the first detailed structural studies of these chemically substituted HBCCO compounds. A brief review of the findings will be given here.

The most successful method to date for shortening the blocking layer has been the full substitution of the smaller Sr ion for Ba.[65] It was found, however, that this substitution could not be achieved unless the Hg site was also partially substituted by Re or Cr and part of the Ca separating CuO_2 layers was substituted by Y.[65, 66, 67, 71] High pressure synthesis methods were used by Yamaura, et al. to synthesize compounds in which this substitution on the Ca site was not required.[68] The structures of these materials, $Hg_{1-x}Re_xSr_2Ca_{n-1}Cu_nO_{2n+2+\delta}$ (for n=2 and 3, and x≈0.25), have been determined by neutron powder diffraction.[72] The portions of the structures containing the Re substitution are shown in Fig. 4.

When Re substitutes at the Hg site, it achieves coordination to six nearby oxygen atoms by incorporating four additional oxygen atoms (O3) into the Hg/Re plane (in addition to the two apical oxygen atoms, O2'). This sixfold coordination of Re to oxygen is typical of Re(VI) compounds. The perovskite ReO_3, where Re has the same coordination, is one of the most metallic oxides. One might expect the Re substitution to make the blocking layer more metallic, but no experiments have been done to characterize how the electronic character of the blocking layer is modified by the Re substitution. One oxygen site per formula unit is available in the Hg/Re plane, setting a conceptual solubility limit for Re of x=0.25, which is observed in synthesis experiments. The substitution of Sr for Ba shortens the blocking layer by about 0.8 Å, as shown in Table 2. The improvement in flux pinning is characterized by measuring the irreversibility field as a function of temperature, as shown in Fig. 5

for Hg-1223. The chemically substituted compound has a lower T_c (by about 15 K), but a higher irreversibility field at any temperature below about 115 K.

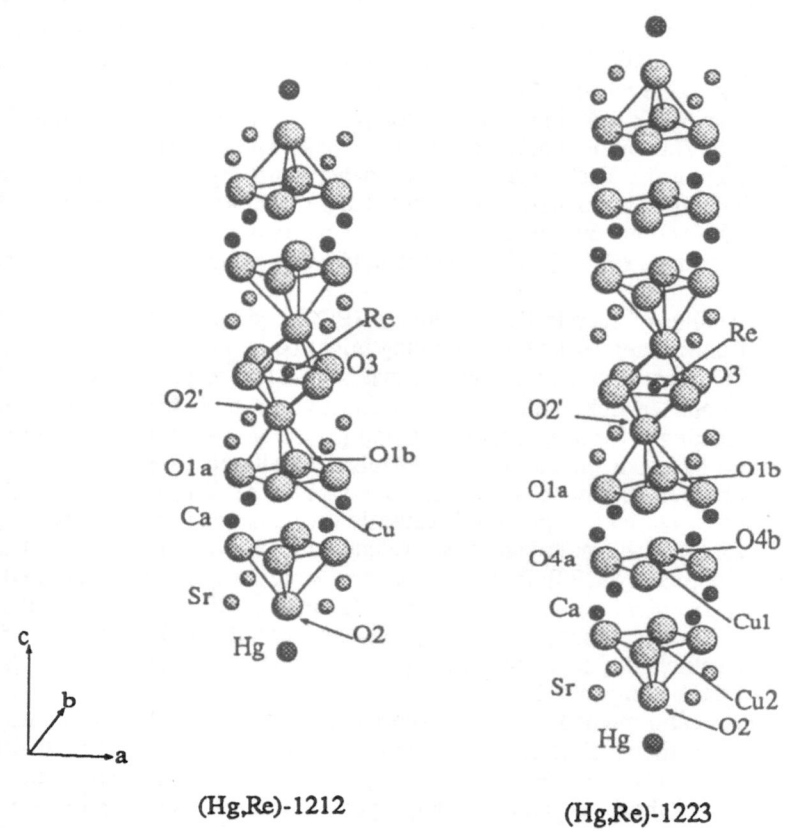

(Hg,Re)-1212   (Hg,Re)-1223

Fig. 4. Portions of the structures of $Hg_{1-x}Re_xSr_2Ca_{n-1}Cu_nO_{2n+2+\delta}$ (for n=1 and 2) showing the local environment around the Re site.[from Ref. 72]

Table 2. Blocking layer distances for ideal and chemically substituted $HgBa_2Ca_{n-1}Cu_nO_{2n+2+x}$ compounds. The chemical substitutions are full substitution of Sr for Ba and partial substitution of Cr for Hg (~40% substitution) for Hg-1201 and Re for Hg (~25% substitution) for Hg-1212 and Hg-1223.[71, 72]

Compound	Ideal	Substituted
Hg-1201	9.52 Å	8.70 Å
Hg-1212	9.58 Å	8.80 Å
Hg-1223	9.43 Å	8.67 Å

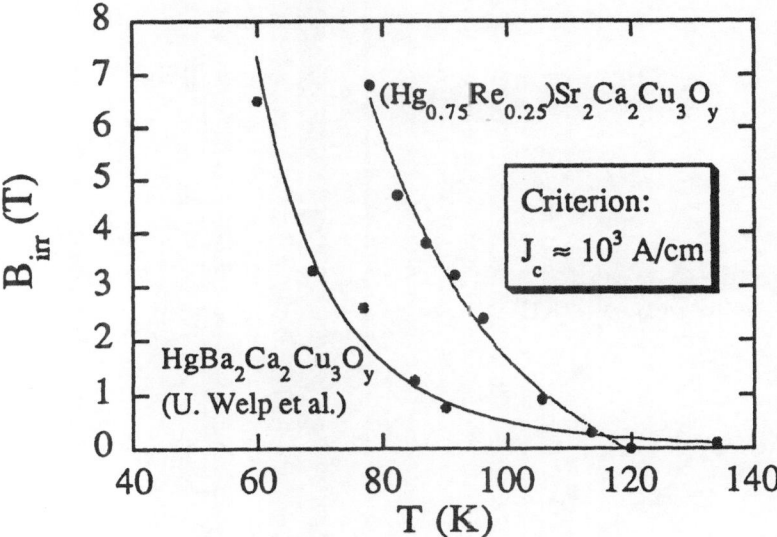

Fig. 5. Irreversibility fields vs. temperature for $HgBa_2Ca_2Cu_3O_{8+\delta}$ and $Hg_{0.75}Re_{0.25}Sr_2Ca_2Cu_3O_{8+\delta}$.[from Ref. 72]

The shortening of the blocking layer is thought to play a significant role in the improvement in flux pinning. Shimoyama, et al. have found a systematic relationship between the blocking layer distance and the irreversibility behavior. However, in other recent experiments, Shimoyama, et al. have used high-pressure synthesis techniques to partially substitute Cr, Mo, or Re on the Hg site without the replacement of Ba by Sr.[65] These compounds also show enhanced flux pinning, even though the blocking layer is not shortened; in fact, it is lengthened in some compounds. These experiments emphasize the need to investigate other structural effects on flux pinning, such as metallization of the blocking layer and the formation of defects that pin flux.

Structural studies of the $Hg_{1-x}Cr_xSr_2CuO_{4+\delta}$ system illustrate how chemical substitution can lead to the formation of extended defects. The structure of this compound, determined by neutron powder diffraction, is shown in Fig. 6.[71] Sr substitution at the Ba site leads to a shortening of the blocking layer distance of about 0.8 Å, as was seen in the two- and three-layer compounds (see Table 2). Cr substitutes for Hg, but is displaced off the ideal site to allow tetrahedral coordination to four oxygen atoms. Two of these are the original apical oxygen atoms (O2) which have been displaced to form an appropriate bond distance to Cr (the O2' site) and two additional oxygen atoms (O3) are incorporated into the Hg/Cr plane. The Cu-O2' distance is 3.13 Å. This apical oxygen atom has essentially been removed from the CuO_6 octahedron. The Cu atom displaces towards the remaining apical oxygen (O2) to form a rather short bond (2.25 Å). Thus, one effect of the Cr substitution is to convert elongated CuO_6 octahedra to CuO_5 pyramids. Energetically, the formation of four-coordinated

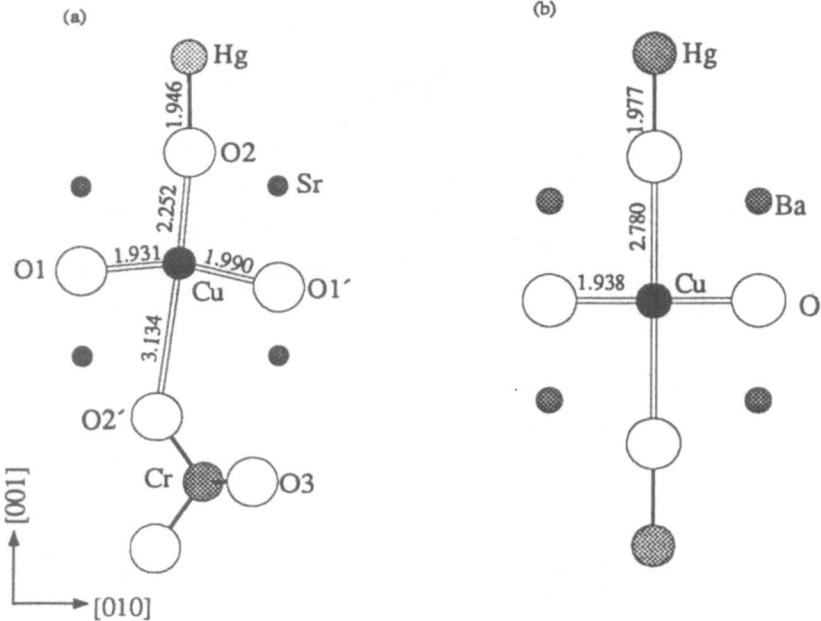

Fig. 6. Portion of the structure of $Hg_{1-x}Cr_xSr_2CuO_{4+\delta}$ showing the local configuration around a Cr site (a) compared to the ideal Hg-1201 structure (b).[from Ref. 71]

Cu^{2+} is unlikely. For this reason, Cr and Hg tend to alternate along the c axis, leading to a doubling of the unit cell in this direction.

All possible O3 oxygen sites in the Hg/Cr plane will be occupied for a conceptual Cr substitution limit of 50%. In experiments, it has been found to be difficult to exceed about 40% substitution. Cr atoms tend to cluster in the Hg/Cr plane. As additional CrO_4 units are added to these clusters, they must orient so that the associated O3 oxygen atoms can occupy available sites as shown in Fig. 7. Ultimately, the size of these clusters is limited by the availability of possible sites for the O3 atoms, leading to a maximum cluster size of about 2.5 unit cells (see Fig. 7). This ordering effect leads to an incommensurate supercell of approximate dimensions 5a x 5a x 2c, which has been observed in electron diffraction measurements. In this supercell, Cr- and Hg-rich regions alternate in all three directions, but the orientational frustration of the CrO_4 units as they cluster (as illustrated in Fig. 7) leads to considerable disorder within the supercell.

The existence of this supercell provides a structural dimension that allows defects of a size that could be effective for flux pinning. It is important to consider what kinds of defects can form for Cr concentrations below the solubility limit of 50%. In such a case, extended Hg-rich clusters will form. A particularly interesting aspect of these clusters is that they give rise to columns of CuO_6 octahedra along the c axis, embedded in a structure that consists mainly of CuO_5 pyramids, as shown in Fig. 8.

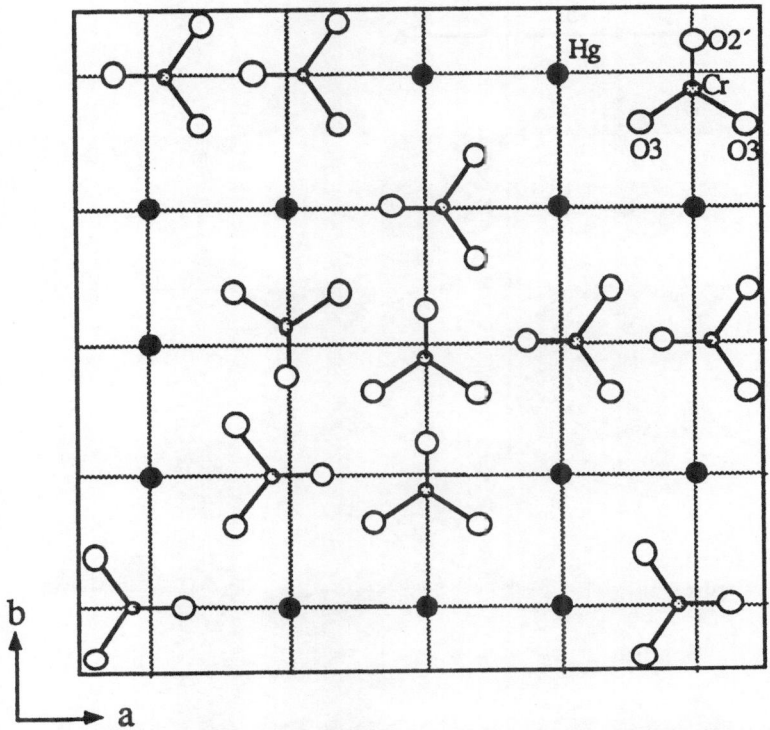

Fig. 7. Representation of one layer of the Hg/Cr plane of $Hg_{1-x}Cr_xSr_2CuO_{4+\delta}$ showing a possible arrangement of CrO_4 units to form a cluster of approximate dimensions 2.5a. Cluster size is limited by the availability of sites for O3 atoms.[from Ref. 71]

Although no measurements have yet been done to determine whether these defects are effective for pinning flux, it is encouraging to observe that defects of a favorable size (comparable to the coherence length) can be produced by chemical substitution. Based on the present observations, it is likely that extended defects will also form in the limit of low Cr concentration. If CrO_4 units tend to cluster, as has been observed for concentrations near 50%, the resulting defects would be islands of Cr in the Hg/Cr plane about 10 Å (2.5a) in diameter. In this region, CuO_6 octahedra with long apical bonds would be converted to CuO_5 pyramids with short apical bonds, locally disrupting the superconducting properties of the CuO_2 plane to form the pinning center.

3 A Strategy for Optimization

To produce the ideal material for applications, one would like to optimize both T_c and J_c. A summary of the methods for doing this is given in Table 3. Shortening the blocking layer will improve flux pinning, but if it is done in a way that leads to a shorter apical Cu-O bond, T_c will likely be lowered. This effect is observed in the chemically substituted HBCCO compounds, where T_c's are typically 15 K or more

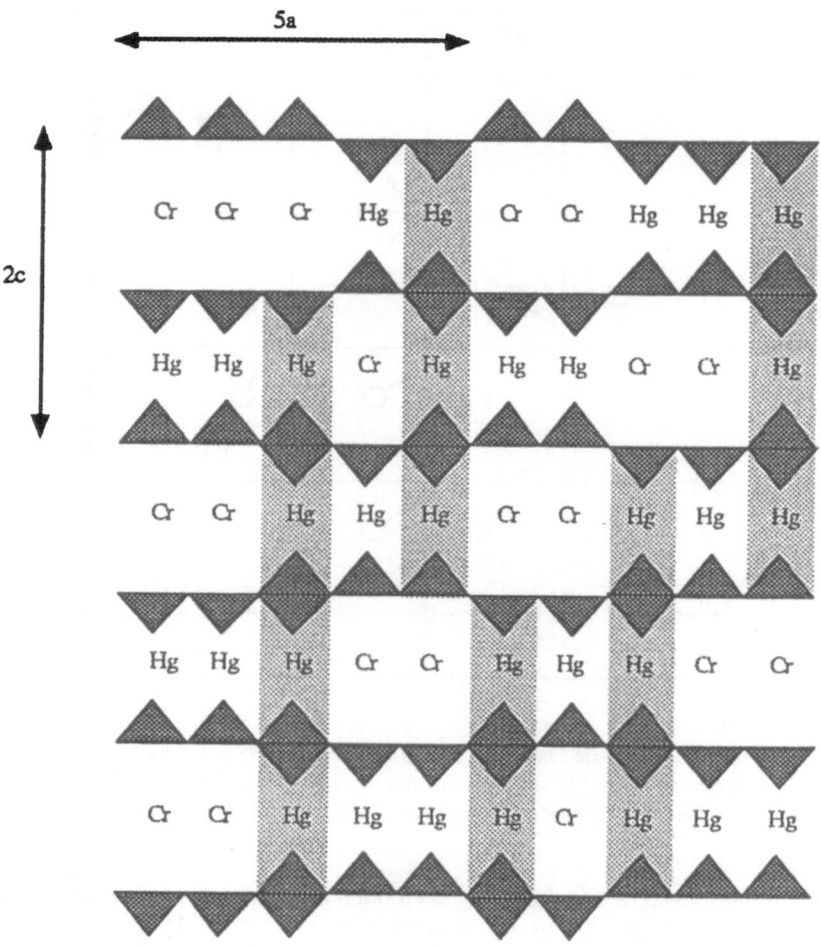

Fig. 8. Representation of the columns of CuO_6 octahedra along the c axis that are expected to form for Cr concentrations below 50%.[from Ref. 71]

lower than for the ideal compounds.[71, 72] We propose that metallizing the blocking layer will raise T_c and improve flux pinning at the same time. Chemical substitutions that produce this effect should be explored to verify whether this hypothesis holds. Additionally, there must be defects that will pin flux. Our work suggests that it may be possible to produce favorable defects (having dimensions approaching the coherence length) by chemical means. If these defects are in the blocking layer rather than in the CuO_2 planes, the effect on T_c should not be too severe, but, clearly, the defects must perturb superconductivity locally to pin flux. It is also important to correctly adjust the concentration of such defects.

Table 3. Possible methods for structurally optimizing T_c and J_c.

	Effect on T_c	Effect on J_c
Shorten blocking layer	Short apical Cu-O bond lowers T_c	Short blocking layer improves flux pinning
Metallize blocking layer	Metallic blocking layer raises T_c	Metallic blocking layer improves flux pinning
Insert pinning defects	Defects in the wrong site lower T_c	Defects of the proper size can pin flux

4 Conclusions

The desire to optimize high-T_c materials for applications has motivated investigations of how both T_c and J_c might be controlled by chemical/structural methods. A considerable data base, coming from years of work on the relationship between T_c and structure, has led to a consensus of what structural features lead to the highest T_c. For compounds where the blocking layer is an insulating charge reservoir, there is a clear correlation between T_c and buckling (or corrugation) of the CuO_2 plane and the apical Cu-O bond length. Flat and square CuO_2 planes and long apical bonds yield the highest T_c's. Compounds like Y-123, Y-124, and PSYCO, which seem to violate these rules, suggest that T_c will be even higher if the blocking layer is metallic. High pressure studies of HBCCO compounds seem to confirm this view. Thus, the highest possible T_c would be observed in a compound with flat and square CuO_2 planes, long apical Cu-O bonds, and a metallic blocking layer. No compound yet discovered (including the HBCCO compounds at high pressure) combines all of these features. Thus, T_c's even higher than 160 K may be possible through creative chemistry if the desired structural features can be achieved.

Work to increase J_c by improving the intrinsic flux pinning properties of the materials is still in its infancy. Initial studies indicate that structural properties can have a significant effect on flux pinning. At least three strategies look promising. The first two focus on improving the coupling along the c axis by shortening the blocking layer or metallizing the blocking layer. The third strategy is to devise chemical methods for creating defects that are effective for pinning flux. Experiments to date have focused on compounds in which more that one of these effects may be present at the same time. It will be important in the future to investigate each effect individually. In the long term, the goal of such work will be to optimize both T_c and J_c at the same time. It is already clear that significant progress toward this goal is likely.

Acknowledgments. The authors wish to thank K. Kishio, Y. Shimoyama, K. Yamaura, Z. Hiroi, and M. Takano for collaboration and discussion on materials properties that enhance flux pinning. This work was supported by the U. S. Department of Energy, Office of Basic Energy Sciences - Materials Sciences, under contract No.

W-31-109-ENG-38 (JDJ, DGH), the National Science Foundation, Office of Science and Technology Centers, under grant No. DMR 91-2000 (OC, DNA, BD), and the U. S. Department of Energy Distinguished Postdoctoral Research Program (JFM).

1. See, for example, the articles in *MRS Bulletin*, Vol. 17, No. 8, (August 1992).
2. J.D. Jorgensen, *Physics Today, June 1991*, p. 34.
3. M.H. Whangbo, D.B. Kang, and C.C. Torardi, *Physica C.* **158**, 371 (1989).
4. Y.J. Uemura, et al., *Phys.Rev.Lett.* **62**, 2317 (1989).
5. J.B. Torrance, Y. Tokura, A.I. Nazzal, A. Bezinge, T.C. Huang, and S.S. Parkin, *Phys. Rev. Lett.* **61**, 1127 (1988).
6. J.L. Tallon, R.G. Buckley, E.M. Haines, M.R. Presland, A. Mawdsley, N.E. Fowler, and J. Loram, *Physica C.* **185-189**, 855 (1991).
7. M.-H. Whangbo and C.C. Torardi, *Science.* **249**, 1143 (1990).
8. J.D. Jorgensen, P. Lightfoot, and S. Pei, *Supercond. Sci. Technol.* **4**, S11 (1991).
9. J.D. Jorgensen, P.G. Radaelli, H. Shaked, J.L. Wagner, B.A. Hunter, J.F. Mitchell, R.L. Hitterman, and D.G. Hinks, *J. Supercond.* **7**, 145 (1994).
10. Y. Tokura, J.B. Torrance, T.C. Huang, and A.I. Nazzal, *Phys.Rev.B.* **38**, 7156 (1988).
11. P.G. Radaelli, D.G. Hinks, A.W. Mitchell, B.A. Hunter, J.L. Wagner, B. Dabrowski, K.G. Vandervoort, H.K. Viswanathan, and J.D. Jorgensen, *Phys. Rev. B.* **49**, 4163 (1994).
12. J.L. Tallon, C. Bernhard, H. Shaked, R.L. Hitterman, and J.D. Jorgensen, *Phys. Rev. B.* **51**, 12911 (1995).
13. Y.J. Uemura, L.P. Le, G.M. Luke, B.J. Sternlieb, W.D. Wu, J.H. Brewer, T.M. Riseman, C.L. Seaman, M.B. Maple, M. Ishikawa, D.G. Hinks, J.D. Jorgensen, G. Saito, and H. Yamochi, *Phys. Rev. Lett.* **66**, 2665 (1991).
14. R.J. Cava, A.W. Hewat, E.A. Hewat, B. Batlogg, M. Marezio, K.M. Rabe, J.J. Krajewski, W.F. Peck Jr., and L.W. Rupp Jr., *Physica C.* **165**, 419 (1990).
15. J.D. Jorgensen, B.W. Veal, A.P. Paulikas, L.J. Nowicki, G.W. Crabtree, H. Claus, and W.K. Kwok, *Phys. Rev. B.* **41**, 1863 (1990).
16. J.L. Tallon and N.E. Fowler, *Physica C.* **204**, 237 (1993).
17. Y. Shimakawa, Y. Kubo, T. Manako, and H. Igarashi, *Phys. Rev. B.* **40**, 11400 (1989).
18. Y. Kubo, Y. Shimakawa, T. Manako, and H. Igarashi, *Phys. Rev. B.* **43**, 7875 (1991).
19. H. Shaked, J.D. Jorgensen, B.A. Hunter, R.L. Hitterman, K. Kinoshita, F. Izumi, and T. Kamiyama, *Phys. Rev. B.* **48**, 12941 (1993).
20. Y. Maeno, T. Tomita, M. Kyogoku, S. Awaji, Y. Aoki, K. Hoshino, A. Minami, and T. Fujita, *Nature.* **328**, 512 (1987).
21. J.M. Tarascon, P. Barboux, P.F. Miceli, L.H. Greene, G.W. Hull, M. Eibschutz, and S.A. Sunshine, *Phys. Rev. B.* **37**, 7458 (1988).
22. R.S. Howland, T.H. Geballe, S.S. Laderman, A. Fischer-Colbrie, M. Scott, J.M. Tarascon, and P. Barboux, *Phys. Rev. B.* **39**, 9017 (1989).
23. J.B. Parise, P.L.Gai, M.K. Crawford, in *High temperature superconductors: relationship between properties, structure and solid-state chemistry*, Edited by J.D. Jorgensen, K. Kitazawa, J.M. Tarascon, M.S. Thompson, and J.B. Torrance,

46. B.A. Hunter, J.D. Jorgensen, J.L. Wagner, P.G. Radaelli, D.G. Hinks, H. Shaked, R.L. Hitterman, and R.B.V. Dreele, *Physica C.* **221**, 1 (1994).
47. D.J. Singh and W.E. Pickett, *Phsyica C.* **233**, 237 (1994).
48. D.L. Novikov, O.N. Myrasov, and A.J. Freeman, *Physica C.* **222**, 38 (1994).
49. D.L. Novikov and A.J. Freeman, *Physica C.* **216**, 273 (1993).
50. D.L. Novikov and A.J. Freeman, *Physica C.* **212**, 233 (1993).
51. R.S. Markiewicz, *Physica C.* **217**, 381 (1993).
52. D.M. Newns, C.C. Tsuei, P.C. Pattnaik, and C.L. Kane, *Comm. Condens. Matt. Phys.* **15**, 273 (1992).
53. D.J. Singh, *Phys. Rev. B.* **48**, 3571 (1993).
54. E.V. Antipov, J.J. Capponi, C. Chaillout, O. Chmaissem, S.M. Loureiro, M. Marezio, S.N. Putilin, A. Santoro, and J.L. Tholence, *Physica C.* **218**, 348 (1993).
55. M. Cantoni, A. Schilling, H.-U. Nissen, and H.R. Ott, *Physica C.* **215**, 11 (1993).
56. S.M. Loureiro, E.V. Antipov, J.L. Tholence, J.J. Capponi, O. Chmaissem, Q. Huang, and M. Marezio, *Physica C.* **217**, 253 (1993).
57. S.N. Putilin, E.V. Antipov, and M. Marezio, *Physica C.* **212**, 266 (1993).
58. D.J. Singh and W.E. Pickett, *Phys. Rev. Lett.* **73**, 476 (1994).
59. M. Itoh, A. Tokiwa-Yamamoto, S. Adachi, and H. Yamauchi, *Physica C.* **212**, 271 (1993).
60. S. Adachi, A. Tokiwa-Yamamoto, M. Itoh, K. Isawa, and H. Yamauchi, *Physica C.* **214**, 313 (1993).
61. A. Umezawa, W. Zhang, A. Gurevich, Y. Feng, E.E. Hellstrom, and D.C. Larbalestier, *Nature.* **364**, 129 (1993).
62. U. Welp, G.W. Crabtree, J.L. Wagner, D.G. Hinks, P.G. Radaelli, J.D. Jorgensen, J.F. Mitchell, and B. Dabrowski, *Appl. Phys. Lett.* **63**, 693 (1993).
63. U. Welp, G.W. Crabtree, J.L. Wagner, and D.G. Hinks, *Physica C.* **218**, 373 (1993).
64. J.A. Lewis, C.A. Platt, M. Weigmann, M. Teepe, J.L. Wagner, and D.H. Hinks, *Phys. Rev. B.* **48**, 7739 (1993).
65. J. Shimoyama, S. Hahakura, R. Kobayashi, K. Kitazawa, K. Yamafuji, and K. Kishio, *Physica C.* **235-240**, 2795 (1994).
66. J. Shimoyama, S. Hahakura, K. Kitazawa, Y. Yamafuji, and K. Kishio, *Physica C.* **224**, 1 (1994).
67. J. Shimoyama, K. Kishio, S. Hahakura, K. Kitazawa, K. Yamaura, Z. Hiroi, and M. Takano, in *Advances in Superconductivity VII, Proc. of the 7th International Symposium on Superconductivity (ISS'94)*, (Edited by K. Yamafuji and T. Morishita), Vol. p. 287. Springer-Verlag, Tokyo (1995).
68. K. Yamaura, J. Shimoyama, S. Hahakura, Z. Hiroi, M. Takano, and K. Kishio, *Physica C.* **246**, 351 (1995).
69. K.E. Gray and D.H. Kim, *Physica C.* **180**, 139 (1991).
70. K.E. Gray, *Appl. Superconductivity.* **2**, 295 (1994).
71. O. Chmaissem, D.N. Argyriou, D.G. Hinks, J.D. Jorgensen, B.G. Storey, H. Zhang, L.D. Marks, Y.Y. Wang, V.P. Dravid, and B. Dabrowski, *Phys. Rev. B.* **52**, 15636 (1995).
72. O. Chmaissem, J.D. Jorgensen, K. Yamaura, Z. Hiroi, M. Takano, J. Shimoyama, and K. Kishio, *Phys. Rev. B.* (in press) (1996).

Role of Doping, Pressure and Van Hove Singularities on Highest T_c Materials

D. L. Novikov and A. J. Freeman

Science and Technology Center for Superconductivity, Department of Physics and Astronomy, Northwestern University, Evanston, IL, 60208-3112, USA

Abstract. A review of some striking predictions of electronic structure theory about the role of van Hove singularities (vHs) on the electronic structure and properties of the newest and highest temperature superconducting cuprates is given. The results provide possible strong evidence for the role of vHs in the superconductivity of quasi-2D high T_c systems. They thus serve to call attention to their role not only in enhancing T_c through large increases in $N(E_F)$ and Fermi surface areas, but also in possibly providing support for vHs based excitonic pairing mechanisms for superconductivity. Further, this information, derived from the understanding - expressed as an "empirical rule" - gained from the related role of doping and pressure, is used to investigate other likely systems for being made into high T_c superconductors.

1 Introduction

Electronic structure theory (EST) has been an active and important part of the high T_c story since late 1986. Unlike exotic theories proposed and long since gone, our EST predictions still remain valid and increasingly demonstrate that the normal state properties of the cuprates can be understood as conventional metals (i.e., Fermi liquids).

It is thus all the more striking that a number of experiments, notably photoemission, de Haas-van Alphen and positron annihilation [1, 2, 3], have confirmed in detail and with excellent agreement the predictions of band theory on the Fermi surfaces and energy bands near E_F. This has served to firmly establish a Fermi liquid picture as the basis for developing the theory of high T_c superconductivity. In addition, the high resolution local density band structure results have provided experimentalists a detailed picture of these complex systems, emphasizing the close relation of the physics (band structure) and chemistry (bonds and valences) to the crystal structure.

More recently, there has developed mounting evidence for the possible important role of van Hove singularities on the superconducting properties of the cuprates. From the first calculations of the electronic structure of the high T_c cuprates [4, 5], van Hove singularities (vHs) were expected and found to play an important role. While most obvious in La 214, where the single $Cu - O_2$ layer yields the now well-known 2D $dp\sigma$ band crossing E_F and a clearly separated vHs, the appearance and role of vHs in the other high T_c cuprates has not been clearly apparent. Nevertheless, a rich literature [6] exists that attempts to relate the vHs, via exotic pairing mechanisms, to the origin of the high T_c and other properties of these layered cuprates.

In this paper, we review some striking predictions about the role of vHs on the electronic structure and properties of the newest and highest temperature superconducting cuprates, and use the knowledge so obtained to investigate other likely systems for being made into high T_c superconductors. All calculations were performed within the local density approximation (LDA) using the full-potential linearized augmented plane wave (FLAPW) [7] and the full-potential linear muffin-tin orbital (FLMTO) [8] methods.

2 vHs in Different Structures

2.1 The infinite layered superconductor $Sr_xCa_{x-1}CuO_2$

Superconductivity at 110 K in the infinite-layer (IL) compound $(Sr_{1-x}Ca_x)_{1-y}$-CuO_2, where x=0.3 and y=0.1, was reported in [9]. The parent compounds are $Ca_{0.86}Sr_{0.14}CuO_2$ [10] and recently synthesized $SrCuO_2$ [11] which has been found to be stable at high pressure. They crystallize in a simple-tetragonal structure (space group P4/mmm) and consist of flat CuO_2 planes that are separated by oxygen-free layers of Ca-Sr atoms; each Cu atom is square coordinated by four oxygen atoms. The electronic structure of tetragonal $CaCuO_2$ has been previously calculated with full-potential LAPW [12, 13], LMTO-ASA [14] and full potential pseudo-function methods [15].

We have investigated [16] in detail the band structure, chemical bonding and density of states and resulting properties of $(Sr_{1-x}Ca_x)CuO_2$ for x=0.0, 0.3 and 1.0, which we correlated with varying composition y of the divalent Ca/Sr vacancies. We showed that a highly important role is played by a strong 2D van Hove saddle-point singularity (vHs) in the electronic band structure for $y > 0$. This singularity strongly affects the Fermi surface topology as discussed in the work of Lifshitz [17], who showed a coexistence of changes in the topology of the Fermi surface and anomalies in the thermal and electronic properties.

The valence band complex spans an energy range of about 9 eV. As is typical of all cuprate superconductors, the valence bands originate from the Cu 3d and O 2p states and exhibit the characteristic feature of cuprate superconductors: the uppermost band (the Cu $d_{x^2-y^2}$ - O p_σ antibonding subband) is exactly half-filled. The total band width along the c-axis (Γ-Z) direction is about 2 eV and is almost five times smaller than that in the planes (Γ-X-Z-Γ) and (Z-R-A-Z). This indicates that, as expected, the electronic interactions in the Cu-O planes are much stronger than those in the c-direction.

The band structure near E_F has a number of interesting features. Only a single free-electron-like band crosses E_F and gives rise to the simple Fermi surface displayed in Fig. 2.1.

An important striking feature is the occurrence of vHs at the X and R points in the Brillouin zone that contributes to the enhancement of the DOS close to E_F. The nearly two-dimensional character of the chemical bonding in these systems is clearly seen from the calculated Fermi surface for the undoped case, Fig. 2.1, shown in the extended zone scheme. The single Cu-O band crossing E_F produces a highly two-dimensional Fermi surface. It has the shape of a rounded square and is centered around the M point in the (ΓXM) plane and around the A point in the (ZRA) plane. A comparison of the shape of the Fermi surface cross sections in both planes displays some differences between them; the Fermi surface shows nesting features in the (100) and (010) directions. Such strong nesting of the Fermi surface can give rise to singularities in the generalized susceptibility and, as has been emphasized previously, may lead to electronically driven instabilities such as charge density or spin density waves. The simple shape of the 2D Fermi surface of $MCuO_2$ is similar to that calculated for La_2CuO_4 [4, 5], Nd_2CuO_4 [18], $Tl_2Ba_2CaCu_2O_8$ [19] and to the predicted [20, 21] and observed [1, 22] Fermi surface of $YBa_2Cu_3O_7$.

Since Ca and Sr atoms act solely as electron donors, we expect that small changes of Ca or Sr concentration would not change any major features of the band structure. Thus, the use of a rigid-band model may be considered as a good first approximation. Our results for $(Sr_{0.7}Ca_{0.3})CuO_2$ confirm this assumption. In order to move E_F to the vHs one has to decrease the electron concentration by about 0.14 electrons for $CaCuO_2$ and about 0.25 electrons for $SrCuO_2$. This can be done by decreasing the Ca (Sr) concentration by 7(12) percent leading to the compound composition $(Ca,Sr)_{1-y}$-CuO_2 where y is about 0.07 - 0.12; this is consistent with the y=0.1 composition found to be superconducting at 110 K [9]. Indeed, for the case of $Sr_{0.7}Ca_{0.3}CuO_2$ calculated using the experimental lattice constants, the calculated y value is 0.11. This is in striking agreement with the experimental value [9] of y=0.1 for which T_c=110 K. Moreover, the total DOS appears

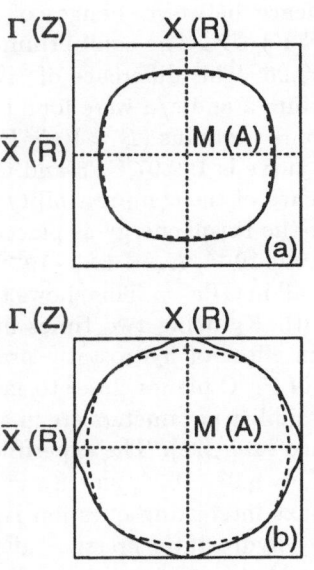

Figure 2.1: The Fermi surface for Hg-1201 in the (ΓXM) plane (dashed line) and (ZRA) plane (solid line) at different hole doping levels: a) undoped case, b) doping for maximum T_c.

to be almost unchanged in a small energy region (about 0.1 eV) below vHs. From the chemical point of view, this fact may reflect the existence of the superconducting phase in a small but finite Ca/Sr vacancy concentration region.

It is interesting to see how the shape of the Fermi surface (FS) would change when doping shifts E_F to the vHs. Again, assuming a rigid-band approximation, we calculated the FS for the non-stoichiometric composition of $(Sr_{0.7}Ca_{0.3})_{0.9}$-CuO_2 which causes E_F to move exactly onto the vHs. The nesting character of the FS in the bottom (ΓXM) plane stays almost unchanged while the top part ((ZRA) plane) changes drastically [16]. At the composition at which E_F touches the vHs, the former rounded square FS becomes pinched and the FS area is maximized. At such a point, this change of the topology of the FS may be accompanied by various anomalies in thermal and electronic properties [17] and possible changes in T_c as well [23].

Further, we investigated [24] the effect of pressure on the electronic structure and properties of the IL superconductor using the FLMTO method. Total energy calculations were performed for a set of different volumes (V/V_0= 0.8, 0.85, 0.9, 0.95, 1.0, 1.05, where V_0 is the experimental volume of $Sr_{0.7}Ca_{0.3}CuO_2$). For each particular volume, a "scan" over c/a was performed in order to find an equilibrium c/a. Knowing the $E(V/V_0)$ dependence, we established a correspondence between changes of volume and external hydrostatic pressure as $P = \partial E(V)/\partial V$. The equilibrium volume for the IL compound was found to be V/V_0=0.96. This difference of -4% is typical for the LDA approximation. The equilibrium a and c/a were found to be equal to 3.882 Å and 0.84 respectively.

The bulk modulus ($B = V(\partial^2 E(V)/\partial V^2)|_{V=V_0}$) calculated from the energy-volume curve is 122.07 GPa and the cohesive energy E=-25.0 eV/unit cell. The dependence of the compressibility on pressure was calculated. For zero pressure (and for the Fermi energy as placed at zero doping in these simulations) we have $K_a = 2.00 \cdot 10^{-3}$, $K_c = 4.23 \cdot 10^{-3}$, and the volume compressibility $K_v = 8.22 \cdot 10^{-3}$ — all in GPa^{-1}. This shows the anisotropic character of the compressibility of IL with K_c being two times higher than K_a. From this we conclude that the main effect of hydrostatic pressure is a "squeezing" of the c-axis and the moving of Cu-O planes closer to each other. These theoretical predictions [24] on compressibility parameters are in an excellent agreement with the measurements performed later [25]. The experimentally found compressibilities are $K_a = 2.2 \cdot 10^{-3}$, $K_c = 4.07 \cdot 10^{-3}$, and $K_v = 8.5 \cdot 10^{-3}$ — all in GPa^{-1}.

The next interesting question is, how do the changes in crystal structure induced by hydrostatic pressure affect its electronic structure. An examination of the band structure corresponding to different volumes of the unit cell shows that the most important difference is the movement of the vHs (at the R-point) with respect to E_F. The decrease of volume, which is mainly due to the c-axis shortening, leads to a raising of the vHs towards the Fermi level. Such a transformation of the band structure may provide a clue to understanding the nontrivial pressure dependence of T_c in high-T_c superconductors.

2.2 HgBa$_2$Ca$_{n-1}$Cu$_n$O$_{2n+2+\delta}$: Apparent importance of the role of van Hove singularities on high T$_c$

The recent exciting discovery of superconductivity in HgBa$_2$CuO$_{4+\delta}$ with T$_c$=94 K, which followed closely the discovery of superconductivity in (Sr$_{1-x}$Ca$_x$)$_{1-y}$-CuO$_2$, turned attention away from the more complex cuprates based on Y, Bi and Tl to these simpler single layer high T$_c$ materials.

We have determined [26, 27] the electronic structure and Fermi surface of HgBa$_2$CuO$_{4+\delta}$ (by the FLMTO method) which shows again that the electronic structure and properties are dominated by a major van Hove saddle point singularity. As for (Sr,Ca)CuO$_2$, the hole Fermi surface has the shape of a rounded square in the undoped case [16]. The role of the measured doping in the 95 K superconductor was found to expand the cross-sectional area of the Fermi surface and to pin the Fermi energy exactly at the position of the vHs — at which point the density of states and the Fermi surface area are maximized. The results show that the vHs may well play an important role in the enhancement of superconductivity of the layered cuprates.

The FLMTO calculations were performed for defect-free HgBa$_2$CuO$_4$ using lattice parameters obtained in Ref. [28] for the "oxygenated" phase (T$_c$=95 K). The modeling of structural defects was performed assuming a "rigid-band" behavior of the band structure. As found from the self-consistent FLMTO energy band results shown in Fig. 2.2, E$_F$ for the undoped system again falls in the region of antibonding Cu 3d - O 2p states primarily from Cu and O(1) states forming Cu-O layers. As expected, the Hg and Ba components are negligible in this region, showing that these constituents are electronically inactive. The same is true for the "apical" oxygen, O(2), and so may be considered as acting solely as "hole" donors. The band structure is very similar to that calculated for the IL materials (Sr$_{1-x}$Ca$_x$)CuO$_2$ [16]. Only a single free-electron-like antibonding dpσ band crosses E$_F$ in the planes parallel to the (XMΓ) plane. There are van Hove saddle points at X and R in the BZ. The total DOS for HgBa$_2$CuO$_4$ shows two peaks derived from the two saddle-points of the uppermost valence band. Since these vHs may play a crucial role in T$_c$ in the layered cuprate compounds [4, 5, 6], we again estimated the hole doping level needed to shift E$_F$ to the vHs. Our estimates show that one needs to introduce about 0.36 "holes"/unit cell. N(E$_F$) in this case is strongly enhanced to 1.46 states/eV-unit cell, compared with 0.745 states/eV-unit cell in the undoped case. Now the highest T$_c$ of 95 K was reached [28] when the neutron measured defects - 0.059 atoms of oxygen/unit cell (located in the center of the Hg squares) [28] and 0.07 Cu atoms/cell situated at the Hg sites and 0.09 oxygen atoms/cell at the sides of Hg squares - were incorporated in the crystal.

It was found later [29] that $\delta = 0.18$ corresponds to the highest (or optimum) T$_c$ value, which turns out to be surprisingly close to our prediction [27] if one assumes that oxygen acts as a divalent ion in doping Hg layers. Further, based on the same simple ionic considerations, we estimate the number of holes in the "oxygen reduced" compound displaying T$_c$=59 K [28] to be 0.31. For E$_F$ corresponding to this doping value, the corresponding value of N(E$_F$) is 1.12

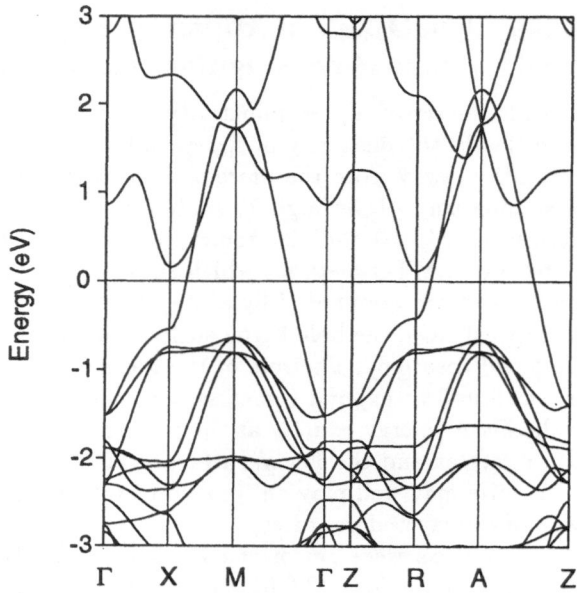

Figure 2.2: FLMTO energy-band results along high symmetry directions in the simple tetragonal BZ for Hg-1201

states/(eV — unit cell), which is considerably less than that of the "oxygenated" crystal. The correlation of $N(E_F)$ with T_c seems clearly established (see also Ref. [30], where the thermodynamic evidence for a DOS peak near E_F is given). It also suggests, as in the Bi and Tl cuprates, that Hg based compounds with more Cu planes would have higher $N(E_F)$ and T_c values.

Consider the specific changes of Fermi surface topology, when the initially ideal crystal is doped by holes. For undoped $HgBa_2CuO_4$ the single free-electron-like band crossing E_F gives rise to the simple Fermi surface, which again has the shape of a rounded square, centered at the M point in the (ΓXM) plane and around the A point in the top (ZRA) plane. Assuming that a rigid-band approximation is valid for the doping region considered, we calculated the FS [26, 27], corresponding to the doping level at which the maximum critical temperature (T_c=95 K) is reached. In this case, E_F hits the van Hove singularity at R in the top (ZRA) plane. As was found for the infinite layered material [16], the diameter of the FS in the (ΓXM) plane is increased but the nesting character stays almost unchanged, while in the (ZRA) plane the former rounded square Fermi surface becomes pinched, touches the BZ boundaries and serves to maximize the FS area. This change in FS may be accompanied by various kind of electronically driven anomalies [17]. It also leads to some kind of "empirical" rule, namely that we can expect the maximum in T_c for hole doped high-T_c superconductors based on layered Cu-O structure, when the area of the FS hits its maximum value.

We have focused on the simple one layer Hg compound, Hg-1201. The results for Hg-1212 and 1223 are very similar [27], but they have "self-doped" band structures, that move the Cu-O $dp\sigma$ antibonding band away from half-filling. This band pattern is reminiscent of the case for the Tl- based superconductors $(Tl_2Ba_2Ca_{n-1}Cu_nO_{4+2n}$, n=1,2,3) [31, 32, 33] — the crystal analog of the Hg-based materials — where the Tl 6s-O 2p band is found to be located always below E_F.

Our results lend strong support to our earlier finding that the vHs may play a dominant role in the superconductivity of these quasi - 2D high T_c systems. Further, they again call attention to vHs based pairing mechanisms for superconductivity - as emphasized in the recent work of Markiewicz [34]. Finally, these findings also may be useful in predicting a synthesis pattern and doping levels for stabilizing large a volume of Hg-1223 and hence its maximum T_c.

2.3 Superconductors without apical oxygen: $Sr_2CuO_2F_2$, $Sr_2CuO_2Cl_2$ and $Ca_2CuO_2Cl_2$

The recent discovery of superconductivity in $Sr_2CuO_2F_{2+\delta}$ [35] and $(Ca_{1-x}Na_x)_2CuO_2Cl_2$ [36] opened a new class of superconducting compounds with the apical oxygen substituted by other elements (F, Cl). Although T_c found in these phases (46 K for $Sr_2CuO_2F_{2+\delta}$ and 26 K for $(Ca,Na)_2CuO_2Cl_2$) is not as high as for other high-T_c layered Cu-O superconductors, the new phases are of great interest because of the expectation that they can help improve our understanding of the nature of high-temperature superconductivity.

We have investigated [37] the role of the lack of apical oxygens on the band structure and density of states of the parent compounds, $(Sr_2CuO_2F_2$, $Sr_2CuO_2Cl_2$ and, $Ca_2CuO_2Cl_2)$ and their electronic properties which we correlate with varying degree of "hole doping". We found that the electronic structures of these compounds display the same features as found in the high-T_c cuprates (strong two-dimensionality, a low density of states and a simple Fermi surface in the form of a rounded square). Since a major 2D van Hove saddle point singularity (vHs) exists near E_F, we again assume, [16, 26, 27], that its position determines the optimum hole (or electron) doping level needed to achieve the highest possible T_c in a specific class of compounds. From our calculations, we argue that $Sr_2CuO_2F_{2+\delta}$ with δ=0.6, which is the estimated composition synthesized by the authors of Ref. [35], turns out to be heavily overdoped and that proper doping may help achieve better superconducting properties, including higher T_c. Since the optimum doping level for $(Ca_{1-x}Na_x)_2CuO_2Cl_2$ [36] is unknown so far, our theoretical investigation might be helpful in estimating this important parameter and in establishing common features of the electronic structure of high-T_c superconductors without apical oxygen in their structures.

The band structures of these compounds are very similar. For all of them, a CuO_2-derived free-electron like $dp\sigma$ antibonding band crosses the Fermi energy. The prominent vHs derived from this band is found at the X and R points in the BZ. For $Sr_2CuO_2Cl_2$ and $Ca_2CuO_2Cl_2$, the vHs at the R-point undoubtedly shows the features of an "extended van Hove singularity", the importance of which was discussed recently by Abrikosov [38].

The densities of states for these compounds also look very similar. The only difference is, of course, in the energy positions of the Cl 3p and F 2p bands, with the former located deeper in energy. For all compounds, the dominant feature near the Fermi level is a peak in the DOS caused by the vHs, which is found at -0.36 eV for $Sr_2CuO_2F_2$, at -0.27 eV for $Sr_2CuO_2Cl_2$ and at -0.28 eV for $Ca_2CuO_2Cl_2$ below the Fermi level. The values of the DOS at E_F and the energy corresponding to the vHs are listed in Table 2.1. For the ideal structures,

Table 2.1: Total and l-decomposed DOS at E_F and at the position of vHs for $Sr_2CuO_2F_2$ and $Sr_2CuO_2Cl_2$ in states/(Ry-unit cell).

	Total DOS	Cu 3d	O 2p	F 2p	Cl 3p
$Sr_2CuO_2F_2$	10.6	6.1	2.7	0.1	—
at the vHs	27.3	16.4	4.6	0.6	—
$Sr_2CuO_2Cl_2$	12.7	7.0	3.3	—	0.4
at the vHs	33.3	18.5	5.8	—	2.7
$Ca_2CuO_2Cl_2$	11.6	6.6	2.8	—	0.3
at the vHs	26.1	15.4	4.6	—	1.7

$N(E_F)$ for these compounds is about 1 state/(eV-unit cell) which is the value (per Cu atom) typical for all high-T_c Cu-O materials. If we assume that the vHs may play an important role in the high-T_c superconductivity, it is important to estimate the total DOS at the energy of the vHs, cf. Table 2.1.

As seen, the main contribution to the total DOS comes from the Cu 3d and O 2p orbitals and is almost three times larger than that at E_F for the undoped compound. Note also that the Cl 2p contribution to the DOS at the vHs for $Sr_2CuO_2Cl_2$ and $Ca_2CuO_2Cl_2$ is considerably higher than that for $Sr_2CuO_2F_2$, but is still much smaller than the Cu and O contributions. The total DOS at the vHs increases from $Ca_2CuO_2Cl_2$ to $Sr_2CuO_2F_2$, which correlates well with the T_c enhancement. Note also that the total DOS reaches its maximum for $Sr_2CuO_2Cl_2$, which could indicate that the best value of T_c might be obtained for $Sr_2CuO_2Cl_2$, if properly doped.

The two-dimensional Cu-O band again gives rise to a simple FS [37]. It has the shape of a rounded square, is practically undispersed along the Γ-Z direction, and drastically changes its shape when doping shifts E_F onto the vHs in which case the FS again becomes pinched and touches the BZ boundaries.

As done previously [16, 26, 27], we correlate the optimum T_c value for a given compound with pinning of the vHs to the Fermi energy by appropriate doping (whether electrons or holes). Thus, we estimate the number of holes needed to hit the vHs to be about 0.38 holes for $Sr_2CuO_2Cl_2$ and $Sr_2CuO_2F_2$ and about 0.35 for $Ca_2CuO_2Cl_2$. It is interesting to note that these values coincide with the "optimum doping level" for the one layer Hg-1201 compound [27]. This suggests that the "optimum doping level" might be an intrinsic property of the band structure of just the single Cu-O layer reflected as a universal value. The transition temperature of 46 K was achieved in $Sr_2CuO_2F_{2+\delta}$ for an estimated composition of $\delta=0.6$ [35]. Providing that F is incorporated into the crystal as assumed [35], we may conclude that the doping level of 0.6 holes, is in a heavily overdoped regime. If correct, we may propose that a substantional increase of T_c is possible by removing the excess fluorine down to the level of $\delta \sim 0.38$. The observed doping level for $(Ca,Na)_2CuO_2Cl_2$ is estimated to be about 0.08 holes per Cu atom [36], which is far below our theoretical optimum value. This may mean that (i) the authors have not reached the optimum doping level and hence that a further increase of T_c in $(Ca,Na)_2CuO_2Cl_2$ is possible, or (ii) their estimate of the number of holes needs to be improved.

Finally, if one considers the possibility of achieving superconductivity by monovalent metal substitutions for Ca in $Ca_2CuO_2Cl_2$ or Sr in $Sr_2CuO_2Cl_2$, we may propose from our results an optimum composition of $M_{0.35-0.38}M'_{1.65-1.62}CuO_2X_2$, where M stands for a monovalent metal such as Na, K, ... and M' is Ca or Sr and X stands for F or Cl. In such a case, from the total DOS at the vHs we can expect a higher T_c for $M_{0.35-0.38}Sr_{1.65-1.62}CuO_2X_2$ rather than for the compounds considered.

2.4 Electronic structure of perovskite related La_2CuSnO_6

Recently, a series of papers have appeared on the synthesis of new compounds containing structural units of Al, Ga, Sn along with the CuO_2 planes that are common for high-T_c superconductors. These compounds are La_2CuSnO_6 [39], $LaSrCuAlO_5$[40], $LaSrCuGaO_5$ [41], and $RESr_2GaCu_2O_7$ [42, 43, 44], RE_2Sr_2M-Cu_2O_9 (M = Al, Co, Ga) [45].

The crystal structure of the first compound, La_2CuSnO_6, is closely related to the structure of the high-T_c copper based superconductors (e.g. Hg-1201, Tl-2201), but contains layers of CuO_2 and SnO_6 octahedra connected by oxygen atoms. However, according to magnetic susceptibility measurements [46], the doped compound appears not to be superconducting down to at least 5 K. The nature of the conductivity of this compound is not understood nor, in particular, why all attempts to drive it into the metallic state have failed. For this reason, we turned to the possible understanding that can be obtained from electronic structure calculations.

The crystal structure of La_2CuSnO_6 includes four formula units per unit cell. The main structural feature that makes the crystal structure complex is the buckling of the CuO_2 planes. Owing to this buckling, there are deviations in the geometry of the crystal from the orthorhombic structure: oxygen octahedra around the Cu and Sn cations are strongly tilted and the La cations are displaced toward the Cu-O layer. The distortions give rise to two copper, two tin, and four lanthanum sites in the unit cell [39].

We performed calculations [47] for both the real (buckled) structure and the ideal one which makes it possible to reduce the size of the unit cell to one formula unit and hence to simplify the band structure analysis. The ideal structure was deduced by flattening the Cu-O planes (i.e., removing the buckling) and removing the tilting of the SnO_6 octahedra, while preserving the length of the most covalent Cu-O bonds

The FLMTO band structure of the *idealized* La_2CuSnO_6 shows, as is typical for all Cu-O based high-T_c superconductors, that a quasi two-dimensional Cu-O derived free-electron-like $dp\sigma$ band crosses E_F. There are also some additional bands, mainly arising from the O(2) 2p (apical oxygen) and O(3) 2p (oxygen in SnO_2 planes) states, that cross E_F around M and A in the BZ. A remarkable feature of the band structure of ideal La_2CuSnO_6, which makes this compound very different from other high-T_c materials, is that the Fermi level is located exactly at the vHs even for the undoped case. The vHs in this structure appears to be strongly dispersed along X-A in the BZ; the overall dispersion of the vHs is about 0.5 eV. E_F is located on the broad peak in the DOS, consisting mainly of Cu 3d, O(2) and O(3) 2p states. This again shows the difference of this compound from the usual band composition of high-T_c Cu-O based materials, where the main contribution to the DOS at E_F comes from Cu 3d and "in-plane" oxygen (O(1) in our case) 2p states.

Consider now the *real* La_2CuSnO_6 structure with four formula units per unit cell. The main changes in the electronic structure in the vicinity of E_F are: *i)* the vHs related peak in the total DOS splits into two peaks; this reduces the high density of states at E_F from - 2.4 states/eV per Cu-atom for the ideal structure to 0.8 states/eV per Cu-atom for the real one; and *ii)* the band character in the vicinity of E_F gets back to what is regarded to be a common high-T_c composition (i.e., Cu 3d - O(1) 2p states with only a small admixture of the apical - O(2) - and O(3) 2p states). It needs to be stressed that E_F is again located exactly in the middle of the vHs, which is now effectively split by the lattice distortions (buckling of Cu-O planes and tilting of the SnO_6 octahedra).

In a view of these results, we can discuss the possibility of achieving super-conducting properties for this material. The first, and probably most important difference between the electronic structure of La_2CuSnO_6 and its idealized counterpart from all "conventional" high-T_c materials is the location of E_F exactly in the energy region of the vHs for the undoped case. This high DOS makes the system unstable, and this is the most likely reason for the crystal distortions observed in the real structure that effectively reduce the value of the total DOS at E_F and make the distorted compound more stable. If one recalls the empirical "recipe" for achieving superconductivity, one must first consider driving the initially semiconducting material into a metallic state. This is usually possible by appropriate electron or hole doping. A second step is the "optimization" of the doping, which usually should lead to moving E_F into close vicinity with the vHs. From the calculated band structures of both the idealized and real structures, both steps cannot be realized simultaneously. Any kind of doping will immediately move E_F away from the vHs and, moreover, for the real structure this doping will lead to the moving of E_F onto DOS peaks of the split vHs, which apparently may make the structure unstable and hence energetically unfavorable. This may be one of the reasons for the failure of attempts [46] to drive the system into a metallic state. Thus, the Mott-Hubbard picture of the electronic structure of this material proposed in [46] should be regarded as an appropriate one.

If one manages to make the ideal crystal of La_2CuSnO_6 (with flat Cu-O planes) by means of high pressure, or suitable chemical additions, one may try to dope this compound with electrons in order to achieve the superconducting onset. And we believe, the electron doping is the only possible way to achieve this goal, since this is the only way to get rid of the $O(2)$, $O(3)$ 2p states and to bring the composition of the states at E_F to be of the Cu 3d - $O(1)$ 2p (in-plane) type, and also to make the states at E_F more two dimensional-like, as in all "conventional" high-T_c copper-based superconductors. But if successful, we can not expect a high value of T_c, because doping moves E_F away from the vHs, which would be unfavorable for reaching high T_c values.

Clearly, further investigations of this compound are very interesting from the fundamental point of view. If a superconducting state (even with low T_c value) can be obtained with this material by taking into account our considerations, it may further enhance the role of the van Hove singularity in achieving the optimum T_c and the ability of LDA band structure approaches to analyze and predict new and "optimal" high-T_c materials.

2.5 Quadruple perovskites $Ln'Ln''Ba_2Cu_2Ti_2O_{11}$ as possible good candidates for HTCS: Role of oxygen defects

In the search for new and better high-temperature superconductors, many new families of layered perovskite structures have been discovered. One of them is the family of the quadruple perovskites $Ln'Ln''Ba_2Cu_2Ti_2O_{11}$ (Ln = lanthanide, Y). et al. [48, 49, 50, 51, 52, 53, 54]. Recently, a comprehensive study of the crystal chemistry of $Ln_2Ba_2Cu_2Ti_2O_{11}$ was carried out and in addition, new layered materials with mixed lanthanide site stoichiometries $Ln'Ln''Ba_2Cu_2Ti_2O_{11}$ were reported [55]. These materials satisfy some general conditions for the occurrence of high-temperature superconductivity: they contain complete CuO_2 planes and a "charge reservoir" consisting of a double perovskite structure composed of Ln', Ti and O. Thus, they may show promise to be superconductors if appropriate doping of the CuO_2 planes is achieved. Since recent attempts to substitute Gd by Ca in $Gd_2Ba_2Ti_2Cu_2O_{11}$ [56] have not been successful, the question of whether it is possible to make these materials superconducting still remains open.

The crystal structure of $LaYBa_2Cu_2Ti_2O_{11}$ [55] is tetragonal and belongs to a P4/mmm space group. It was found [55] that Y almost exclusively occupies the 8-coordinate site between Cu-O planes. We label the oxygen positions in the following way: $O(1)$ is located in La-O plane, $O(2)$ in the Ti-O plane, $O(3)$ is an "apical" oxygen with respect to Cu-O planes and $O(4)$ together with Cu atoms form the Cu-O layers. We have performed calculations [57] for (i) the defect-free $LaYBa_2Cu_2Ti_2O_{11}$ structure, (ii) for the structure containing oxygen between the Cu-O planes and (iii) for a structure containing 50% $O(2)$ vacancies.

Surprisingly enough, the band structure in the vicinity of E_F is very simple and looks like the band structure typical to all Cu-O based high-T_c superconductors. Only the quasi two-dimensional Cu-O derived free-electron-like $dp\sigma$ bands cross E_F. Again, as for all Cu-O based high-T_c superconductors, the only contribution to the DOS at E_F comes from the Cu $d-$ and O(4) $p-$ (in-plane oxygen) states. The prominent van Hove singularity is clearly seen on the DOS at 0.33 eV below E_F. The DOS composition at E_F is also typical for high-T_c superconductors — it mainly consists of Cu 3d and O(4) 2p states. Thus, from these features of the band structure we may expect $LaYBa_2Cu_2Ti_2O_{11}$ and most probably other members of the $Ln'Ln''Ba_2Cu_2Ti_2O_{11}$ family (since their crystal structure and valence electron concentration are practically the same) to be able to display superconductivity, if properly doped.

If we associate the "optimum doping" with the position of the Fermi energy in the close vicinity of the vHs, which seems to be quite relevant [34, 27], we predict the hole doping to be about ~ 0.6 holes per unit cell, or ~ 0.3 holes per Cu-O layer. Most likely, such a doping can be achieved by substituting mono- or divalent atoms for Y and/or lanthanides.

Next we investigated the effect of these imperfections on the electronic structure of $Ln'Ln''Ba_2Cu_2Ti_2O_{11}$. The main differences in the results from those for ideal $LaYBa_2Cu_2Ti_2O_{11}$ are: *i)* the lowering of the antibonding part of the Ti 3d states down to E_F because of the change in electrostatic potential of the Cu-O layers and *ii)* the appearance of new inter-layer oxygen states at E_F. Since these oxygen states contribute heavily to the former Cu-O $pd\sigma$ band, they change its character. It is interesting to note that this one extra oxygen per unit cell, which one can think would heavily dope the Cu-O $pd\sigma$ band by holes, does not change much the energy distance between the vHs and E_F. This suggests that the extra oxygen in between the Cu-O layers is mostly in its atomic-like state. Consequently, any doping coming from the "charge reservoir" would most probably first localize on that interlayer oxygen and prevent the $pd\sigma$ band from being optimally doped. Also, the incorporation of this extra oxygen apparently changes the two-dimensional character of the $pd\sigma$ band, which is thought to be responsible for the high-T_c properties of Cu-O layered ceramics. Thus, we can speculate on why the attempts to make $Gd_2Ba_2Ti_2Cu_2O_{11}$ superconducting have failed. Most probably, in the process of doping some additional oxygen is incorporated between the Cu-O layers and this alters the character of the states at E_F and destroys the main condition for high-T_c superconductivity.

Finally, using a supercell approach, we examined [57] the electronic structure of $LaYBa_2Cu_2Ti_2O_{11}$ containing 50% oxygen vacancies at the O(2) site. The band structure shows that additional bands appear in the vicinity of E_F; the Ti d states shift down in energy and now contribute to the DOS at E_F. The nature of those states can be understood from their specific space localization. A detailed analysis of the charge distribution coming from these states shows that they are mainly localized in the region of the Ti-O plane.

The appearance of "vacancy states" below E_F is caused by rearrangement of electron states around oxygen vacancies with a decisive role played by changes of wave functions of atoms closest to the vacancy Ti atoms. Specifically, the appearance of an O(2) vacancy leads to the breakdown of direct Ti-O bonding and to the creation new Ti-Ti bonds through the vacancy region. This was also found to be a common feature of the electronic structure of carbides and nitrides of transition metals containing metalloid vacancies [58]. Such an analysis suggests that "extra" electrons created in the Ti-O layers by oxygen vacancies will not be able to move to the Cu-O layers and change the carrier concentration since they stay in the vicinity of O-vacancies taking part in Ti-Ti bonding.

Thus, we should conclude that oxygen vacancies in the Ti-O plane are also not good for superconductivity since they (i) do not contribute to the doping ability of the "charge reservoir" and (ii) destroy the two-dimensional character of the Cu-O states at E_F.

3 Possible physical consequences of vHs existing close to E_F

3.1 Effects of pressure on the A_{1g} phonon frequency

Since the only available experimental data on the pressure dependence of the phonon modes in Hg-1201 are for the A_{1g} Raman active mode [59], we concentrate our attention on it [60]. As follows from a symmetry analysis, the A_{1g} phonon represents O(2) and Ba vibrations along the z direction. We have calculated the total energy (frozen phonon approximation) for a sufficient number of independent O(2) and Ba displacements to make a fit by a sixth-order polynomial. Such a high order polynomial was needed to describe satisfactorily the strongly anharmonic total energy surface.

The calculated O(2) and Ba A_{1g} phonon frequencies for different pressure, $\omega_{O(2)}(P)$ and $\omega_{Ba}(P)$, are plotted in Fig. 3.1. Frequencies for both atoms start increasing linearly with pressure. Both display a striking feature — a dip that occurs around the pressure at which the vHs passes through E_F. After that, they proceed linearly with further increase of pressure. Since both $\omega_{O(2)}(P)$ and $\omega_{Ba}(P)$ show two distinct linear regions (I and II), we fit the calculated points with different linear functions in those regions. The fitting coefficients are listed in the inset in Fig. 3.1. From this fit we estimate the frequencies corresponding to zero pressure to be $\omega_{O(2)} = 586$ cm^{-1} and $\omega_{Ba} = 158$ cm^{-1} and their rate of increase with pressure as $d\omega_{O(2)}/dP = 3.9$ cm^{-1}/GPa and $d\omega_{Ba}/dP = 1.2$ cm^{-1}/GPa for pressure region I and 1.68 cm^{-1}/GPa and 0.23 cm^{-1}/GPa for pressure region II. These zero pressure frequencies appear to be in excellent agreement with experiment [61, 62, 63] and in good agreement with those calculated previously [64] at the experimental lattice parameters (cf. Table 3.1).

In order to further compare our results on the pressure dependence of the phonon modes with experiment [59], we have calculated the square of the normalized frequencies $(\omega/\omega_0)^2$. The values of $\Delta(\omega^2/\omega_0^2)/\Delta(P)$ were found from a linear least square fit (again in two distinct pressure regions, I and II). For O(2), the A_{1g} mode $\Delta(\omega^2/\omega_0^2)_{O(2)}/\Delta(P)$ is about 0.014 GPa^{-1} which is in strikingly

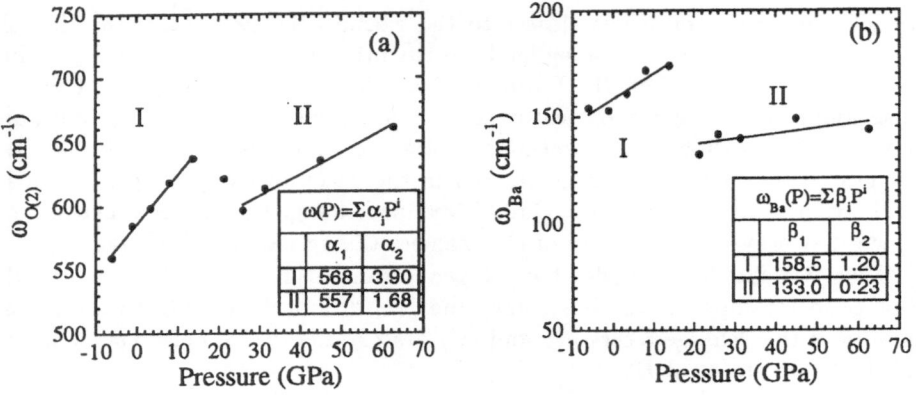

Figure 3.1: The calculated (a) O(2) and (b) Ba A_{1g} phonon frequency ($\omega_{O(2)}(P)$ and $\omega_{Ba}(P)$) for different pressures and coefficients of a linear least square fit to them in different pressure regions (I and II).

good agreement with the experimental value of 0.015 GPa^{-1} [59] for Hg-1201 over the pressure range from 0 to 10 GPa. The corresponding value in the "high pressure region" (II) is 0.006 GPa^{-1}. For $\Delta(\omega^2/\omega_0^2)_{Ba}/\Delta(P)$ we have 0.016 GPa^{-1} and 0.003 GPa^{-1} for the I and II pressure regions, respectively. Since we reproduce the experimental trend for the low pressure part of $\omega(P)$, we may be quite confident in our results for higher pressures, including the softening of the A_{1g} mode in the pressure region where the vHs comes close to and passes through E_F.

Despite our initial hope to interpret the results of phonon frequency measurements for Hg-1223 [59] on the basis of our calculations for Hg-1201, we can not clearly see the reason for the kink in $(\omega/\omega_0)^2$ at pressures around 5 GPa for Hg-1212. It was previously supposed [59] that this kink originates from the fact that Hg-O(2) derived band (which is located above E_F in Hg-1201 and slightly below E_F in Hg-1212 and Hg-1223 [27]) moves down with pressure [65], becomes populated and changes the strength of Hg-O(2) bond, which in turn

Table 3.1: Comparison of experimental and theoretical frequencies of the Raman active A_{1g} phonon mode. FLAPW, SM and FLMTO refer to full-potential linearized augmented plane wave method, shell model and full-potential LMTO method, respectively.

A_{1g} mode	exp. [61]	exp. [62]	exp. [63]	FLAPW [59]	SM [64]	FLMTO [64]	FLMTO [this work]
O(2)	592	592	591	587	591	540	586
Ba	161	158	—	—	124	161	158

affects the frequency of O(2) vibrations. But our calculations do not support this idea. Moreover, since this Hg-O(2) derived band is *anti-bonding*, one might expect a *softening* of the Hg-O(2) bond and thus a decrease in the O(2) phonon frequency. Further, the effect of filling the Hg-O(2) derived band on the calculated O(2) phonon frequency was not observed in our simulations. Hence, we have to conclude that the change in this band population is too small to produce an observable effect and so this particular result of the phonon frequency measurements [59] still remains unresolved.

3.2 Possible experiments on vHs affected properties

As already mentioned, vHs based models of high-T_c superconductivity have recently attracted considerable attention [66]. The models try to establish the connection between anomalies in the physical properties of the high-T_c materials and the existence of a vHs close to E_F. As we have shown previously for the infinite layer compound [16] and for Hg-1201 and Hg-1212 [27, 26], there is an apparent correlation between T_c as a function of doping and/or pressure and a maximum in the density of states, located close to E_F. Or in other words, the behavior of T_c depends strongly on the vHs passing close to or through E_F; the latter case is known as an electronic topological transition (ETT) [17]. The direct proof of this hypothesis would be experimental observations of those properties of high-T_c materials that are sensitive to this ETT. These include thermopower [67], bulk modulus, phonon frequencies and thermal expansion coefficients [68, 69, 70, 71].

The thermopower α (in the non-superconducting state) must display very sharp anomalies of the type $\delta\alpha \sim T\partial N(E_F)/\partial E_F$ [67] as a function of pressure or doping. Unfortunately, these anomalies can be seen clearly only at very low temperatures (see e.g. [72, 73]) and so they might be difficult to measure in superconductors with high critical temperatures. One possible way to detect an ETT in such systems is to conduct low temperature measurements with the superconductivity suppressed by a high magnetic field. However, it might be more convenient to detect anomalies in the lattice properties at a not too low temperature. Corresponding expressions for ETT in the two-dimensional case (*i.e.* the logarithmic anomaly in $N(E_F)$ caused by a saddle point) were derived in an almost-free-electron approximation in Ref. [74] in connection with a discussion of magneto-volume and magneto-elastic effects in itinerant ferromagnets.

We propose therefore [60] the following "tests" to detect ETT in high-T_c materials when subjected to either pressure or doping. One should look: (1) For a minimum in the phonon frequencies of the A_{1g} mode. It appears that the position of the vHs is especially sensitive to the deformation along the c axis, and thus anomalies in this phonon mode should be especially strong. (2) For a minimum in the thermal expansion coefficient. (3) For strong anomalies in the thermopower at low temperatures when superconductivity is suppressed by high magnetic fields.

We have to mention also that it is quite possible that the first type of anomaly has already been observed, although indirectly. It was reported [29] that the maximum in T_c correlates with the maximum in the ratio of mean square amplitudes of Ba displacement along the c axis and in the ab plane. At first glance, this looks puzzling since Ba states do not contribute to the DOS at E_F and apparently have nothing to do with superconductivity. By symmetry, however, Ba vibrations are connected to O(2) vibrations and their softening by ETT causes the amplitude enhancement of the thermal vibrations in the c direction. At the same time, Ba vibrations in the ab plane are practically insensitive to the position of the vHs.

4 Acknowledgments

This work was supported by the National Science Foundation (through the Northwestern University Science and Technology Center for Superconductivity, Grant No. DMR 91-20000, and by a grant of computer time at the Pittsburgh Supercomputing Center supported by its Division of Advanced Scientific Computing).

References

[1] L. Smedskjaer et al., Physica C **156**, 269 (1988).

[2] H. Haghighi et al., in *The University of Miami Workshop on: Electronic Structure and Mechanisms for High Temperature Superconductivity* (University of Miami, Miami, 2-9 January 1991).

[3] F. Mueller, J. Phys. Chem. Solids **52**, 1457 (1991).

[4] J. Yu, A. Freeman, and J. Xu, Phys. Rev. Lett. **58**, 1035 (1987).

[5] J. Xu, T. Watson-Yang, J. Yu, and A. Freeman, Phys. Rev. Lett. **120**, 489 (1987).

[6] This literature is so vast that we cite only some recent references and refer to these for earlier work; J.E. Hirsh and D.J. Scalapino, Phys. Rev. Lett. **56** (1986) 2735; J. Friedel, J. Phys. Condens. Matter **1** (1989) 7757; R.S. Markiewicz, Physica C **200** (1992) 65; D.M. Newns, et al., Phys. Rev. Lett. **69** (1992) 1264; K. Levin, et al., in "Electronic Structure and Mechanisms for High Temperature Superconductivity", edited by J. Ashkenazi and G. Vezzoli (Plenum, New York) 1992 p. 481.

[7] E. Wimmer, H. Krakauer, M. Weinert, and A. J. Freeman, Phys. Rev. B **24**, 864 (1981), and references therein.

[8] M. Methfessel, Phys. Rev. B **38**, 1537 (1988).

[9] M. Azuma et al., Nature **356**, 775 (1992).

[10] T. Siegrist, S. M. Zahurak, D. W. Murphy, and R.S.Roth, Nature (London) **334**, 231 (1988).

[11] Z. Hirori, M. Azuma, M. Takano, and Y. Bando, J. Solid State Chem. **95**, 230 (1991).

[12] L. Mattheiss and D. Hamman, Phys. Rev. B **40**, 2217 (1989).

[13] D. Singh et al., Physica B **163**, 470 (1990).

[14] M. A. Korotin and V. I. Anisimov, Mat. Lett. **10**, 28 (1990).

[15] S. Hatta, R. V. Kasowski, and W. Y. Hsu, Appl. Phys. **A55**, 508 (1992).

[16] D. L. Novikov, V. A. Gubanov, and A. J. Freeman, Physica C **210**, 301 (1993).

[17] I. M. Lifshitz, Sov. Phys. - JETP **11**, 1130 (1960).

[18] S. Massidda, N. Hamada, J. Yu, and A. Freeman, Physica C **157**, 571 (1989).

[19] J. Yu, S. Massidda, and A. Freeman, Physica C **152**, 273 (1988).

[20] S. Massidda, J. Yu, A. Freeman, and D. Koelling, Phys. Lett. A **122**, 198 (1987).

[21] J. Yu, S. Massidda, A. Freeman, and D. Koelling, Phys. Lett. A **122**, 203 (1987).

[22] A. A. Manuel and et al., Europhys. Lett. **6**, 61 (1987).

[23] L. Dagens, J. Phys. F **8**, 2093 (19987).

[24] D. L. Novikov and A. J. Freeman, Physica C **219**, 246 (1993).

[25] H. Shaked et al., Phys. Rev. B **50**, 12752 (1994).

[26] D. L. Novikov and A. J. Freeman, Physica C **212**, 233 (1993).

[27] D. L. Novikov and A. J. Freeman, Physica C **216**, 273 (1993).

[28] J. L. Wagner et al., Physica C **210**, 447 (1993).

[29] Q. Huang, J. W. Lynn, Q. Xiong, and C. W. Chu, Phys. Rev. B **52**, 462 (1995).

[30] C. Tsuei et al., Phys. Rev. Lett. **69**, 2134 (1992).

[31] J. Yu, S. Massidda, and A. J. Freeman, Physica C **152**, 273 (1988).

[32] D. R. Hamann and L. F. Mattheiss, Phys. Rev. B **38**, 5138 (1988).

[33] D. J. Singh and W. E. Pickett, Physica C **203**, 193 (1992).

[34] R. S. Markiewicz, Physica C **217**, 381 (1993).

[35] M. Al-Mamouri, P. P. Edvards, C. Greaves, and M. Slaski, Nature **369**, 382 (1994).

[36] Z. Hiriri, N. Kobayashi, and M. Takano, Nature **371**, 139 (1994).

[37] D. L. Novikov, A.J.Freeman, and J.D.Jorgensen, Phys. Rev. B **51**, 6675 (1994).

[38] A. A. Abrikosov, J. C. Campuzano, and K. Gofron, Physica C **214**, 73 (1993).

[39] M. T. Andersen and K. R. Poeppelmeier, Chem. Mater. **3**, 476 (1991).

[40] J. B. Wiley *et al.*, J. Solid State Chem. **87**, 250 (1990).

[41] J. T. Vaughey, J. B. Wiley, and K. R. Poeppelmeier, Z. Anorg. Allg. Chem. **598/599**, 372 (1991).

[42] J. T. Vaughey *et al.*, Chem. Mater. **3**, 935 (1991).

[43] G. Roth *et al.*, J. Phys. **1**, 721 (1991).

[44] R. J. Cava *et al.*, Physica C **185-198**, 180 (1991).

[45] T. Krekels *et al.*, J. Solid State Chem. **105**, 313 (1993).

[46] M. T. Andersen, K. R. Poeppelmeier, S. A. Gramash, and J. K. Burdett, J. Solid State Chem. **102**, 164 (1993).

[47] D. L. Novikov, A. J. Freeman, K. R. Poeppelmeier, and V. P. Zhukov, Physica C **252**, 7 (1995).

[48] A. Gomezano and M. T. Weller, J. Mater. Chem. **3**, 771 (1993).

[49] A. Gomezano and M. T. Weller, J. Mater. Chem. **3**, 979 (1993).

[50] K. B. Greenwood *et al.*, Physica C **235-240**, 349 (1994).

[51] P. A. Salvador *et al.*, J. Solid State Chem. (1995), in press.

[52] M. R. Palacin, A. Feurtes, N. Casan-Pastor, and P. Gómez-Romero, Adv. Mater. **6**, 54 (1994).

[53] P. Gómez-Romero, M. R. Palacin, and J. Rodrigues-Carvajal, J. Chem. Mater. **6**, 2118 (1994).

[54] M. R. Palacin, F. Krumeich, M. T. Caldés, and P. Gómez-Romero, J. Solid State Chem. (1995), in press.

[55] K. B. Greenwood *et al.*, J. Chem. Mater. **7**, 1355 (1995).

[56] A. Fukuoka *et al.*, Physcia C **231**, 372 (1994).

[57] D. L. Novikov, A. J. Freeman, and K. R. Poeppelmeier, .

[58] A. L. Ivanovsky, V. P. Zhukov, and V. A. Gubanov, *Electronic Structure of Refractory Carbides and Nitrides* (University Press, Cambridge, 1994).

[59] I.-S. Yang *et al.*, Phys. Rev. B **51**, 644 (1995).

[60] D. L. Novikov *et al.*, Phys. Rev. B , submitted.

[61] M. C. Krantz, C. Tompsen, H. Mattausch, and M. Cardona, Phys. Rev. B **50**, 1165 (1995).

[62] N. H. Hur *et al.*, Physica C **218**, 365 (1993).

[63] Y. T. Ren *et al.*, Physica C **226**, 209 (1994).

[64] M. G. Stachiotti *et al.*, Physica C **243**, 207 (1995).

[65] D. L. Novikov and A. J. Freeman, Physica C **222**, 38 (1994).

[66] A. A. Abrikosov, Physica C **244**, 243 (1995).

[67] V. G. Vaks, A. V. Trefilov, and S. V. Fomichev, Sov. Phys. - JETP **53**, 830 (1981).

[68] V. G. Vaks and A. V. Trefilov, J. Phys. F **18**, 213 (1988).

[69] V. G. Vaks and A. V. Trefilov, J. Phys: Cond. Matter **3**, 1389 (1991).

[70] M. I. Katsnelson, I. I. Naumov, and A. V. Trefilov, Phase Transitions B **49**, 143 (1994).

[71] V. I. Nizhankovski, M. I. Katsnelson, G. V. Peschanskikh, and A. V. Trefilov, Sov. Phys. - JETP Lett. **59**, 733 (1994).

[72] V. S. Egorov and A. I. Fedorov, Sov. Phys. - JETP Lett. **58**, 959 (1983).

[73] N. V. Bashkatov and N. L. Sorokin, Sov. Phys. Sol. State **31**, 910 (1989).

[74] V. Y. Irkhin, M. I. Katsnelson, and A. V. Trefilov, J. Magn. Magn. Mater. **117**, 210 (1992).

Localized–Itinerant Electronic Transitions in Perovskites

John B. Goodenough

Center for Materials Science & Engineering, ETC 9.102 University of Texas at Austin, Austin, TX 78712-1063

Introduction

Single–valent transition–metal (M) oxides with perovskite–related structures have localized or itinerant d^n electron configurations depending on the strength of $ca.$ 180^0 metal–oxygen–metal (M–O–M) interactions relative to the effective on–site electrostatic Coulomb energy U_{eff} separating the empty d^{n+1} configuration from the highest occupied orbital. A relatively small U_{eff} is found where addition of a d electron does not require overcoming either a cubic crystal–field splitting Δ_c or an interatomic exchange energy Δ_{ex}; it may also be small where the highest occupied orbital has a large O–$2p$ component, which makes U_{eff} a charge–transfer gap. The conditions for a small U_{eff} are found for the $3d^1$ configurations of Ti^{3+} and V^{4+}, for the octahedral–site high–spin $3d^4$ configurations of Mn^{3+} and Fe^{4+}, and for the $3d^9$ configuration of Cu^{2+} in four, five, or six-fold oxygen coordination. Accordingly, the mixed–valent systems $La_{1-x}Sr_xTiO_3$, $La_{1-x}Sr_xVO_3$, $La_{1-x}Sr_xMnO_3$, $La_{1-x}Sr_xFeO_3$, and $La_{2-x}Sr_xCuO_4$ all exhibit transitions from magnetic insulators at $x = 0$ to itinerant–electron behavior with increasing x; but each transition proceeds in a different manner. U_{eff} is also small for octahedral–site Co(III), and the system $La_{1-x}Sr_xCoO_3$ is further complicated by a low–spin to high–spin transition in the parent $LaCoO_3$ that transforms to an intermediate–spin configuration on the cobalt in the mixed-valent compositions. Single–valent $NdNiO_3$ exhibits a global first–order transition from magnetic insulator to Pauli paramagnetic metal whereas strongly and weakly correlated electrons coexist in the system $Sr_{1-x}Ca_xVO_3$, a finding that suggests the presence of correlation fluctuations. It is argued that the superconductive system $La_{2-x}Sr_xCuO_4$ is distinguished by the presence of a vibronic coupling. In the superconductive compositions, the vibronic coupling appears to stabilize large ($ca.$ 5 Cu centers), nonadiabatic polarons that condense into a polaron liquid below room temperature and form an ordered polaron array within a CuO_2 sheet below a $T_d \geq T_c$; ordering between sheets may be required for superconductivity.

1   Structural Considerations

The cubic AMO_3 structure of Fig.1(a) contains 180^0 M–O–M bonds that dominate the interactions between transition–metal cations M; each oxygen atom has four coplanar A–cation near neighbors and two apical M cations. A measure of the matching of A–O and M–O equilibrium bond lengths is the tolerance factor

$$t = (A - O)/\sqrt{2}\,(M - O) \qquad (1)$$

which may be calculated with empirical ionic radii obtained from oxide lattice parameters at ambient pressure and temperature. A different compressibility and thermal expansion of each equilibrium bond length makes $t = t(P, T)$. In all known cases, t increases with T because of a larger thermal expansion of the A–O bond. Normally a more compressible A–O bond makes t decrease with increasing pressure (1); but in the special case that the oxygen atom resides in a double–well potential between the two M atoms, an unusually large compressibility of the M–O–M bonds may make $dt/dP > 0$ [2]. This situation is found, for example, in the superconductive copper oxides.

A $t < 1$ places the M–O bonds under compression, the A–O bonds under tension; a $t > 1$ does the inverse. The structure can relieve the internal stresses created by a $t < 1$ with a cooperative distortion of either the MO_6 octahedra to rhombohedral symmetry, as in the $LaNiO_3$ structure of Fig.1(b), or a cooperative rotation of the MO_6 octahedra as in the tetragonal distortion of low–temperature $SrTiO_3$, Fig.1(c), or the orthorhombic distortion of $GdFeO_3$, Fig.1(d); in each case the M–O–M bond angles bend from 180^0. Accommodation to a $t > 1$ is made by a progressive, ordered change from all–cubic to all–hexagonal stacking of the AO_3 close–packed planes in a sequence of hexagonal polytypes [1]; but this structural change occurs at the expense of creating face–shared MO_6 octahedra, an energetic cost that increases with the formal charge on the M cation.

The structure of Fig.1(a) may also be viewed as an alternate stacking along an [001] axis of MO_2 and rock–salt AO planes. From this perspective, it should be possible to stabilize more complex intergrowth structures in which an AO rock–salt plane of one layer interfaces an MO_2 sheet of another layer. The T–tetragonal K_2NiF-4 structure of high–temperature La_2CuO_4, Fig.2(a), is the simplest such intergrowth; it is the $n = 1$ end member of the Ruddlesden–Popper sequence of alternating rock–salt and perovskite layers $AO \cdot (AMO_3)_n$.

An A_2MO_4 compound can also accommodate a $t < 1$ by bending the M–O–M bond angle of the MO_2 layers from 180^0 in a cooperative rotation of the MO_6 octahedra, arrows in Fig.2(b), as occurs in La_2CuO_4 below an orthorhombic–tetragonal transition temperature $T_t \approx 540$ K. A different cooperative rotation of the CuO_6 octahedra results in a low–temperature tetragonal phase with a larger a–axis lattice parameter in $La_{1.88}Ba_{0.12}CuO_4$ [3].

As illustrated by Nd_2CuO_4, an A_2MO_4 compound accommodates a $t > 1$ by a change of the $(AO)_2$ rock–salt layer to an $A-O_2-A$ fluorite layer to form the T'–tetragonal structure of Fig.2(c). In practice, some oxygen atoms of an $A-O_2-A$ fluorite layer tend to be displaced to the A–atom planes above an M atom, a position also available to excess oxygen; the presence of these apical oxygen at the M atoms creates fivefold instead of fourfold coordination at the M atoms, which perturbs the periodic potential of an MO_2 plane. Perturbations of the periodic potential by two types of M–atom oxygen coordination suppresses

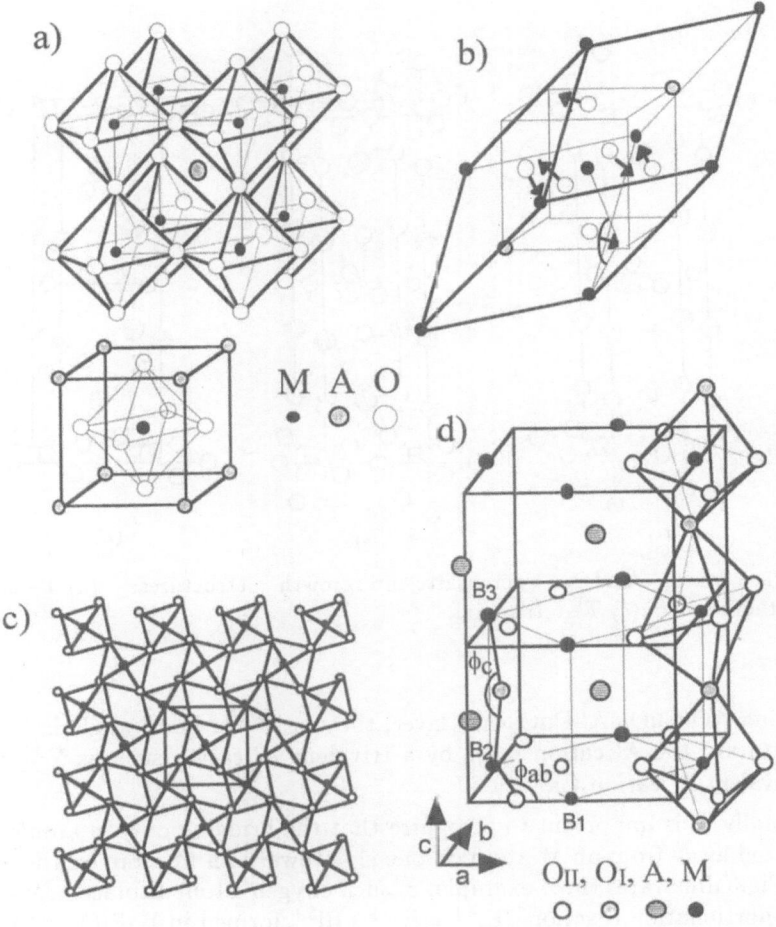

Fig. 1. Some AMO_3 perovskite structures: (a) cubic, (b) rhombohedral ($R3c$), (c) tetragonal, and (d) orthorhombic ($Pbnm$).

superconductivity. The Aurivillius intergrowth structures $Bi_2O_3 \cdot A_{n-1}(MO_3)_n$ contain oxygen in both sites in the $BiO \cdot O_2 \cdot BiO$ layers.

The perovskite structure can also accommodate either cation or anion vacancies. Of particular interest for the copper–oxide superconductors is the preferential removal of oxygen from planes of smaller A cations stable in eightfold oxygen coordination. This type of oxygen–vacancy ordering is found where the M cation has a d^9 configuration as at Cu^{2+} or Ni^+; in this case, a cooperative ordering of the d-configuration hole into $x^2 - y^2$ σ-antibonding orbitals of an MO_2 plane introduces a global tetragonal component to the structure that is compatible with a cooperative removal of an apical oxygen on the z-axis. In the

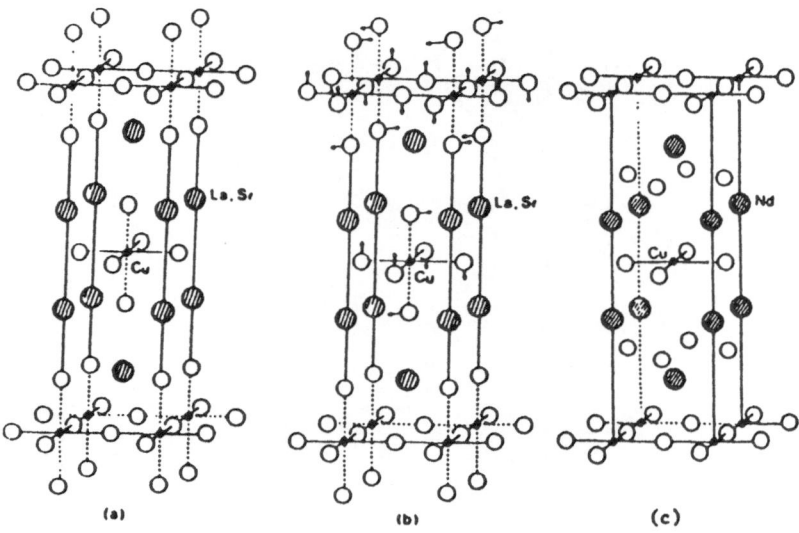

Fig. 2. Some A_2MO_4 perovskite–intergrowth structures: (a) T–tetragonal, (b) orthorhombic, (c) T'–tetragonal.

resulting $AO\cdot CuO_2\cdot A'\cdot CuO_2\cdot AO$ layer, the Cu–O–Cu bond angle is bent from 180^0 toward the A' cation more by a trivalent A' cation such as Y^{3+} than by the divalent A' cation Ca^{2+}.

Finally, it is important to recognize that the bridging oxygen atoms may be displaced away from an M atom on one side toward an M atom on the opposite side. Fig. 3 illustrates three examples of such oxygen–atom displacements: (a) the disproportionation reaction $2Bi^{4+} = Bi^{3+} + Bi^{5+}$ formed in $BaBiO_3$, (b) the ferroelectric transition in tetragonal $BaTiO_3$, and (c) the cooperative Jahn–Teller distortion of $LaMnO_3$. In each case, the oxygen atom sits in a double–well potential between a longer equilibrium M–O bond that is more ionic and a shorter equilibrium M–O bond that is more covalent. In these three examples, the cooperativity of the oxygen–atom displacements is long–range and static, which makes them directly detectable by diffraction techniques. In this paper, indirect evidence is presented for cooperative oxygen–atom displacements that are short–range and may be dynamic. It is further argued that a vibronic coupling associated with a pseudo Jahn–Teller deformation of a Cu center may stabilize, in the presence of a double–well potential for the oxygen atoms, non–adiabatic polarons containing about 5 Cu centers in a CuO_2 sheet of the high–T_c copper–oxide superconductors. In the superconductive compositions, elastic coupling between polarons condenses a polaron gas above room temperature into a polaron liquid below room temperature and, below T_c, into a polaron solid. Evidence is also presented for a vibronic dispersion curve $\varepsilon(\mathbf{k})$ that is flattened in the direction of the Cu–O bonds of a CuO_2 sheet; straightening the Cu–O–Cu bond angle enhances the vibronic coupling, which flattens the $\varepsilon(\mathbf{k})$ curve further, thus raising T_c and the thermopower.

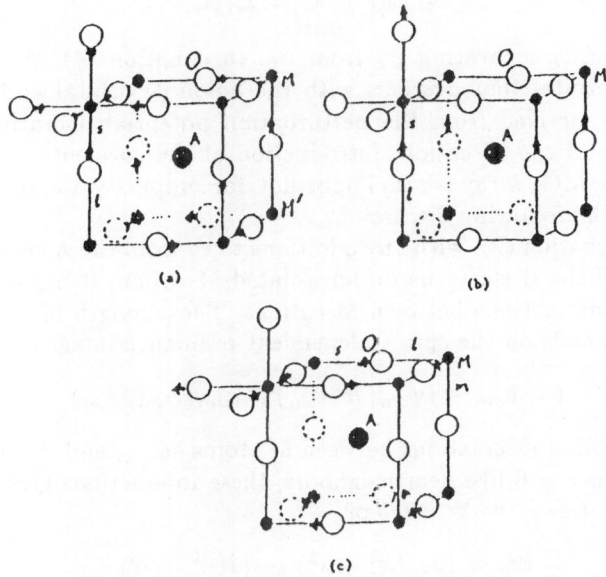

(a)   (b)

(c)

Fig. 3. Illustrative cooperative displacements of oxygen towards one neighboring M atom and away from another found in (a) monoclinic $BaBiO_3$, (b) tetragonal $BaTiO_3$, and (c) $LaMnO_3$.

2 Electronic Considerations

The d–electron orbitals of a transition–metal cation M are split by an octahedral–site cubic field into twofold–degenerate e orbitals x^2-y^2 and $[(z^2-x^2)+(z^2-y^2)]$ (abbreviated as z^2) and threefold–degenerate t_2 orbitals xy and $yz\pm izx$; choice of the local M–O z–axis is determined where there is an ordering of electrons among partially occupied d orbitals that further removes any residual configuration degeneracy of the ground state.

In an ionic model, covalent mixing of O–$2s$, $2p$ and M–nd states is introduced *via* second–order perturbation theory. The resulting ligand–field e and t_2 orbitals remain localized at the M atoms and are described by the wave functions:

$$\psi_e = N_\sigma(f_e + \lambda_\sigma\phi_\sigma + \lambda_s\phi_s)\,,$$
$$\psi_t = N_\pi(f_t + \lambda_\pi\phi_\pi)\,, \tag{2}$$

where ϕ_s, ϕ_σ, and ϕ_π are appropriately symmetrized O–$2s$, O–$2p_\sigma$, and O–$2p_\pi$ wave functions of neighboring oxygen atoms. The covalent-mixing parameters

$$\lambda_i = b^{ca}/\Delta E \tag{3}$$

are the ratio of the M–O resonance integral

$$b^{ca} \equiv (\psi_d, H'\phi_i) \approx \varepsilon_i(\psi_d, \phi_i) \tag{4}$$

to the energy ΔE_i separating ψ_d from ψ_i. In equation (4), Ψ_d is the lowest unoccupied d orbital that overlaps with the anion ϕ_i orbital and ε_i is a one–electron energy derived from the perturbation potential operating on ϕ_i due to the presence of the M cation. Introduction of the covalent mixing of the A cations with the MO$_3$ array is also important; for simplicity, we introduce it only as required in the following discussion.

Although equation (2), with any additions to ψ_t from the A cations, provides a description of the d electrons on an isolated M cation, it neglects the dominant M–O–M interactions between M cations. The strength of these neglected interactions depends on the spin–independent resonance integrals

$$b = b_{mn} \equiv (\psi_{dm}, H'\psi_{dn}) \approx \varepsilon_{mn}(\psi_{dm}, \psi_{dn}) \tag{5}$$

for nearest–neighbor interactions between M atoms at R_m and R_n. For a simple–cubic array with $z = 6$ like near neighbors, these interactions give independent tight–binding bandwidths $W \approx 2zb$ of

$$W_\sigma \approx 12\varepsilon_\sigma(\lambda_\sigma^2 + \lambda_s^2) \cos(180^0 - 2\theta),$$
$$W_\pi \approx 12\varepsilon_\pi \lambda_\pi^2. \tag{6}$$

The parameter b_σ decreases to zero as the M–O–M bond angle $(180^0 - 2\theta)$ decreases from 180^0 to 90^0; at 90^0 the O–$2p_\sigma$ orbital of the ψ_e for M at R_m is orthogonal to that for the M at $R-n$ whereas at 180^0 the same O–$2p_\sigma$ orbital enters ψ_e for each M atom. Although the M–O–M bond angle does not enter explicitly into W_π, substitution of a smaller isovalent A cation for a larger one reduces the tolerance factor t so as to create a greater bending of the M–O–M bond angle. A smaller isovalent cation is more acidic, which means that the A–O bonding competes more strongly for the O–$2p$ orbital that π bonds with the M atom, thereby reducing λ_π in equation (5) by an "inductive effect;" this effect makes W_π also decrease as the bond bending increases. However, if a double–well potential for the oxygen atom results in a large compressibility of the M–O–M bond, then the increased pressure on this bond imposed by a smaller A cation may induce a transition from the more ionic to the more covalent equilibrium M–O bond length so as to increase $\lambda - \pi$ more than the inductive effect decreases it. As discussed below, the rare–earth orthovanadates RVO$_3$ appear to illustrate this latter situation.

Two limiting descriptions of the M–O–M interactions are available for single–valent MO$_3$ arrays: one applies where the on–site electrostatic energy U required to add another electron to the M–atom d^n configuration is greater than the corresponding d–electron tight–binding bandwidth W_π or W_σ, i.e. $U > W$, whereas the other applies where a $W > U$ renders the electrons itinerant. In the strong correlation Mott–Hubbard limit $U > W$; successive d^{n-1}/d^n and d^n/d^{n+1} redox energies are separated by an energy gap $E_g \approx (U - W)$; the localized d^n configurations give a Curie-Wiess paramagnetism; and the spin-spin interactions between localized d^n configurations can be treated by the superexchange perturbation theory. In the itinerant–electron limit $W > U$, successive redox energies are generally not split; partially filled bands give metallic conduction and a Pauli paramagnetism that is progressively enhanced as the conduction band becomes narrower. Electron correlations are treated by a renormalization to give itinerant quasiparticles of a Fermi liquid rather than itinerant electrons of a Fermi gas. These correlations may be strong enough to just split successive redox energies, rendering a single–valent MO_3 array semiconductive and introducing a net spin density at an M atom, but the splitting is not large enough for the interactions between neighboring M–atom spins to be treated by the superexchange perturbation theory. Of particular interest is whether the transition from one limiting description to the other is smooth or first–order.

It is logical to anticipate that a change from localized to itinerant electronic behavior involves a discontinuous change in the mean electron kinetic energy $\langle T \rangle$. In this case, it follows from the virial theorem

$$2\langle T \rangle + \langle V \rangle = 0 \qquad (7)$$

that such a transition would result in a first–order phase change since a corresponding discontinuous change in the mean electronic potential energy $\langle V \rangle$ would require a discontinuous change in the unit–cell volume across the transition. Recent theoretical calculations [4,5] support this conclusion, which has two important consequences: first, a discontinuous change in the on–site electrostatic energy U must occur across the transition and, second, a discontinuous change in the equilibrium M–O bond length implies a double–well potential at the transition.

3   Transitions in Single–Valent Systems

The instability of an electronic system with bandwidth $W \approx U$ generally leads to a classical phase segregation to stabilize phases with $W > U$ and/or $W < U$. However, where it proves possible to stabilize a compound having a $W \approx 2zb \approx U$, a structural transition may signal either a global transition from the localized to the itinerant electronic regime or an internal segregation into molecular clusters of shorter — or of longer — near–neighbor bonding. In the AMO_3 perovskites, excluding the hexagonal polytypes, the M–O–M bonding dominates within the MO_3 array, and a greater electronic potential energy $\langle V \rangle$ for localized, antibonding electrons is created by an expansion of the M–O bonds. The itinerant–electron phase $W > U$ appears to have not only the greater mean kinetic energy $\langle T \rangle$, but the greater entropy as well. Consequently, global transitions are marked by a lattice dilatation on lowering the temperature through a transition from a Pauli paramagnetic metal to a magnetically ordered, insulating phase. An internal segregation is marked by the trapping of holes in shorter M–O bonds within a molecular cluster described by molecular electronic orbitals and the trapping of electrons in longer M–O bonds that can be described with an ionic model. Where the internal segregation is static, it can be detected by diffraction and/or spectroscopic techniques; where an internal segregation is only quasistatic or dynamic, more indirect techniques may be required to establish a cooperative segregation into domains of shorter and longer equilibrium M–O bond lengths.

The $RNiO_3$ perovskite family, where R is a rare–earth ion, appears to illustrate a global transition from a Pauli paramagnetic metal at high temperatures to an antiferromagnet at lower temperatures. We begin our discussion of this family with the electronic energies of NiO.

Antiferromagnetic NiO has an unoccupied Ni^{2+}/Ni^+ redox couple, the upper Hubbard band of the half–filled e–orbital manifold of the localized Ni^{2+} : $t_2^6 e^2$ $(^4A_{2g})$ configuration, located about 3 eV above the lower Hubbard band as illustrated in the schematic energy diagram of Fig. 4. The ionic model of the Zaanen–Sawatzky–Allen [7] scheme would assign the highest occupied orbitals to an O^{2-} : $2p^6$ band and distinguish the gap as a charge–transfer gap Δ. However, the introduction of covalent mixing between the O–$2p_\sigma$ and Ni–e orbitals places at the top of the occupied bands the most strongly antibonding states, which are states having a ligand–field e parentage. Therefore $\Delta = U_{\text{eff}}$, and at issue is only the relative weights of the Ni–e and O–$2p_\sigma$ states in the lower Hubbard band at the top of the occupied states. On the other hand, it is apparent from Fig.4 that oxidation of Ni^{2+} to Ni^{3+} results in a small ΔE in equation (3) even though a larger Madelung energy must raise the empty state of the Ni^{3+}/Ni^{2+} couple relative to the occupied states. Consequently a $W_\sigma < U$ in NiO may become a $W_\sigma \geq U$ in the perovskite family $RNiO_3$. In fact, it was demonstrated several decades ago [8,9] that the covalent mixing parameter λ_σ of equation (8) is large enough in $LaNiO_3$ to make this end member of the $RNiO_3$ family a Stoner–enhanced, Pauli–paramagnetic metal with low–spin Ni(III): $t_2^6 \sigma_*^1$ ions participating in a quarter–field, antibonding σ_* conduction band of e–orbital parentage in the NiO_3 array.

Substitution of a progressively smaller rare–earth ion in the $RNiO_3$ family bends the Ni–O–Ni bond angle ever further from 180^0, which systematically narrows the width W_σ of the conduction band according to equation (6). As W_σ narrows, synthesis requires hydrostatic pressure, so the $RNiO_3$ family was not studied systematically until recently [6]. Whereas $LaNiO_3$ remains a Pauli paramagnetic metal to lowest temperatures, substitution of a smaller rare–earth atom R for La introduces at lower temperatures a first–order dilatation transition to an antiferromagnetic phase that is a poorer electronic conductor. As the size of the R^{3+} ion decreases, the transition temperature T_t increases, see Ref.[6]. Moreover, where a $T_t > T_N$ is found, the Néel temperature T_N increases with the size of the $R^{3}+$ ion, i.e. with increasing W_σ, even though the electrons are apparently not sufficiently localized to give a cooperative Jahn–Teller deformation

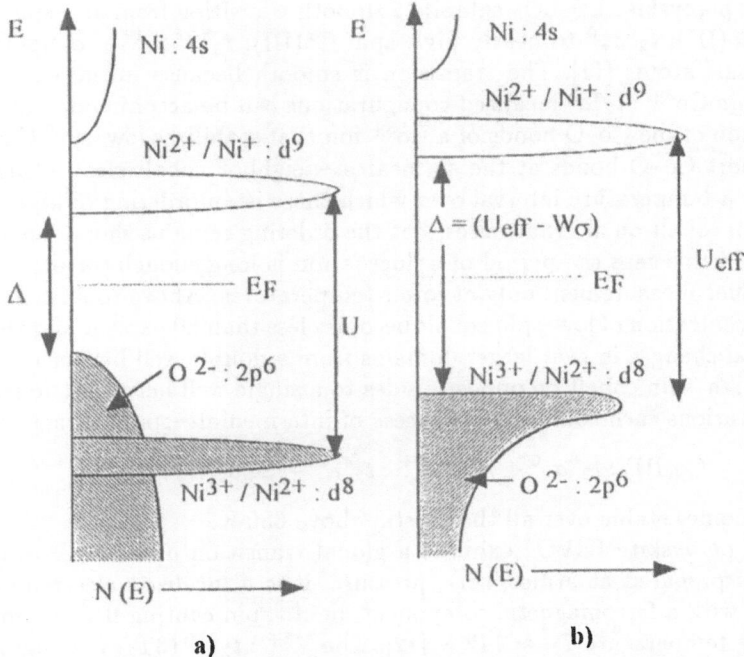

Fig. 4. Schematic electron–energy density of states of NiO: (a) without and (b) with Ni-O covalent mixing.

or to exhibit the ferromagnetism expected of a narrow, quarter–filled band. The intraband–exchange interactions are not strong enough to create parallel spins of all the conduction–band electrons, and a commensurate spin–density wave (SDW) is stabilized instead of ferromagnetism. This observation leaves open the possibility that there could be a second first–order transition to a localized configuration on further narrowing of the σ_* band.

The metallic perovskite $CaFeO_3$ illustrates an internal segregation in which the high–spin $Fe^{4+} : t_2^3\sigma_*^1$ configuration has been found, from Mössbauer measurements [10], to undergo below room temperature a disproportionation reaction

$$2Fe^{4+} : t_2^3\sigma_*^1 = Fe^{3+} : t_2^3e^2 + Fe(v)t_2^3\sigma_*^0 , \qquad (8)$$

in which the localized Fe^{3+}–ion configurations alternate with $Fe(v)O_6$ clusters containing localized t_2^3 configurations, but molecular orbitals of ψ_e parentage. On the other hand, neutron diffraction experiments have not revealed the cooperative oxygen displacements of Fig.3(a) required of a static disproportionation. Therefore it appears that the smooth "negative–U" charge–density–wave transition signaled by the Mössbauer data may be dynamic or quasistatic with a fluctuation period longer than the 10^{-8} sec of a Mössbauer measurement to quite low temperatures.

The perovskite $LaCoO_3$ exhibits a smooth transition from low–spin, diamagnetic Co(III): $t_2{}^6\sigma_*{}^0$ to nearly high–spin Co(III): $t_2{}^{5-\delta}\sigma_*{}^{1+\delta}$ configurations at the cobalt atoms [11]. The transition is smooth because initial excitations to high–spin $Co^{3+} : t_2{}^4e^2$ localized configurations can be accommodated by a local expansion of the Co–O bonds of a Co^{3+} ion that stabilizes low–spin Co(III) ions with short Co–O bonds at the six nearest–neighbor cobalt sites. Consequently there is a temperature interval over which there is an ordering of high–spin and low–spin cobalt on alternate sites, but the ordering remains short–range and dynamic. In this case the period of a fluctuation is long enough for detection by a Mössbauer measurement only at lower temperatures. Above room temperature, the concentration of low–spin cobalt becomes less than 50 percent and the oxygen potential changes in ever larger domains from a double well between a low–spin and a high–spin cobalt on opposite sides to a single well between two like cobalt configurations each somewhat in excess of intermediate–spin character:

$$Co(III) : t_2{}^6\sigma_*{}^0 + rmCo^{3+} : t_2{}^4e^2 \rightarrow 2Co(III) : t_2{}^{5-\delta}\sigma_*{}^{1+\delta} \tag{9}$$

that becomes stable over all the crystal above 650 K.

The perovskite $LaVO_3$ exhibits a global transition under hydrostatic pressure. As prepared at atmospheric pressure, it is a localized–electron antiferromagnet with a ferromagnetic component due to spin canting below a magnetic–ordering temperature $T_N \approx 142$ K [12]. The $V^{3+} : t_2{}^2e^0 (3T_1g)$ configuration in a cubic octahedral field retains an unquenched orbital angular momentum, and ordering of the spins below T_N introduces, via spin–orbit coupling, a cooperative ordering of the localized–electron orbital angular moment below a first–order magnetostrictive transition at $T_t \approx 137$ K $< T_N$ that lowers the symmetry from the orthorhombic structure of Fig.1(d) to monoclinic [13]. A giant magnetocrystalline anisotropy accompanies the magnetostrictive distortion, and below T_t the magnetization versus applied field exhibits no hysteresis to $H = 50$ kOe [14]. More striking is the observation [14-16] of an "anomalous diamagnetism" on cooling from room temperature to 4 K in an $H = 1$ kOe, a phenomenon that was shown [14,17] to be due to a reversal of the magnetization on traversing T_t. This reversal appears to be caused by a discontinuous change in the orbital angular momentum in crossing T_t, a change that induces a reversal of the persistent atomic currents responsible for the orbital angular momentum so as to give an orbital moment in opposition to the magnetizing field in accordance with Lenz's law. For a T_t close to T_N, this change dominates the interaction of the spins with the magnetic field, and a spin–orbit coupling reverses the spin component also. Significantly, the first–order character of the magnetostrictive transition and its associated remarkable magnetic behavior is suppressed in $LaVO_3$ prepared under a hydrostatic pressure $p > 8$ kbar; but the high-pressure sample reverts in a first-order transition back to the ambient–pressure phase on heating [17]. Moreover, substitution for La or Pr or a smaller rare–earth atom in the RVO_3 family also suppresses the first–order character of the transition and the "anomalous diamagnetism." Since introduction of a smaller R^{3+} ion reduces the tolerance factor t of equation (1), it introduces a greater pressure on the V–O bonds that can be only partially relieved by a greater bond bending. In the presence of a double–well potential at a transition from localized to itinerant electronic behavior, the V–O bond is susceptible to a first–order contraction to relieve the bond–length mismatch. Although smaller R^{3+} ions normally decrease λ_π [18], the high compressibility of the localized–electron V–O bond of the VO_3 array dominates the "inductive effect" associated with the greater acidity of the smaller R^{3+} ions.

The coexistence of electrons in a band of incoherent electronic states, precursor of a lower Hubbard band located well below the Fermi energy e_F, and a coherent band of states overlapping e_F has been identified by photoemission spectroscopy in the perovskite system $Sr_{1-x}Ca_xVO_3$ [19] as well as in $LaTiO_3$ and VO_2 [20]. Since these compositions are all Pauli paramagnetic over their metallic temperature range [21,22] where there is no evidence for a static segregation of electronic states, it is reasonable to consider the incoherent band to represent electrons of a dynamic Mott–Hubbard fluctuation within a Fermi liquid phase of coherent states. Particularly instructive are the data for $Sr_{1-x}Ca_xVO_3$; they reveal a monatomic transfer of spectral weight from the coherent to the incoherent band with increasing x, as expected if the smaller Ca^{2+} ion reduces λ_π, and a maximum in the effective mass m^* at an intermediate value of x.

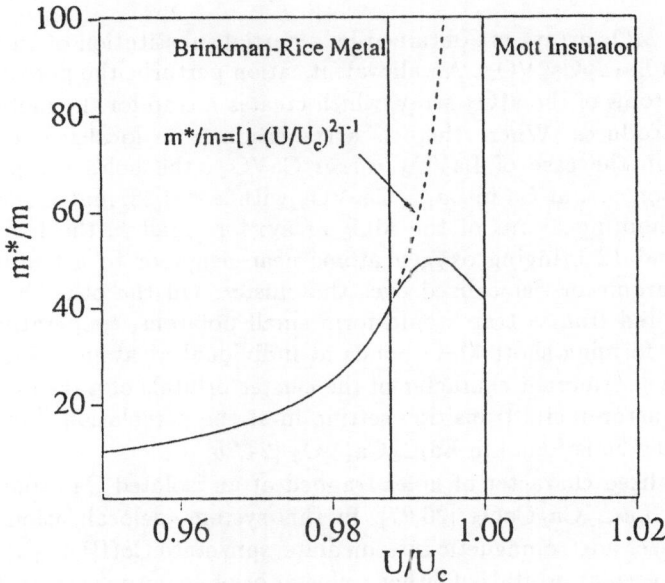

Fig. 5. Schematic variation of normalized effective mass versus normalized Hubbard U for a mass–enhanced metal.

Fig.5 shows a curve of the mass enhancement m^*/m, where m is the lattice mass without enhancement, *versus* the on–site electrostatic Hubbard energy

U normalized to the critical value U_c for the transition to a Mott–Hubbard insulator. As U/U_c increases to 1.0, the on–set of correlation fluctuations between Fermi–liquid (Brinkman–Rice) and Hubbard states, which transfers spectral weight from e_F to the lower Hubbard band, reduces the mass enhancement from the Brinkman–Rice value (dotted line), introducing instead a maximum in m^*/m before U/U_c reaches 1.0. The end member $CaVO_3$ approaches the limit $U = U_c$, and the band narrows with increasing x in the system $Sr_{1-x}Ca_xVO_3$ because the inductive effect is dominant in the itinerant–electron limit of a $(VO_3)^{2-}$ array where the V–O bonds are not so compressible.

Since hydrostatic pressure increases the bandwidth of an MO_3 array in a perovskite, it follows that m^* should increase with pressure in $CaVO_3$ whereas it decreases in a normal metal. The predicted anomalous increase in m^* with hydrostatic pressure has been verified by a measurement of the temperature dependence of the Seebeck coefficient as a function of pressure [23]. Thus the phenomenon of "correlation fluctuations" associated with cooperative oxygen displacements appears to be firmly established for a single–valent MO_3 array of an AMO_3 perovskite.

4   Transitions in Mixed–Valent Systems

Mixed–valent MO_3 arrays are obtained by a partial substitution of an aliovalent A cation as in $La_{1-x}Ca_xVO_3$. An aliovalent cation perturbs the potential at the neighboring atoms of the MO_3 array, which creates a trap for the mobile charge carriers it introduces. Where there is a transition from localized to itinerant electrons, as in the case of $LaVO_3$ *versus* $CaVO_3$, the holes trapped at the aliovalent cation, *i.e.* at Ca in $La_{1-x}Ca_xVO_3$ with $x < 0.25$, may be delocalized over the neighboring atoms of the MO_3 array; for small x, the hole trapped at the 8 M and 12 bridging oxygen atoms near–neighbor to a Ca atom may be either polaronic or delocalized over the cluster. On the other hand, holes thermally excited from a trap would form small polarons, cooperative atomic displacements forming short M–O bonds at individual M atoms removed from the clusters. The itinerant character of the cluster orbitals of a trap is manifest as a semiconductor–metal transition setting in at the percolation threshold x_c, where an $x_c \approx 0.26$ is found in $La_{1-x}Ca_xVO_3$ [24,25].

The delocalized character of holes trapped at an isolated Ca atom is nicely illustrated by $La_{1-x}Ca_xCoO_3$ [26,27]. In this system, delocalization of the e orbitals stabilizes a ferromagnetic intermediate–spin state Co(III): $t_2^5\sigma_*^1$, where σ_* refers to molecular orbitals of either a cluster or of an entire crystal. For small values of x, the clusters are superparamagnetic, and at the percolation threshold x_c they form a ferromagnetic matrix that is interpenetrated by trivalent–cobalt domains exhibiting the low–spin to high–spin transition of the parent compound $LaCoO_3$.

The systems $La_{1-x}A_xMnO_3$, A = Ca or Sr, have provided a series of surprises. The initial surprise was the observation of an anisotropic Mn–O–Mn magnetic coupling in the parent compound LaMnO₃ [28]; ferromagnetic (001) planes of the orthorhombic structure of Fig. 1 were coupled antiferromagnetically along the c–axis. This finding was rationalized by two postulates [29], subsequently confirmed: (1) stabilization of a cooperative displacement of the oxygen within the ferromagnetic planes away from one Mn toward the other to create short Mn–O bonds with an empty e orbital and longer Mn–O bonds stabilizing a localized e electron to give a cooperative non–collinear Jahn–Teller deformation at each Mn^{3+} ion, Fig.4(c), and (2) the rules governing the sign of the superexchange interaction.

At that time it was also recognized that preparation of LaMnO₃ by conventional ceramic methods required an inert atmosphere; in air, the product is $La_{1-\delta}Mn_{1-\delta}O_3$. However, the driving force for the oxidation reaction was not understood. Moreover, the mixed–valent systems $La_{1-x}A_xMnO_3$, A = Ca or Sr, show a decreasing tendency to oxidation with increasing x, and they each exhibit a wide ferromagnetic domain with an abrupt change from semiconductive to metallic behavior on cooling through the Curie temperature T_C in the interval $x_c < x < 0.5$, where x_c is smaller for A = Sr [18]. The metallic conductivity was thought to signal a lack of polaron formation within a majority–spin lower Hubbard band; the ferromagnetism could then be described by double exchange [30].

The next surprise came from a study [31] motivated by the use of $La_{1-x}Sr_xMnO_3$ as a cathode material for a solid–oxide fuel cell. In a part of that study, van Roosmalen and Cordfunke used thermogravimetric analysis to argue, convincingly in my opinion, for oxidation of Mn^{2+} ions resulting from disproportionation fluctuations of the type

$$2Mn(III) : t_2{}^3\sigma_*{}^1 = Mn^{2+} : t_2{}^3e^2 + Mn(IV) : t_2{}^3\sigma_*{}^0 \qquad (10)$$

In my view, the fluctuations occur at a cross–over from itinerant–σ_* to localized–e electronic behavior; they are analogous to the high–spin $Co^{3+} : t_2{}^4e^2$ fluctuations found at low temperatures within a low–spin Co(III): $t_2{}^6\sigma_*^0$ matrix of LaCoO₃, but in this case the cooperative motion of bridging oxygen atoms away from a central Mn atom toward six neighboring Mn atoms causes an electron transfer from the six neighboring Mn atoms to create a central Mn^{2+} ion having longer Mn–O bonds and a localized $t_2{}^3e^2 : {}^6A_{1g}$ configuration. The disproportionation reaction then becomes another example where the oxygen displacements manifest a double–well oxygen potential at a cross–over from localized to itinerant electronic behavior.

A further surprise was the realization that the ferromagnetic transition at T_c may be first–order [32] and T_c can be increased by an applied magnetic field [33]. The resulting giant negative magnetoresistance has stimulated an intensive and extensive ongoing reinvestigation of the $R_{1-x}A_xMnO_3$ family of compounds, where R = La or Pr and A = Ca or Sr. As should be expected for a transition at the cross–over from localized–e to itinerant–σ_* electrons, the transition temperature T_c and the critical value x_c for stabilization of metallic conductivity below T_c vary sensitively with the width of the σ_* band and therefore with the magnitude of the tolerance factor t of equation (1). Since t decreases with temperature, metallic conductivity below T_c reverts to semiconductive behavior on further cooling at $x \approx x_c$. The magnetoresistance appears to have its maximum value for x just above x_c. At $x = 0.5$, ordering of Mn^{3+} and Mn^{4+} on alternate Mn atoms as a result of static cooperative oxygen displacements competes with the ferromagnetic metallic phase; but Tomioka et $al.$ [34] have shown that in $Pr_{0.5}Sr_{0.5}MnO_3$ this ordering transition occurs below T_c and the first–order transition temperature can be shifted by an applied magnetic field to give an alternative dramatic negative magnetoresistance at temperatures a little below T_t.

It should be appreciated that the $t_2{}^3$ configuration remains localized at the transition from localized–e to itinerant–σ_* electrons because of the relation $W_\pi < U \approx W_\sigma$. Moreover, stabilization of itinerant σ_* electrons below T_c reflects an electron–transfer resonance integral b of equation (5) that must be modified by a spin dependence. A mixed–valent system has its maximum spin–independent value for ferromagnetic alignment; it is reduced by spin–disorder scattering as the spins become disordered or aligned antiferromagnetically. Above T_c, the electrons become trapped in regions of short–range ferromagnetic fluctuations, and the electron mobility becomes activated, which reduces the resonance integral exponentially with the activation energy for hopping, so the double–exchange coupling becomes sharply reduced. However, a vibronic coupling of empty and occupied e orbitals at neighboring Mn resulting from the oxygen double–well potential can give a ferromagnetic superexchange component to the interatomic exchange.

5   The Copper–Oxide Superconductors

The copper–oxide p–type superconductors all have intergrowth structures containing superconductive layers of $n \geq 1$ CuO_2 sheets separated by $(n-1)$ planes of smaller A cations Ca^{2+} or R^{3+} in eightfold oxygen coordination; these layers alternate with non–superconductive layers bounded by (001) AO rock–salt sheets containing larger A cations. The superconductive CuO_2 sheets may contain Cu in fourfold planar, fivefold square–pyramidal, or sixfold octahedral oxygen coordination, but every copper of a sheet has the same oxygen coordination. Doping to make the CuO_2 sheets mixed–valent and superconductive may be achieved either by aliovalent A–cation substitution or by anion insertion/substitution within (excluding the AO boundary sheets) the non–superconductive layers. The non–superconductive layers may act as charge reservoirs for the superconductive layers.

The single–valent end member of a copper–oxide superconductor system is an antiferromagnetic insulator with only Cu^{2+} ions in the CuO_2 sheets; the single hole in the $Cu^{2+} : 3d^9$ configuration is localized and occupies an $x^2 - y^2$ orbital directed toward the bridging oxygens of a CuO_2 sheet. A charge–transfer gap $\Delta = (U_{eff} - W_\sigma) \approx 2$ eV separates the empty $x^2 - y^2$ Hubbard band from the

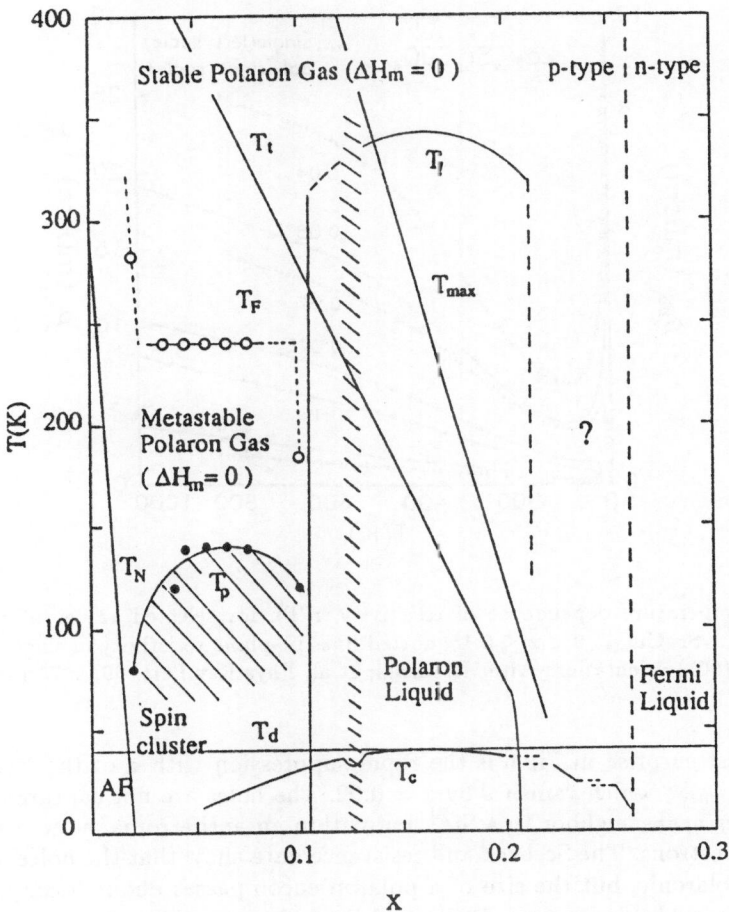

Fig. 6. Preliminary phase diagram for the system $La_{2-x}Sr_xCuO_4$.

highest occupied band, which is also an antibonding $x^2 - y^2$ band of strongly hybridized $O-2p_x$, $2p_y$ and $Cu-3d_{x^2-y^2}$ states. A Cu(III) ion in an oxide is always low–spin and diamagnetic forming four coplanar, strongly covalent Cu–O bonds; holes introduced into the lower $x^2 - y^2$ band occupy molecular–orbital as opposed to ligand–field $x^2 - y^2$ states, and a transition from an antiferromagnetic insulator to a Pauli paramagnetic metal can be anticipated with increased p-type doping. The simplest superconductive system for monitoring this transition is $La_{2-x}Sr_xCuO_4$, which has $n = 1$ and a non–superconductive $(LaO)_2$ rock-salt layer that does not act as a charge reservoir, Fig.2(a). Moreover, a solid-solution range $0 \leq x \leq 0.33$ above room temperature spans the transition from antiferromagnetic insulator to superconductor to overdoped Pauli–paramagnetic metal. Fig.6 is a preliminary phase diagram for this system [35].

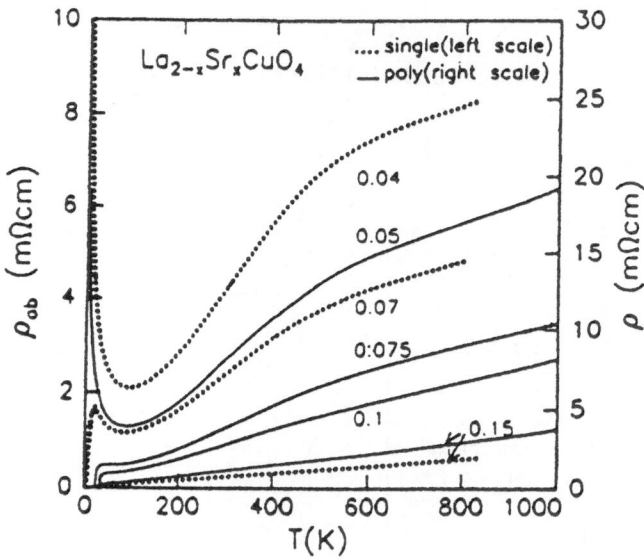

Fig. 7. Temperature dependence of resistivity, r(T), for selected values of x in the system $La_{2-x}Sr_xCuO_4$, $0 < x \leq 0.15$; dotted lines in–plane resistivity of single–crystal films with (001) orientation. After H.Takagi; et al, Phys.Rev.Lett. 69, 2975 (1992).

The first surprise in Fig.6 is the rapid suppression with x of the Néel temperature T_{rmN}, which vanishes by $x = 0.02$. The holes are not captured at the four copper near–neighbor to a Sr^{2+} ion within an antiferromagnetic matrix of localized electrons. The Seebeck and resistance data show that the holes are mobile and polaronic, but the size of a polaron encompasses about 5 copper sites with their neighboring oxygen atoms and the polaron mobility is activated only at lowest temperatures, Fig.7. Holes trapped within mobile large polarons would occupy molecular orbitals just as holes trapped in stationary clusters neighboring Cu^{2+} ions in the case of $La_{1-x}Ca_xVO_3$. The first–order phase change between the region inside a polaron and the surrounding matrix would introduce a discontinuity of the Hubbard on–site electrostatic energy U at the polaron–matrix interface. Therefore we identify these non–adiabatic polarons as "correlation" polarons [35].

An immediate problem raised by this model is that non–adiabatic polarons are generally small, confined to a single atomic site, and move with an activated mobility to highest temperatures. However, the temperature dependence of the resistivity $\rho(T)$, Fig.7, shows a semiconductive character only at lowest temperatures; above a temperature T_ρ, the $\rho(T)$ curve increases nearly linearly with temperature. The minimum in the $\rho(T)$ curve decreases with increasing x, falling below the critical temperature T_c in the optimally doped superconductive compositions. In order to address these problems, we [36] have invoked a dynamic pseudo Jahn–Teller deformation known to occur at copper sites. In an isolated Cu(III) complex, mixing of an excited state into the ground state introduces a Jahn–Teller deformation in which the square oxygen coordination is transformed into a rhombus by a stretching of two opposing Cu–O bonds. Such a deformation would be stabilized by cooperativity among neighboring Cu sites as illustrated in Fig.8; this stabilization favors enlargement of the non–adiabatic polaron to include more than one Cu site. Calculation [36] shows that a dynamic pseudo Jahn–Teller deformation at the Cu sites within a polaron would stabilize a non-adiabatic polaron containing 5 to 7 Cu sites, which is in good agreement with experiment. Moreover, the large polaron can move a fraction of itself in amoeba fashion one Cu–O bond at a time, and calculation of this type of motion gives only a small activation energy for the mobility with an R(T) curve like that of the lowest–doped sample of Fig.7. Indirect evidence for a vibronic state within a polaron comes from thermopower data for the superconductive and overdoped compositions, see below. Finally, such a polaron is too large to be trapped at a Sr^{2+} ion.

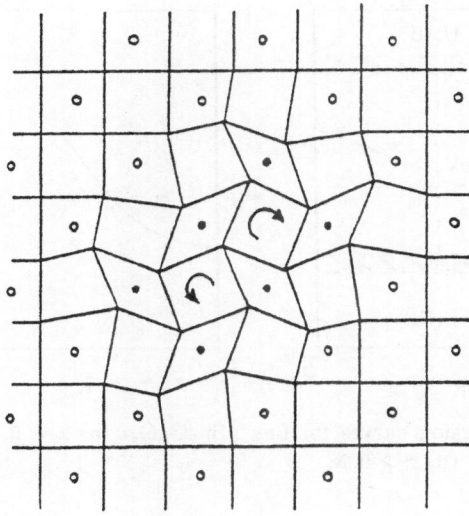

Fig. 8. Cooperative pseudo Jahn–Teller deformations within a correlation polaron.

Another surprise was evidence [37] for a dynamic phase segregation below T_ρ into an antiferromagnetic phase and a superconductive phase in the compositional range $0.02 < x < 0.10$. A classical phase segregation into oxygen–rich and oxygen–poor domains occurs in $La_2CuO_{4+\delta}$ in the corresponding range of hole concentrations $0.01 < \delta < 0.05$ [38,39]. This observation led to the deduction [40] that the superconductive phase is a thermodynamic state distinguishable from both the underdoped and the overdoped compositions; it is stable in a narrow range of hole concentrations $0.10 < p < 0.22$ per Cu atom of a CuO_2 sheet, optimum doping occurring at $p \approx 0.16 \pm 0.02$. However, a competitive, non–superconductive phase may be stabilized near $p = 0.125$ [3,41].

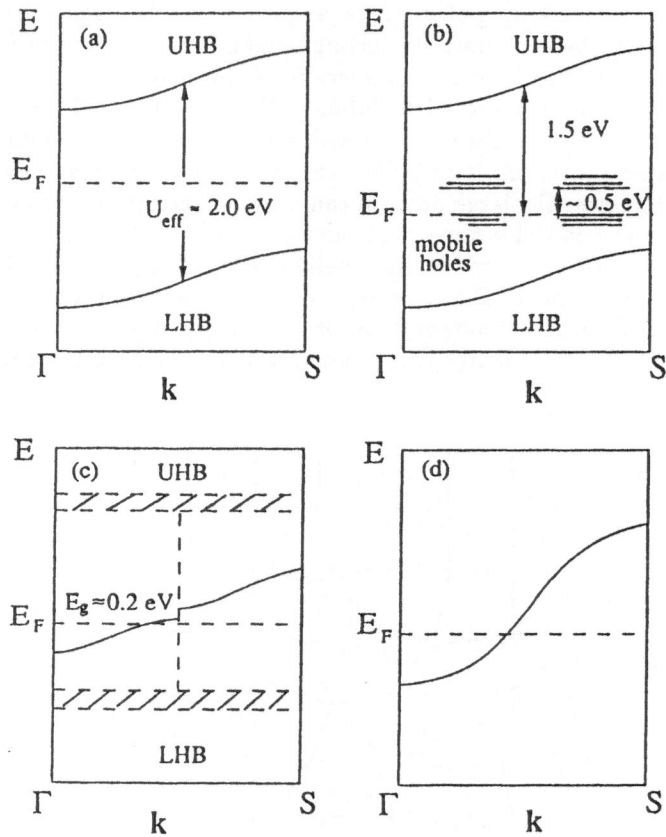

Fig. 9. Schematic dispersion curves for $La_{2-x}Sr_xCuO_4$: (a) $x = 0$, (b) $0 < x < 0.10$, (c) $0.10 \leq x \leq 0.22$, and (d) $x \geq 0.28$.

A systematic study [2] of the evolution of T_c with x from the optimally doped composition $x = 0.15$ to the overdoped composition $x = 0.30$ showed a step–wise transition to the overdoped regime. Moreover, the observation of the pressure dependence of $R(T)$ in this study revealed that T_c increases with a straightening of the Cu–O–Cu bond, reaching a maximum value independent of pressure on passing from the orthorhombic to the tetragonal phase. Stabilization of the tetragonal relative to the orthorhombic phase with hydrostatic pressure demonstrates that the tolerance factor t of equation (1) increases with pressure, which means that the Cu–O bonds have an unusually high compressibility; a compressibility of an M–O bond that is higher than that of the A–O bond is only found where there is a double–well oxygen potential at a cross–over from localized to itinerant electronic behavior of the M–O antibonding electrons.

In addition, the $R(T)$ data [2,37] showed the presence of a composition-independent transition temperature Td that sets an upper limit to T_c; we suggested that this transition marks the temperature below which superconductive fluctuations in a CuO_2 sheet can be coupled along the c–axis to give a three–dimensional superconductivity in this system. The mechanism of inter-layer c–axis coupling may differ significantly from one copper–oxide system to another, but the mechanism of superconductive–pair formation within a CuO_2 plane should be the same in all.

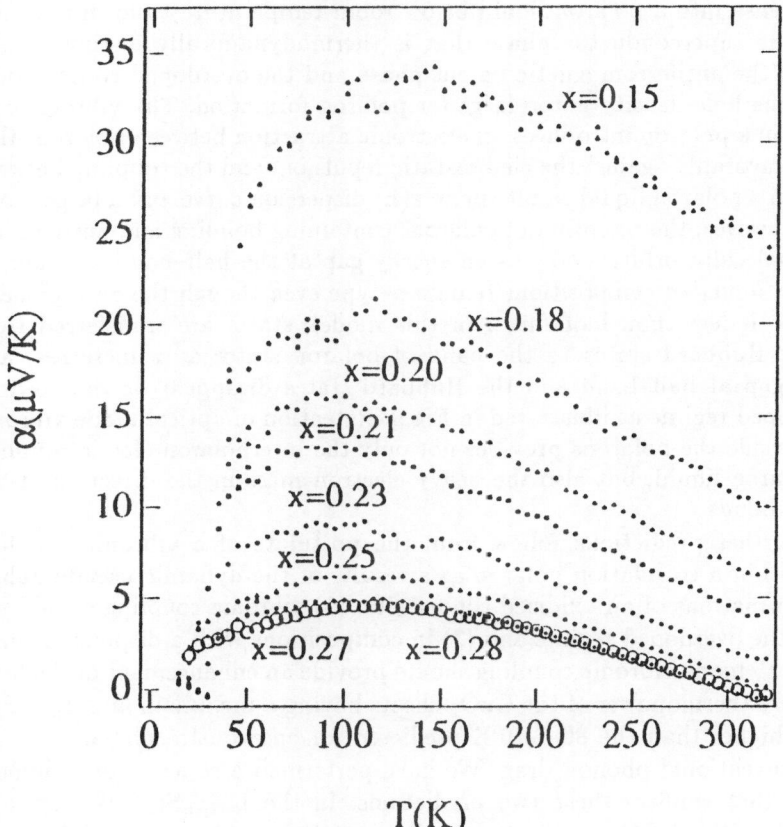

Fig. 10. Temperature dependence of the Seebeck coefficient, $\alpha(T)$, for the system $La_{2-x}Sr_xCuO_4$, $0.15 \leq x \leq 0.28$.

An observation of fundamental importance comes from angle–resolved photoemission spectroscopy (ARPES) [42-44]; it has shown the existence of an electronic dispersion curve $\varepsilon(\mathbf{k})$ in the normal state of the superconductive compositions and a large Fermi surface located close to the position in the Brillouin zone predicted by band theory [45]. It also shows a remarkable flattening of the dispersion curves in the Cu–O–Cu bond directions. Moreover, NMR data [46] record the presence of spin fluctuations persisting into the superconductive compositions, which indicates retention of strong–correlation fluctuations. Such fluctuations bound a correlation polaron and would be retained in the superconductive compositions if the polarons are also retained. However, an $\varepsilon(\mathbf{k})$ dispersion curve would then require a strong coupling between the polarons. We were thus led to the conclusion [35, 36] that the correlation polarons are present as a polaron gas down to T_r in the underdoped compositions $0 < x < 0.10$ of Fig.6 and also in the superconductive compositions above room temperature; they condense into a polaron liquid below room temperature to form a normal state of the superconductor phase that is thermodynamically distinguishable from both the antiferromagnetic parent phase and the overdoped compositions in which the hole density is too large for polaron formation. The vibronic coupling within a polaron introduces an electronic attraction between polarons that competes favorably against the electrostatic repulsion, and the coupling between polarons in a polaron liquid results in an $\varepsilon(\mathbf{k})$ dispersion curve and a large Fermi surface. However, the retention of polarons containing bonding and antibonding M-O-M molecular orbitals creates an energy gap at the half–band position, so the superconductive compositions remain p–type even though the $x^2 - y^2$ band as a whole is less than half–filled. In this model, states are transferred from incoherent Hubbard states to the band of polaron states as x increases, and both the gap at half–band and the Hubbard states disappear on doping into the overdoped regime as illustrated in Fig.9. Retention of optical mode vibronic coupling inside the polarons provides not only the interpolaron elastic coupling in the polaron liquid, but also the heavy electron mass in the direction of the Cu–O–Cu bonds.

Two further predictions follow from the postulate of a vibronic coupling present within a correlation polaron as a results of the dynamic pseudo Jahn–Teller deformations of an oxidized copper site: (1) vibronic coupling should extend into the overdoped regime and (2) in compositions with a dispersion curve $\varepsilon(\mathbf{k})$, optical–mode vibronic coupling should provide an enhancement of the low–temperature thermopower of the CuO_2 sheets having a maximum at a T_{max} significantly higher than the 80 ± 10 K predicted for an acoustic–phonon process such as conventional phonon drag. We have performed a series of experiments [36,47–50] that confirm these two predictions. In the $La_{2-x}Sr_xCuO_4$ system, for example, only the CuO_2 sheets contribute to the thermopower, and a low–temperature enhancement indicative of an $\varepsilon(\mathbf{k})$ dispersion curve is found from the optimally doped sample $x = 0.15$ to the overdoped sample $x = 0.30$. As shown in Fig.10 for different values of x in the range $0.15 \leq x \leq 0.30$, the thermopower for all samples can be described by

$$\alpha(T) = \alpha_0 + \delta\alpha(T) \tag{11}$$

in which α_0 remains nearly temperature–independent to temperatures above 600^0C (not shown) and decreases monotonically with increasing hole concentration x; the magnitude of ao is a measure of the asymmetry of e vs $\mathbf{k}$ in the neighborhood of ε_F [36]. The enhancement term $\delta\alpha(T)$ appears below room temperature and reaches a maximum value at $T_{max} \approx 140$ K. The enhancement term, though diminished, persists into the overdoped regime.

In the $YBa_2Cu_3O_{6.96}$ 90 K–superconductor, the Cu–O chains of the non–superconductive layers are metallic and contribute a competitive negative term to the thermopower that has a similar enhancement term to that of the CuO_2 sheets, but of opposite sign [48]; however, isolated Cu–O chains in $Sr_2CuO_{3-\delta}$ were shown to exhibit an enhancement term with $T_{max} \approx 75$ K typical of a phonon drag [49]. The $T_{max} \approx 140$ K appears to be a signature of superconductive and overdoped CuO_2 sheets or of metallic non–superconductive layers that are coupled by a bridging oxygen to a superconductive CuO_2 sheet; in no other situation has a $T_{max} \approx 140$ K been found.

Measurements of $\alpha(T)$ under different hydrostatic pressures have also been made [50]. For $La_{1.85}Sr_{0.15}CuO_4$, we found $dT_c/dP > 0$, $d\alpha_0/dP > 0$, and $d(\delta\alpha)/dP > 0$. At room temperature, where $\alpha \approx \alpha_0$, a high value of α indicates a large curvature of $\varepsilon(\mathbf{k})$ at ε_F, which is consistent with the $\varepsilon(\mathbf{k})$ curves obtained by ARPES. Since there is no charge reservoir in $La_{2-x}Sr_xCuO_4$, a $d\alpha_0/dP > 0$ can only reflect an increase in the curvature of $e(\mathbf{k})$ at e_F with pressure. Since $d\delta\alpha/dp$ and T_c also increase with hydrostatic pressure P, the data demonstrate that the phenomenon responsible for flattening further the $\varepsilon(\mathbf{k})$ curve at ε_F also underlies the enhancement of both $\delta\alpha$ and T_c. Moreover, a $dT_c/dP > 0$ in the orthorhombic, but not the tetragonal, phase reflects an increase in T_c with straightening of the Cu–O–Cu bond angle. In view of these deductions, it is significant that pseudo Jahn–Teller deformations would have their maximum stability for 180^0 Cu–O–Cu bonds and would also generate vibronic states having a flat $\varepsilon(\mathbf{k})$ in the direction of the Cu–O bonds in a CuO_2 sheet. Straightening of the Cu–O–Cu bond angle under pressure would enhance the vibronic coupling, thereby flattening further $\varepsilon(\mathbf{k})$. A flatter dispersion curve would increase the maximum value of the superconductive energy gap, thereby increasing T_c. The correlation between $dT_c/dP > 0$, $d\alpha_0/dP > 0$, and $d(\delta\alpha)/dP > 0$ is therefore consistent with dynamic pseudo Jahn–Teller deformations inducing a flattening of a vibronic dispersion $\varepsilon(\mathbf{k})$ as the common underlying physical phenomenon.

6 Summary

This paper has emphasized the first–order character of the transition from localized to itinerant and/or Curie–Wiess to Pauli paramagnetic electronic behavior in solids. A larger unit–cell volume is associated with the more localized electrons, a higher entropy with the more delocalized electrons. At the transition, the oxygen atom of an AMO_3 perovskite is located in a double–well potential; the shorter equilibrium M–O bond length is associated with molecular orbitals or itinerant electrons, the longer M–O bond length with localized electrons.

Reference to particular illustrative compounds was made to show that, in single–valent oxides with the perovskite structure, such a transition may be either global or characterized by a coexistence of two electronic phases in a single crystallographic phase. The latter situation was illustrated by (1) a disproportionation into different valence states, (2) a disproportionation into different spin states, and (3) correlation fluctuations between Hubbard states and Fermi–liquid states. The coexistence of two distinguishable electronic states is accomplished by cooperative oxygen–atom displacements that may be either static or dynamic.

In mixed–valent oxides, segregation into hole–rich and electron–rich regions by cooperative atomic displacements commonly results in small-polaron formation; but larger atomic clusters may be stabilized as non-adiabatic polarons or as stationary charge–carrier traps at near neighbors of aliovalent A cations that introduced them. In stationary traps, the carriers are attracted to the aliovalent cations by electrostatic forces, but deep traps are formed by a first–order change in the M–O bond length on passing from the molecular–orbital (or itinerant-electron) domain of the hole-rich electronic phase to the localized–electron domain of the electron-rich phase. Excitation of holes from a hole–rich cluster to a hole–poor matrix results in small–polaron formation at lower concentrations, but in an extended transformation to itinerant–electron behavior at concentrations that exceed a percolation threshold.

In the case of itinerant σ_* electrons in the presence of localized t_{2^n} configurations ($1 \leq n \leq 5$), a spin–dependent resonance integral has its maximum value where the atomic spins are ferromagnetically aligned. Consequently, at a transition from localized to itinerant electronic behavior, a mixed–valent ferromagnet may form magnetic polarons in the paramagnetic state that become larger with the domains of short–range ferromagnetic order on lowering the temperature toward T_c; and a first–order transition to a global itinerant–electron phase below T_c can give a giant magnetoresistance. Similarly, stabilization of a ferromagnetic phase by an applied magnetic field relative to an antiferromagnetic phase of ordered small polarons has been shown to give a first–order insulator–metal transition below T_c of the ferromagnetic phase.

Application of this concept to the transition from localized to itinerant electron behavior in the mixed–valent copper–oxide superconductors requires another phenomenon than a spin–dependent resonance integral to rationalize the stabilization of a non–adiabatic polaron containing about 5 copper centers. A Cu(III) center supports a dynamic pseudo Jahn–Teller deformation of its square–coplanar coordination to introduce the formation of vibronic states. Cooperativity between deforming centers stabilizes a large non–adiabatic polaron and introduces an elastic coupling between polarons that induces condensation of a polaron gas into a polaron liquid in the normal state of the superconductive compositions and into a polaron solid below the critical temperature T_c. Initial experiments were shown to confirm predictions based on the formation of vibronic states in the hole–rich regions within a polaron or globally in the overdoped compositions.

We gratefully acknowledge support from the Robert A.Welch Foundation, Houston, TX, the National Science Foundation, and the Texas Advanced Research Program.

References

1. J.B.Goodenough, J.A.Kafalas, and J.M.Longo, in: *Preparative Methods in Solid State Chemistry*, ed.by P.Hagenmuller (Academic Press, New York, 1972) pp.1-69.

2. J.-S.Zhou, H.Chen, and J.B.Goodenough, *Phys.Rev.*B **49**, 9084 (1994).

3. J.D.Axe, A.H.Moudden, D.Hohlwein, D.E.Cox, K.M.Mohanty, A.R.Moudenbaugh, and Youwen Xu, *Phys.Rev.Lett.* **62**, 2751 (1989).

4. M.J.Rozenberg, G.Kotliar, and X.Y.Zhang, *Phys.Rev.*B **49**, 10181 (1994).

5. P.Majundar and H.R.Krishnamurthy, *Phys.Rev.Lett.* **73**, 1525 (1994).

6. J.B.Torrance, P.Lacorre, D.I.Nazzal, E.J.Ansaldo, and Ch.Niedermayer, *Phys.Rev.*B **45**, 8209 (1992).

7. J.Zaanen, G.A.Sawatsky, and J.W.Allen, *Phys.Rev.Lett.* **55**, 418 (1985).

8. J.B.Goodenough, *J.Appl.Phys.* **37**, 1415 (1966).

9. J.B.Goodenough, N.F.Mott, M.Pouchard, G.Demazeau, and P.Hagenmuller, *Mat.Res.Bull.* **8**, 647 (1973).

10. M.Takano, N.Nakanishi, Y.Takeda, S.Naka, and T.Takada, *Mat.Res.Bull.* **12**, 923 (1977).

11. M.A.Señarís-Rodríguez, and J.B.Goodenough, *J.Solid State Chem.* **116**, 224 (1995).

12. A.S.Borukhovich, G.V.Bazuev, and G.P.Shveiken, *Sov.Phys.- Solid State* **16**, 181 (1974); **15**, 1467 (1974).

13. P.Bordet, C.Chaillout, M.Marezio, Q.Huang, A.Sontoro, S.-W.Cheong, H.Takagi, C.S.Oglesby, and B.Batlogg, *J.Solid State Chem.* **106**, 253 (1993).

14. J.B.Goodenough and H.C.Nguyen, *C.R.Acad.Sci.* **319**, 1285 (1994).

15. N.Shirakawa and M.Ishikawa, *Jpn.J.Appl.Phys.* **30**, L755 (1991).

16. A.V.Mahajan, D.C.Johnston, D.R.Torgeson, and F.Borsa, *Physica C* **185-189**, 1095 (1991); *Phys.Rev.*B **46**, 10966 (1992).

17. H.C.Nguyen and J.B.Goodenough, *Phys.Rev.*B **52**, 324 (1995).

18. J.B.Goodenough and J.M.Longo, in: *Landolt-Börnstein Tabellen, New Series* III/4a, ed.by K.H.Hellwege (Springer-Verlag, Berlin 1970) pp.126-314.

19. I.H.Inoue, I.Hase, I.Aura, A.Fujimori, Y.Haruyama, T.Maruyama, and Y.Nishihara, *Phys.Rev.Lett.* **74**, 2539 (1995).

20. A.Fujimori, I.Hase, H.Namatame, Y.Fujishima, and Y.Tokura, *Phys.Rev.Lett.* **69**, 1796 (1992).

21. M.Onoda, H.Ohta and H.Nagasawa, *Solid State Commun.* **79**, 281 (1991).

22. Hoan C.Nguyen and J.B.Goodenough, *Phys.Rev.*B (in press 1995).

23. J.-S.Zhou and J.B.Goodenough (unpublished).

24. P.Dougier, D.Deglane, and P.Hagenmuller, *J.Solid State Chem.* **19**, 135 (1976).

25. Hoan C.Nguyen and J.B. Goodenough (unpublished).

26. M.Itoh, I.Natori, S.Kubota, and K.Motoya, *J.Phys.Soc.Japan* **63**, 1486 (1994).

27. M.A.Señarís-Rodríguez and J.B.Goodenough, *J.Solid State Chem.* **118**, 323 (1995).

28. E.O.Wollan and W.C.Koehler, *Phys.Rev.* **100**, 545 (1955).

29. J.B.Goodenough, *Phys.Rev.* **100**, 564 (1955).

30. C.Zener, *Phys.Rev.* **82**, 403 (1951);   P.W.Anderson and H.Hasegawa, *Phys.Rev.* **100**, 675 (1955);   P.G.de Gennes, *Phys.Rev.* **118**, 141 (1960).

31. J.van Roosmalen, *Thesis*, Delft Univ.of Technology (1993); J.A.M.van Roosmalen and E.H.P.Cordfunke, *J.Solid State Chem.* **110**, 109 (1994).

32. H.Y.Hwang, S.-W.Cheong, P.G.Radaelli, M.Marezio, and B.Batlogg, *Phys.Rev.Lett.* **75**, 914 (1995);   A.Asamitsu, Y.Moritomo, Y.Tomioka, T.Arima, and Y.Tokura, *Nature* (London) **373**, 407 (1995).

33. S.Jin, T.H.Tiefel, M.McCormack, R.A.Fastnacht, R.Ramesh, and L.H.Chen, *Science* **264**, 413 (1994).

34. Y.Tomioka, A.Asamitsu, Y.Moritomo, H.Kuwahara, and Y.Tokura, *Phys.Rev.Lett.* **74**, 5108 (1995).

35. J.B.Goodenough and J.-S.Zhou, *Phys.Rev.*B **49**, 4251 (1994).

36. G.I.Bersuker and J.B.Goodenough (unpublished);   J.S.Zhou, G.I.Bersuker and J.B.Goodenough, *J.Supercond.* **8**, No.1 (1995);   J.B.Goodenough, J.-S.-Zhou and G.I.Bersuker, *Proc.Int.Worksh.on the "Anharmonic Properties of High-T_c Cuprates"*, Bled, Slovenia, Sept.1–6, 1994.

37. J.B.Goodenough, J.-S.Zhou, and K.Allan, *J.Mater.Chem.* **1**, 715 (1991).

38. J.D.Jorgensen, B.Dabrowski, S.Pei, D.G.Hinks, L.Soderholm, B.Morosin, J.E.Schirber, E.L.Venturini, and D.S.Ginley, *Phys.Rev.*B **38**, 11 337 (1988).

39. J.Ryder, P.A.Midgley, R.Exley, R.J.Beynon, D.L.Yates, L.Afalfliz, and J.A.Wilson, *Physica* C **173**, 9 (1991).

40. J.G.Goodenough, J.-S.Zhou, and J.Chan, *Phys.Rev.*B **47**, 5275 (1993).

41. J.-S.Zhou, H.Chen, and J.B.Goodenough, *Phys.Rev.*B **50**, 4168 (1994).

42. D.M.King, Z-X.Shen, D.S.Dessau, D.S.Marshall, C.H.Park, W.E.Spicer, J.L.Peng, Z.-Y.Li, and R.L.Greene, *Phys.Rev.Lett.* **73**, 3298 (1994).

43. K.Gofron, J.C.Campuzano, A.A.Abrikosov, M.Lindroos, A.Bansil, H.Ding, D.Koelling, and B.Dabrowski, *Phys.Rev.Lett.* **73**, 3302 (1994).

44. Jian Ma, C.Quitmann, R.J.Kelley, P.Alméras, H.Berger, G.Margaritondo, and M.Onellion, *Phys.Rev.*B **51**, 3832 (1995).

45. W.E.Pickett, *Revs.Mod.Phys.* **61**, 433 (1989);   W.E.Pickett, H.Krakauer, R.E.Cohen, and D.I.Singh, *Science* **255**, 46 (1992).

46. A.Sokol and D.Pines, *Phys.Rev.Lett.* **71**, 2813 (1993) and refs. therein.

47. J.-S.Zhou and J.B.Goodenough, *Phys.Rev.*B **51**, 3104 (1995).

48. J.-S.Zhou, J.-P.Zhou, J.B.Goodenough, and J.T.McDevitt, *Phys.Rev.*B **51**, 325 (1995); *J.Supercond.* **8**, No.1 (1995).

49. W.B.Archibald, J.-S.Zhou and J.B.Goodenough, *Phys.Rev.*B **52** (1995).

50. J.-S.Zhou and J.B.Goodenough, (unpublished).

Overview of Recent Magnetic Studies of High T_c Cuprate Parent Compounds and Related Materials

D. C. Johnston[1], T. Ami[2*], F. Borsa[1,3], P. C. Canfield[1], P. Carretta[3], B. K. Cho[1**], J. H. Cho[1***], F. C. Chou[1†], M. Corti[3], M. K. Crawford[2], P. Dervenagas[1], R. W. Erwin[4], J. A. Fernandez-Baca[5], A. I. Goldman[1], R. J. Gooding[6], Q. Huang[4,7], M. F. Hundley[8], R. L. Harlow[2], B. N. Harmon[1], A. Lascialfari[3], L. L. Miller[1], J. E. Ostenson[1], N. M. Salem[6], C. Stassis[1], B. Sternlieb[9], B. J. Suh[1], D. R. Torgeson[1], D. Vaknin[1], K. J. E. Vos[6], X.-L. Wang[5], Z. R. Wang[1‡], M. Xu[1], and J. Zarestky[1]

[1] Ames Laboratory and Department of Physics and Astronomy, Iowa State University, Ames, Iowa 50011, U.S.A.
[2] Du Pont, Wilmington, Delaware 19880, U.S.A.
[3] Dipartimento de Fisica Generale "A. Volta", Universita' di Pavia, 27100 Pavia, Italy
[4] National Institute of Standards and Technology, Bldg. 235, Gaithersburg, Maryland 20899, U.S.A.
[5] Oak Ridge National Laboratory, Oak Ridge, Tennessee 37831, U.S.A.
[6] Department of Physics, Queen's University, Kingston, Ontario, Canada K7L 3N6
[7] Department of Materials and Nuclear Engineering, University of Maryland, College Park, Maryland 20742, U.S.A.
[8] Los Alamos National Laboratory, Los Alamos, New Mexico 87545, U.S.A.
[9] Physics Department, Brookhaven National Laboratory, Upton, New York 11973, U.S.A.

Abstract. Recent studies of the magnetic properties of several high superconducting transition temperature (T_c) cuprate parent compounds and related materials will be reviewed. The observations of a Heisenberg to XY-like crossover upon cooling below $\sim 300\,\mathrm{K}$ towards the Néel temperature $T_N = 257\,\mathrm{K}$ and a subsequent magnetic field-induced XY-like to Ising-like crossover near T_N in single crystals of the K_2NiF_4-type spin 1/2 model compound $Sr_2CuO_2Cl_2$ will be described. The spin 1/2 linear chain compound Sr_2CuO_3, the parent of the $Sr_2CuO_{3+\delta}$ oxygen-doped superconductors, is found to exhibit classic Bonner-Fisher magnetic behavior, with a large antiferromagnetic Cu-Cu superexchange coupling constant. Studies of the evolution of

* Permanent Address: Sony Research Center, 174, Fujitsukacho, Hodogaya-ku, Yokohama 240, Japan
** Present address: Department of Chemistry, Cornell University, Ithaca, NY 14853, U.S.A.
*** Present Address: Superconductivity Technology Center, Los Alamos National Laboratory, Los Alamos, NM 87545, U.S.A.
† Present Address: Center for Materials Science and Engineering, Massachusetts Institute of Technology, Cambridge, MA 02139, U.S.A.
‡ Present address: Department of Physics, University of California at Irvine, Irvine, CA 92717-4575, U.S.A.

$La_{2-x}Sr_xCuO_4$ with Sr doping in the insulating regime ($x < 0.05$) will be summarized, which indicate that the doped holes reside in walls separating undoped domains. We have found that $BaCuO_{2.1}$, a copper-oxygen cluster compound, exhibits ferromagnetic rather than antiferromagnetic Cu-Cu superexchange interactions. Finally, a summary of the magnetic properties of single crystals of the recently discovered RNi_2B_2C layered structure superconductors will be given.

1 Introduction

The discovery in 1986 of high temperature superconductivity (HTSC) in the layered cuprates with the K_2NiF_4 structure (Bednorz and Müller 1986) and subsequently in many related layered cuprate structures (Vanderah 1992) has stimulated extensive investigations of their normal and superconducting state properties (for reviews, see Ginsberg 1989). It was soon found that La_2CuO_4, the so-called "parent" of the first HTSC cuprate system $La_{2-x}M_xCuO_4$ (M = Ca,Sr,Ba), is an antiferromagnetic (AF) insulator with a Néel temperature $T_N \sim 300$ K (Vaknin et al. 1987). In an ionic picture, the La, Cu and O ions have oxidation states of $+3$, -2 and $+2$, respectively, so that the Cu ions have a d^9 electronic configuration with spins 1/2. The AF superexchange coupling strength between adjacent Cu^{+2} ions in the CuO_2 layers was found to be very strong ($J \approx 1500$ K), and strong short range dynamic AF order was found to persist far above T_N, as expected for the two-dimensional (2D) Heisenberg model with $T_N \ll J$ (for reviews, see Birgeneau and Shirane 1989, Johnston 1991). An orthorhombic distortion of the structure occurs at 530 K, which introduces a Dzyaloshinskii-Moriya (DM) interaction between the near-neighbor Cu ions, which is then the largest anisotropy term in the spin Hamiltonian. This term in turn causes a pronounced cusp in $\chi(T)$ at T_N, and a slight canting of the AF structure and a ferromagnetic canted moment perpendicular to each CuO_2 layer below T_N. Since the canted moments in adjacent CuO_2 layers are antiferromagnetically aligned, the weak ferromagnetism below T_N is "hidden", but becomes observable at high magnetic fields H perpendicular to the CuO_2 layers when the canted moments in all CuO_2 layers are forced to align parallel to the field.

Bulk superconductivity in $La_{2-x}M_xCuO_4$ only appears at appreciable doping concentrations $x \gtrsim 0.1$, with the optimim superconducting transition temperature $T_c \approx 38$ K occurring for $M =$ Sr and $x \approx 0.15$, and at these doping levels the long-range AF order is suppressed (Johnston 1991). However, magnetic susceptibility $\chi(T)$ and especially inelastic neutron scattering measurements showed that strong dynamic AF correlations persisted in the metallic and superconducting compositions (Birgeneau and Shirane 1989, Johnston 1991). From these results, the superconducting layered cuprates may be viewed as doped antiferromagnets. All of the known classes of cuprate high temperature superconductors contain the same type of CuO_2 layers as in La_2CuO_4, and where there exists an insulating composition of a given class, it generally exhibits antiferromagnetic ordering. With regard to possible mechanisms for the HTSC, it is therefore of great interest to understand from both experimental and theoretical viewpoints the detailed magnetic behaviors of the parent compounds, as well as how these behaviors change as the compounds are doped into the metallic and superconducting regimes. In this paper, we review our recent results on several classes of HTSC cuprate parent compounds and related materials. In addition, we discuss similarities and differences of the properties of the recently discovered RNi_2B_2C layered structure intermetallic compound superconductors (Nagarajan et al. 1994, Cava et al. 1994, Siegrist et al. 1994) with those of the layered cuprate HTSCs.

2 $Sr_2CuO_2Cl_2$

This compound has the tetragonal K_2NiF_4 crystal structure, containing the same CuO_2 layers as in La_2CuO_4 but with the out-of-plane oxygens replaced by chlorine and the lanthanum by strontium (Miller et al. 1990, and references therein). This leaves the Cu in the +2 oxidation state as in La_2CuO_4, so that the Cu ions still have a d^9 electronic configuration and spins 1/2. $Sr_2CuO_2Cl_2$ is found to be an AF insulator with $T_N = 257$ K as determined from neutron diffraction measurements on single crystals (Vaknin et al. 1990, Greven et al. 1994, Greven et al. 1995). However, an important structural feature is that down to 10 K, $Sr_2CuO_2Cl_2$ remains tetragonal and does not show the orthorhombic distortion found for La_2CuO_4 (Miller et al. 1990, Vaknin et al. 1990). Thus, there is no DM anisotropy term in the spin Hamiltonian. This in turn allows spin exhange and other anisotropies which are much weaker than the DM term in La_2CuO_4 to be more easily identified and studied. The anisotropic magnetic susceptibility $\chi(T)$ from 5 to 400 K of a single crystal is shown in Fig. 1. Consistent with the lack of an orthorhombic distortion and DM interaction, there is no peak in $\chi(T)$ at T_N. In fact, there is no obvious feature identifying T_N in the data. The minimum in χ_c at about 310 K would normally be, and has mistakenly been (Vaknin et al. 1990), attributed to an AF transition, but the neutron diffraction measurements clearly show that T_N is at a much lower temperature. We will return to this question below.

Inelastic neutron scattering measurements (Greven et al. 1994, Greven et al. 1995) have demonstrated that the temperature dependence of the in-plane AF correlation length ξ from about 280 K to 600 K is well-described by theoretical predictions (Hasenfratz and Niedermayer 1991, Makivić and Ding 1991) for the spin 1/2 Heisenberg antiferromagnet on a square lattice. On the other hand, measurements below T_N show that the largest anisotropy present is an (easy-plane) XY anisotropy of magnitude $\alpha_{XY}J$, where $\alpha_{XY} = 1.4 \times 10^{-4}$ and $J = 125$ meV (Greven et al. 1994, Greven et al. 1995). Any in-plane easy-axis anisotropy was too small to observe in the neutron measurements. Since $\xi(300$ K$)$ is already ~ 100 lattice constants, one would expect a crossover from Heisenberg to XY-like behavior of $\xi(T)$ upon further cooling towards T_N. Due to finite resolution of the neutron measurements, this crossover was not observed.

Detailed $\chi(T)$ and ^{35}Cl nuclear magnetic resonance (NMR) measurements on single crystals have recently been carried out up to 400 K (Suh et al. 1995a).

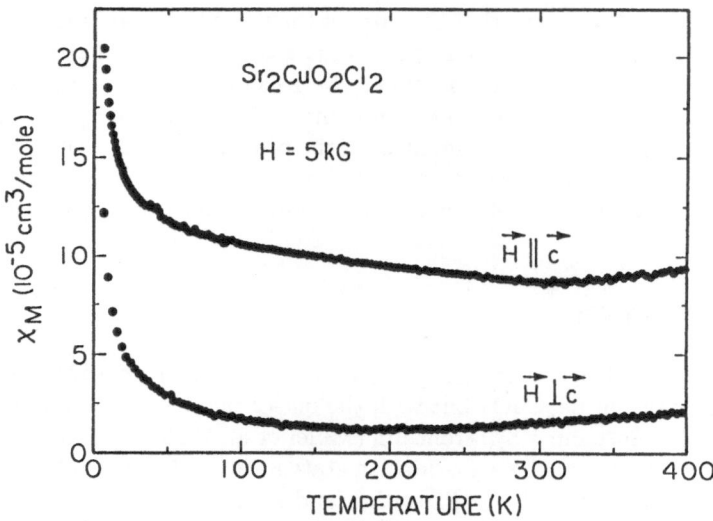

Fig. 1. Magnetic susceptibility χ_M of $Sr_2CuO_2Cl_2$ vs temperature for applied fields H perpendicular and parallel to the c-axis. The c-axis is perpendicular to the CuO_2 planes

From these data, it is found that T_N increases with H when $H \perp c$, and that T_N is independent of H when $H \parallel c$, as shown in Fig. 2. From the NSLRs for $H \parallel c$ and for $H \perp c$, one obtains the ratio of the spectral densities for spin fluctuations out-of-plane to those in-plane. This ratio was found to decrease rapidly as $T \rightarrow T_N^+$, strongly indicating a crossover from Heisenberg to XY-like behavior upon cooling towards T_N. Additional evidence for this crossover is seen in $2W(T)$ for $H \parallel c$ at $H = 4.7$ T as shown in Fig. 3. Whereas the data above 300 K show an exponential $1/T$ behavior as expected for a 2D Heisenberg system, upon further cooling the data diverge according to $2W = A\exp[B/(T - T_{KT})]^{1/2}$ (where $T_{KT} = 231$ K) as expected for an XY-like system (Makivić and Ding 1991, and references therein). The minimum in $\chi(T)$ at about 300 K in Fig. 1 was attributed (Suh et al. 1995a) to the same Heisenberg to XY-like crossover observed in the NMR measurements. The increase in T_N with $H \perp c$ was found to be semiquantitatively consistent with a field-induced crossover from XY-like to Ising-like behavior due to a field-induced in-plane Ising-like anisotropy. Consistent with this interpretation, no change in T_N with $H \parallel c$ was found, since such a field orientation does not change the magnetic anisotropy within the ab-plane. The work of Suh et al. (1995) clearly demonstrates the existence of spin dimensionality crossover from Heisenberg to XY-like (easy plane) behavior upon cooling towards T_N and indicates a further crossover to an Ising-like (easy axis in the ab-plane) behavior near T_N when a sufficiently strong (> 2 T) field is applied in the ab-plane.

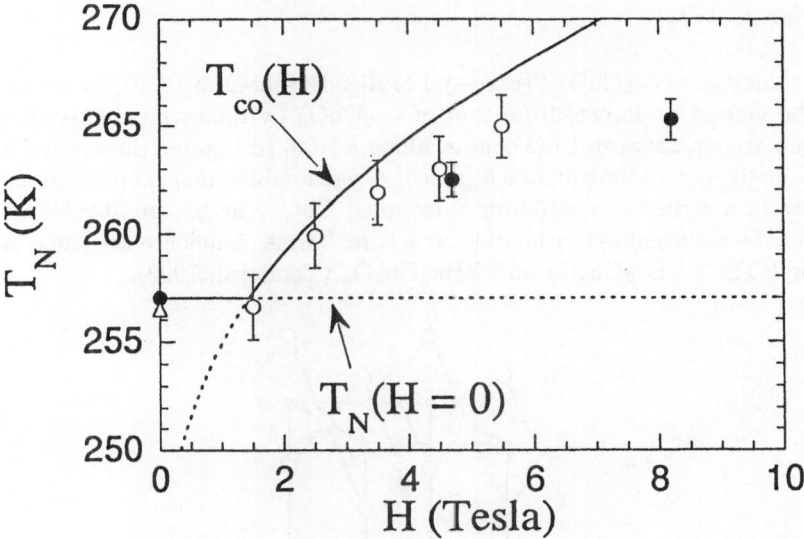

Fig. 2. Néel temperature T_N of $Sr_2CuO_2Cl_2$ vs applied magnetic field H in the ab-plane. The solid curve gives the calculated temperature of the XY to Ising crossover vs H

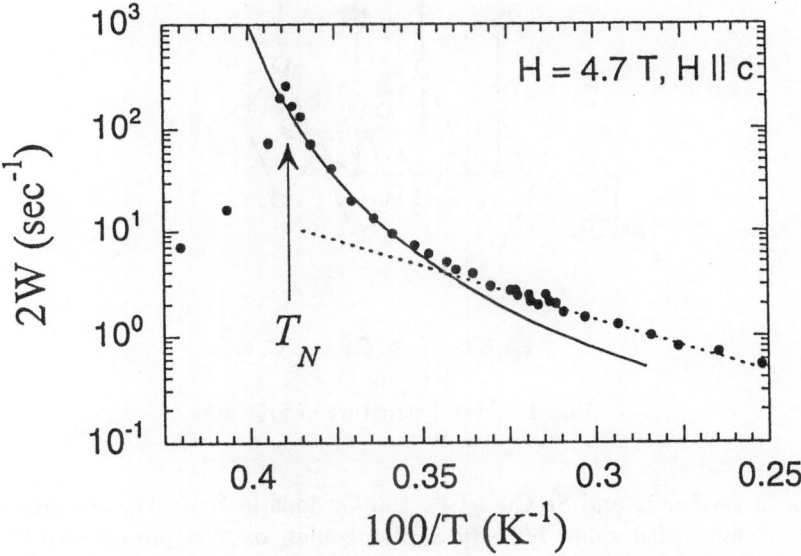

Fig. 3. ^{35}Cl nuclear spin–lattice relaxation rate $2W$ in $Sr_2CuO_2Cl_2$ vs inverse temperature $1/T$ for magnetic field H applied parallel to the c-axis

3 Sr₂CuO₃

The structure of Sr_2CuO_3 (Teske and Muller-Büschbaum 1969), shown in Fig. 4, can be viewed as derived from that of La_2CuO_4 by removing chains of in-plane oxygen atoms, between CuO chains, along a [100] tetragonal direction. The out-of-plane oxygen atom sublattice (apical oxygen sublattice) remains intact. This results in a structure containing uncoupled CuO_3 chains running along the a-axis of the resultant orthorhombic structure. These chains are the same as those in the HTSCs $YBa_2Cu_3O_7$ and $YBa_2Cu_4O_8$ (Vanderah 1992).

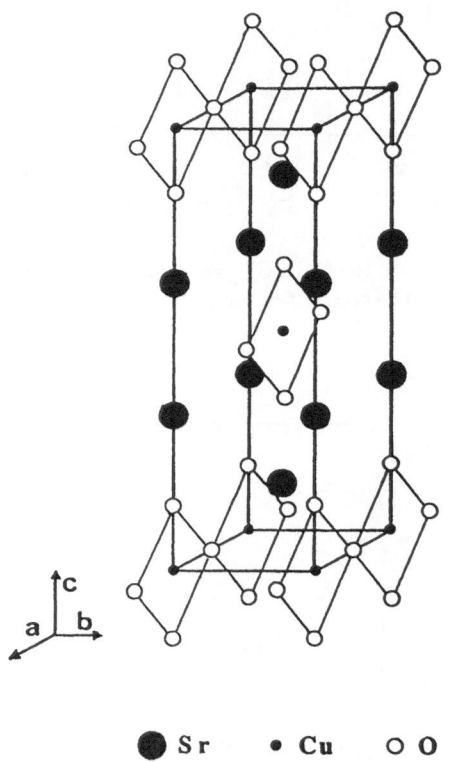

● Sr   • Cu   ○ O

Fig. 4. Crystal structure of Sr_2CuO_3

As in La_2CuO_4 and $Sr_2CuO_2Cl_2$, the Cu ions in Sr_2CuO_3 are presumably Cu^{+2} d^9 ions with spins 1/2. By applying high oxygen pressure (6 GPa) at high temperature (900 °C) to Sr_2CuO_3, one obtains oxygen-doped $Sr_2CuO_{3+\delta}$ with $\delta \approx 0.1$ (Hiroi et al. 1993), which becomes superconducting from 70 K to 94 K depending on the subsequent heat treatment (Hiroi et al. 1993, Han et al. 1994). The superconducting material is, on average, tetragonal with the K_2NiF_4 structure, but with an in-plane superlattice (Hiroi et al. 1993). Neutron diffraction measurements of superconducting samples showed, surprisingly, that the large fraction ($\approx 1/4$) of oxygen vacancies in the K_2NiF_4 structure were *all still located within the Cu planes* (Shimakawa et al. 1994), in contrast to all other known cuprate HTSCs where in-plane disorder invariably rapidly reduces T_c. The nature of the oxygen ordering which presumably gives rise to the in-plane superlattice is not yet known but would certainly be of great interest in this regard.

Because the Cu ions in Sr_2CuO_3 are evidently spin 1/2 ions and the CuO_3 chains are weakly coupled, the magnetic properties of this compound would be expected to correspond well to that of the spin 1/2 Heisenberg chain. Thus, the oxygen doping of the Sr_2CuO_3 parent compound to form the $Sr_2CuO_{3+\delta}$ high temperature superconductor would constitute a novel transformation of a 1D antiferromagnet to a 3D high T_c superconductor, which contrasts with the layered cuprates in which the undoped parent compounds are 2D antiferromagnets. Therefore, studies of the magnetic properties of the undoped Sr_2CuO_3 are of great interest.

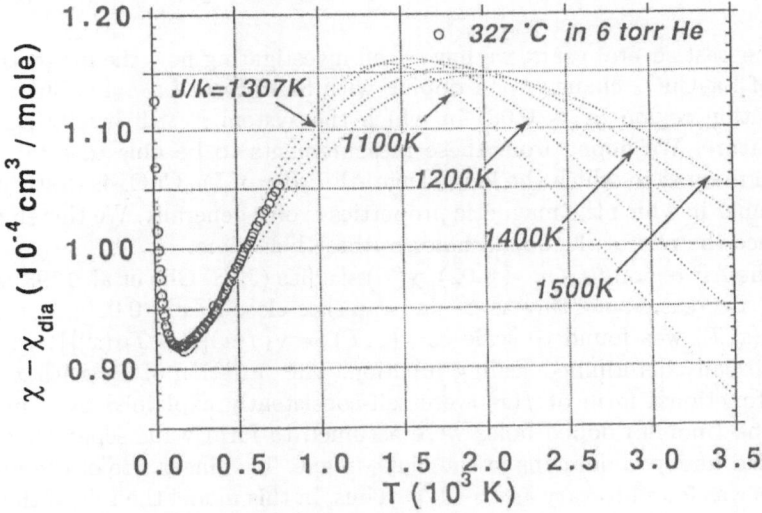

Fig. 5. Magnetic susceptibility corrected for the core diamagnetism, $\chi - \chi_{dia}$, vs temperature T for Sr_2CuO_3. Values calculated using the Bonner-Fisher theory for the respective listed values of AF exchange coupling constant are shown by the dotted lines. In this figure, the Cu-Cu coupling constant is defined as $2J$

Magnetic susceptibility studies were carried out on Sr_2CuO_3 samples with slight differences in oxygen content (Ami et al. 1995). Small amounts of excess oxygen were found to introduce a Curie-like impurity contribution to $\chi(T)$. The $\chi(T)$ for the (presumably) most stoichiometric sample with the smallest Curie term is shown in Fig. 5. The data could be fitted well with the Bonner-Fisher prediction for the spin 1/2 AF Heisenberg chain (Bonner and Fisher 1964), with a Cu-Cu superexchange interaction $J \approx 2600\,K$, significantly larger than the value of about 1500 K found in the layered cuprate HTSCs. A reanalysis of the data in Fig. 5 based on a recent recalculation (Eggert et al. 1994) of the theoretical susceptibility yielded (Eggert 1995) a significantly smaller value of J. Muon spin rotation measurements revealed a magnetic transition occurring at a temperature $T_M \approx 5\,K$ (Keren et al. 1993), but magnetic neutron diffraction measurements failed to reveal any indication of long-range magnetic order down to 1.5 K (Ami et al. 1995). In any case, the maximum value of the ratio T_M/J is apparently smaller than for any other linear chain antiferromagnet, which indicates that Sr_2CuO_3 is the most ideal AF chain system to date.

It would now clearly be of great interest to study how the magnetic properties of Sr_2CuO_3 evolve with oxygen doping, as the material is transformed into a high temperature superconductor. This would reveal similarities and/or differences with the layered cuprate systems.

4 $La_{2-x}Sr_xCuO_4$ in the Insulating Composition Region

Over the past several years, we have been investigating how the magnetic properties of La_2CuO_4 change upon doping with Sr to form $La_{2-x}Sr_xCuO_4$, in the composition region ($x \lesssim 0.08$) in which the system is still insulating at low temperature. We hoped from these measurements to be able to establish the detailed manner in which the long-range AF order of La_2CuO_4 is destroyed and the manner in which the magnetic properties evolve generally. We thereby hoped to deduce the role of the doped holes in these behaviors.

In the AF region ($0 \leq x < 0.02$), $\chi(T)$ studies (J. H. Cho et al. 1993) showed that T_N decreases smoothly as x^2 and vanishes close to $x = 0.020$. At the same time, $\chi(x, T)$ was found to scale as $\chi(x, T) = \chi\{f(x)[T - T_N(x)]\}$, with $f(x)$ an empirically determined scaling function. The variation of T_N with x as well as the functional form of $f(x)$ were self-consistently explained by a model in which the (mobile) doped holes were assumed to form walls separating nearly uncoupled undoped domains in the CuO_2 plane. The linear size of the undoped domains was found to vary as $L \sim 1/x$. Thus, in this model the role of the doped holes is effectively to break up the system up into smaller and smaller pieces with increasing doping level, a finite-size effect.

The same model was used and the same conclusion was reached based on [139]La nuclear quadrupole resonance (NQR) measurements in the so-called spin-glass regime $0.02 < x < 0.08$ (J. H. Cho et al. 1992), but where in that regime $L \sim 1/\sqrt{x}$ was found. The spin-glass transition temperature was found to vary as $T_g \sim 1/x$, also consistent with the model. Thus, in this composition regime the spin-glass was concluded to be a "cluster spin glass" in which locally dynamically AF ordered domains freeze due to their (small) mutual interaction.

Subsequent [139]La NQR measurements in the AF regime ($0 < x < 0.02$) (Chou et al. 1993) revealed a magnetic transition in the AF state at a temperature $T_f = (815\,K)x \ll T_N$. This transition was deduced to be due to freezing of the doped holes' effective spin degrees of freedom below T_f into a spin-glass-like state which is superimposed on the Cu^{+2} antiferromagnetic background. The magnetic phase diagram for $x < 0.05$ constructed on the basis of the above studies is shown in Fig. 6. From the figure, a distinct crossover is seen at $x \approx 0.02$ in the spin-glass transition temperature, implying a drastically different nature of the spin-glass transition in the two regimes $x < 0.02$ and $x > 0.02$.

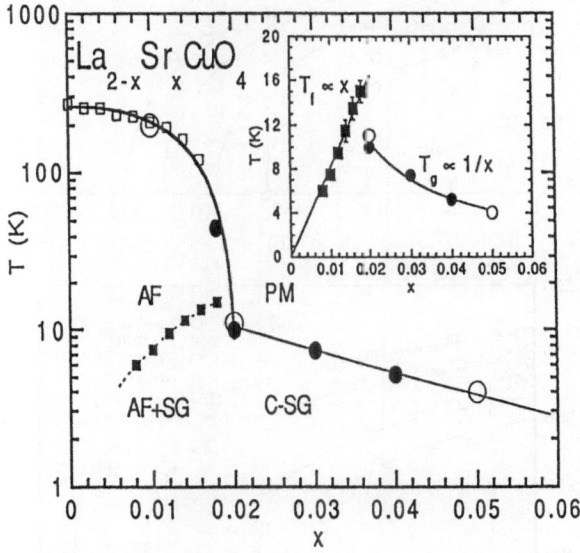

Fig. 6. Magnetic phase diagram of $La_{2-x}Sr_xCuO_4$ in the doping regime $x < 0.06$. Abbreviations: PM, paramagnetic; AF, antiferromagnetic; SG, spin-glass; CSG, cluster spin-glass; T_f and T_g, spin glass freezing temperatures for $x < 0.02$ and $x > 0.02$ respectively

The nature of the spin-glass transition at T_f, in the AF region, was investigated theoretically by Gooding et al. (1994). They assumed that in the temperature regime near T_f, the doped holes are localized on a plaquette of four oxygen ions neighboring a Cu ion in the CuO_2 plane. The effective spin degrees of freedom of the doped holes were found to correspond to the transverse (to the ordered longitudinal Cu moment) components of the nearby Cu spins. These magnetic polarons are the entities which couple and which freeze below T_f. Gooding et al. found a $T_f(x)$ in very good agreement with the data. To be consistent with the model above based on mobile holes, this interpretation implies that at some temperature below T_N, the doped holes become localized. This is consistent with transport measurements which indicate that the doped holes indeed become localized below $\sim 50\,K$ (Chen et al. 1991, Chen et al. 1995).

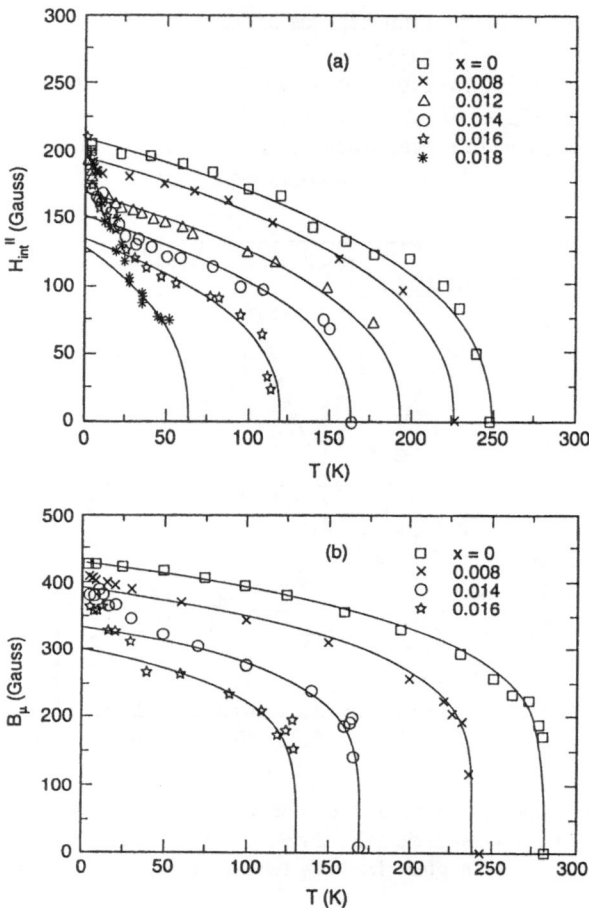

Fig. 7. Local internal magnetic fields in La$_{2-x}$Sr$_x$CuO$_4$ vs temperature T: (a) component of the local field at the La site parallel to the principal axis of the La site crystalline electric field $H_{int}^{\parallel}$, and (b) magnitude of the local field at the muon site B_μ in μSR measurements

Most recently, coordinated ^{139}La NQR and muon spin rotation (μSR) measurements were carried out on the same set of La$_{2-x}$Sr$_x$CuO$_4$ samples in the AF regime ($0 < x < 0.02$) (Borsa et al. 1995). From these measurements, the internal magnetic fields were determined versus temperature, as shown in Fig. 7. In the NQR measurements, the local internal field at the La site $H_{int}^{\parallel}$ is the component of the internal field parallel to the local crystalline electric field axis. In the μSR measurements, the magnitude of the local field at the muon site B_μ is measured. Both types of measurements give very similar temperature dependences, as seen in Fig. 7; i.e., $H_{int}^{\parallel}(T)$ is nearly proportional to $B_\mu(T)$. One concludes that both measurements are probing the temperature dependence of the staggered magnetization $M^\dagger(T)$ below T_N. Thus, the anomalous increases in the local fields in Fig. 7 below $\sim 30\,\mathrm{K}$ must be due to corresponding changes in $M^\dagger$. Indeed, $M^\dagger(x, T = 0)$ is seen to be independent of x! The anomalous increases in $M^\dagger$ below $30\,\mathrm{K}$ are inferred to be due to localization of the doped holes below this T. Consistent with this inference, the linewidth of the ^{139}La NQR is found to be proportional to x and independent of T for $x < 0.02$ below $4.2\,\mathrm{K}$, indicative of localized doped holes in this T range (Borsa et al. 1995).

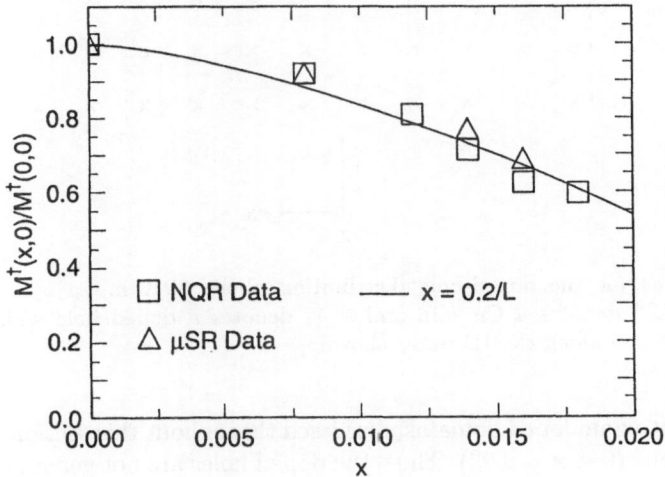

Fig. 8. Staggered magnetization $M^\dagger(x, T = 0)$, normalized to $M^\dagger(0, T = 0)$, vs Sr doping x in $La_{2-x}Sr_xCuO_4$. The values for $x > 0$ were obtained by extrapolating to $T = 0$ from above $30\,\mathrm{K}$. The solid curve is a theoretical fit to the data based on the model discussed in the text

In order to determine the values that $M^\dagger(x, T = 0)$ would have attained in the absence of hole localization, the data in Fig. 7 above $30\,\mathrm{K}$ were fitted to a power law in reduced temperature $T/T_N(x)$, as shown by the solid curves in Fig. 7, and the fits were extrapolated to $T = 0$. The extrapolated values of $M^\dagger(x, 0)/M^\dagger(0, 0)$ are plotted versus x in Fig. 8. These data were modeled assuming a doped hole distribution in the CuO_2 plane shown in Fig. 9 (Borsa et al. 1995). The doped holes are assumed to form 1D walls separating uncoupled undoped domains as in the model discussed above, where the average distance between holes in a wall is S_h and the distance between walls is denoted L. For a density of x holes in a given CuO_2 plane, one has $S_h L = 1/x$. An infinite stack of such planes is then assumed. With a ratio of interplanar to intraplanar Cu-Cu AF superexchange coupling constants given by 5×10^{-5} (Keimer et al. 1992), and with $S_h = 5a$ where a is the intraplanar Cu-Cu distance, conventional spin-wave theory then yields the $M^\dagger(x, 0)/M^\dagger(0, 0)$ versus x behavior shown as the solid curve in Fig. 8. The agreement with the data is seen to be excellent. This agreement supports the model of mobile (above $30\,\mathrm{K}$) doped holes situated in

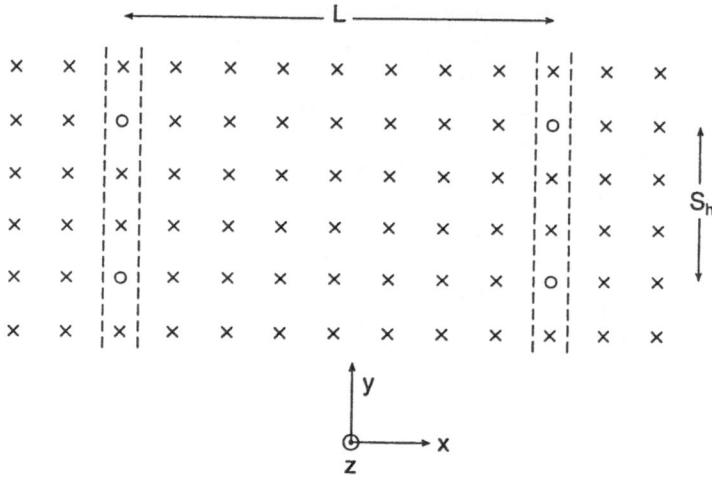

Fig. 9. Model for the doped-hole distribution above $30\,\mathrm{K}$ in $La_{2-x}Sr_xCuO_4$ with $x < 0.02$. "X" denotes a Cu spin and "$\bigcirc$" denotes a doped hole which is mobile in the y-direction along the 1D paths shown

walls separating undoped domains, discussed throughout this section, at least for the AF regime ($0 < x < 0.02$). Thus, the doped holes are not generally randomly distributed, at least not in the $La_{2-x}Sr_xCuO_4$ system, contrary to what is often implicitly assumed.

5 Ferromagnetic Interactions in $BaCuO_{2+x}$

The body-centered-cubic compound $BaCuO_{2+x}$ has a very interesting structure, shown schematically in Fig. 10 (Kipka and Müller-Buschbaum 1977, Weller and Lines 1989). The large unit cell, with lattice parameter $a = 18.3$ Å, contains 90 formula units. The cell consists of two $Cu_{18}O_{24}$ sphere clusters at the origin and body center positions, eight Cu_6O_{12} ring clusters with axes on the body diagonals, and six lone CuO_4 units located on the cell edges (after accounting for 1/2 random occupation). The Cu-O-Cu superexchange pathway within either type of cluster is close to 90°, in contrast to the approximately 180° pathway in all of the cuprate compounds discussed so far. Therefore, one expects the magnetic interaction between nearest-neighbor Cu ions within each cluster to be *ferromagnetic*, as opposed to the AF interaction for the 180° pathway.

Extensive magnetization and magnetic neutron diffraction measurements were carried out to study the magnetic interactions in a sample of He-annealed $BaCuO_{2+x}$ with $x = 0.1\pm0.1$ (Z. R. Wang et al. 1994, Z. R. Wang et al. 1995, X.-L. Wang et al. 1995). The neutron measurements showed the occurrence of some type of AF order below $T_N \approx 15\,\mathrm{K}$ (Z. R. Wang et al. 1994, X.-L. Wang et al. 1995).

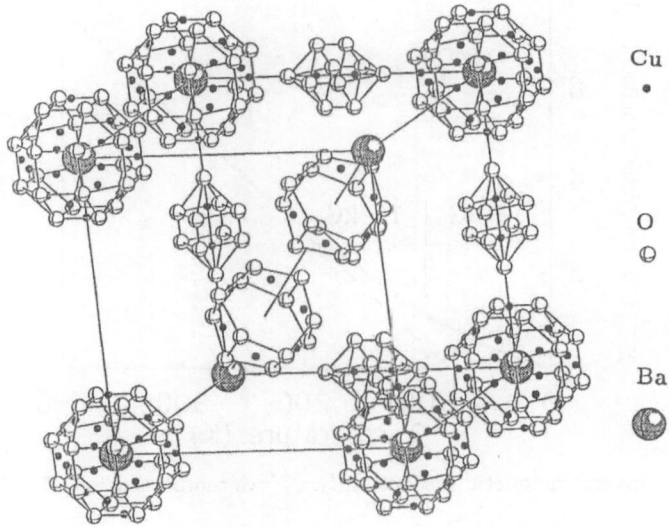

Fig. 10. Schematic illustration of the crystal structure of BaCuO$_{2+x}$. For clarity, only a subset of the atoms in the unit cell is shown

Cu

•

O

☺

Ba

The (low-field) $\chi^{-1}(T)$ data are shown in Fig. 11 (Z. R. Wang et al. 1994). Above ~ 250 K, the data approach a Curie-Weiss law with a Curie constant close to that for free spin 1/2 ions, and with a Weiss temperature $\theta = 81$ K. Thus, at high T, the Cu spins act as if they were all equivalent and interacting ferromagnetically. However, contrary to the prediction of molecular field theory, long-range ferromagnetic (FM) ordering is not observed at (or below) $T = \theta$. Rather, $\chi^{-1}(T)$ exhibits positive curvature below ~ 200 K and then approaches a Curie law below 6 K. Magnetization M versus H isotherms with H up to 5.5 T begin to exhibit negative curvature below ~ 50 K; the $M(H)$ data at 2 K are shown in Fig. 12 (Z. R. Wang et al. 1994, Z. R. Wang et al. 1995). The data can be decomposed into a part M_r linear in H and a part M_s which saturates above ~ 2 T. Analysis of the curvature of $M_s(H)$ reveals that this component must be coming from a species with a large spin $S \sim 10$. Since the only species in the unit cell which could have about that spin value is the Cu$_{18}$ sphere cluster, the $M_s(H)$ was concluded to arise from the sphere clusters with spin $S_s = 18(1/2) = 9$, which therefore remain paramagnetic down to 2 K and do not participate in the AF ordering seen by neutron diffraction. Thus, the Cu ions in the Cu$_{18}$ sphere clusters are ferromagnetically coupled and the sphere cluster have maximal spin 9 ground states. Further analysis shows that three lone Cu ions are antiferromagnetically coupled to each sphere cluster, and that these lone Cu ions also remain paramagnetic to 2 K. Due to the similarity of the internal geometries and bond angles in the Cu$_{18}$ and Cu$_6$ clusters, the Cu spins in the latter cluster were also inferred to be ferromagnetically coupled, yielding a maximal ground state spin $S_r = 3$ for the Cu$_6$ clusters. From Fig. 11, both

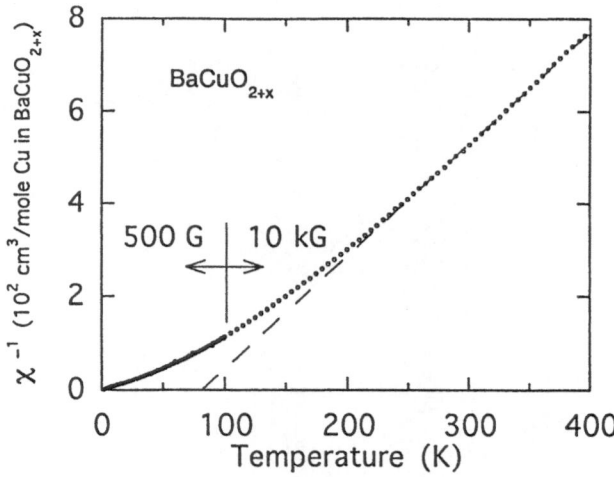

Fig. 11. Inverse magnetic susceptibility χ^{-1} vs temperature for $BaCuO_{2+x}$

types of clusters appear to be in their grouwnd states below $\sim 40\,K$.

The high-field slope of data such as in Fig. 12, plotted versus temperature, show a maximum at $\sim 15\,K$, consistent with the AF ordering at about this T seen by neutron diffraction. Since the only magnetic species in the unit cell not accounted for above are the Cu_6 ring clusters, these must be the species which exhibit AF *intercluster* order below $T_N \sim 15\,K$. The magnetic neutron diffraction peaks below T_N were successfully modeled assuming that only the Cu_6 clusters exhibit AF intercluster order and that the spin on each cluster is $S_r = 3$ (Z. R. Wang et al. 1994, X.-L. Wang et al. 1995). Analysis and fitting of the $\chi^{-1}(T)$ data above $70\,K$ in Fig. 11 allowed the FM Cu-Cu intracluster superexchange constants to be determined, $J_r = (290 \pm 60)\,K$ and $J_s = (80 \pm 16)\,K$ for the ring and sphere clusters, respectively; an excellent fit to the data above $70\,K$ in Fig. 11 was obtained using these parameters.

It is remarkable that despite its complexity, the spin system in $BaCuO_{2+x}$ can be adequately described by a simple cluster model, where the populations of the internal magnetic energy levels of the clusters are temperature dependent. In the field of high-temperature superconductivity, the conditions under which the Cu-O-Cu superexchange coupling is either ferromagnetic or antiferromagnetic are of prime importance. The above studies provide a clear demonstration of ferromagnetic Cu-O-Cu superexchange coupling within the Cu_6 and Cu_{18} clusters in $BaCuO_{2+x}$. This compound is one of very few cuprates in which such ferromagnetic superexchange has been proven to occur.

6 RNi_2B_2C (R = Y, Gd-Tm, Lu) Single Crystals

Although this review emphasizes the magnetic properties of cuprates, we depart briefly in this section to discuss the magnetic properties of the RNi_2B_2C

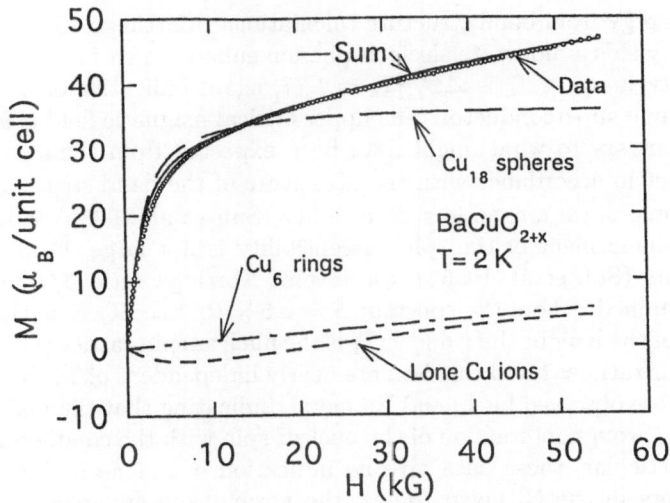

Fig. 12. Magnetization M of BaCuO$_{2+x}$ vs applied magnetic field H at 2 K. The solid curve labeled "Sum" is a theoretical fit to the data. The contributions to the Sum from the Cu$_{18}$ sphere clusters, the Cu$_6$ ring clusters and the lone Cu ions are also shown

compounds noted in the Introduction. These materials have high T_c compared with most other intermetallic compounds, reaching nearly 17 K for the $R =$ Lu member (Cava et al. 1994). One would like to know whether the layered atomic structure leads to strong anisotropies in the superconducting properties in applied magnetic fields. Even the compounds containing the magnetic rare earths $R =$ Ho, Er, Tm (Cava et al. 1994) and Dy (B. K. Cho et al. 1995a, Lin et al. 1995) exhibit superconductivity, albeit at lower temperatures. The Ni atoms form square lattice layers bounded by B layers, and these Ni$_2$B$_2$ layer blocks are separated and bonded together with RC layers (Siegrist et al. 1994). Since Ni compounds are often magnetic, one wonders whether the Ni layers have magnetic character similar to the Cu layers in the layered cuprates, and if so, what role that magnetic character plays in the superconducting mechanism. It is also of interest to investigate how the superconducting electrons interact with the magnetic R ions. All of these issues can be attacked most successfully by studying single crystals. A concerted effort has been underway at Ames Laboratory for the past 1 1/2 years to grow single crystals of the RNi$_2$B$_2$C compounds and to study their properties. Here, we briefly describe some of the salient results. The crystals are grown from a Ni$_2$B flux at 1200–1500 °C, yielding crystal masses up to 700 mg (B. K. Cho et al. 1995b).

Superconducting state magnetization (Xu et al. 1994, Xu et al. 1995, Johnston-Halperin et al. 1995) and heat capacity (Movshovich et al. 1994) measurements of single crystal YNi$_2$B$_2$C show that this material is a bulk type II superconductor with a Ginzburg-Landau parameter $\kappa \sim$ 13–15. The normal state linear heat capacity coefficient is $\gamma = 19$ mJ/mole K^2. Using the bare density of states at the Fermi energy from band structure calculations (Mattheiss 1994, Pickett and Singh 1994) yields a moderate electron-phonon enhancement factor $\lambda = 0.7$. The heat capacity jump at T_c is $\Delta C/\gamma T_c = 1.77$, again indicative of a moderately strong-coupling superconductor. The upper critical magnetic field is found to be isotropic, contrary to what might have been expected from the layered crystal structure, but in accordance with the 3D nature of the band structure.

The normal state magnetic susceptibility (Suh et al. 1995b) indicates that exchange enhancement of the spin susceptibility is not large. From ^{11}B NMR measurements (Suh et al. 1995b), the inverse Korringa ratio $S/K^2 T_1 T$ versus T was determined, where the constant $S = 2.5 \times 10^{-6}$ sec K, K is the isotropic component of the Knight shift, and $1/T_1$ is the nuclear spin-lattice relaxation rate (NSLR). This ratio ≈ 1 and the data are nearly independent of T; the results are similar to those observed for Li and Na metal, indicating that the main source of the NSLR is through interaction of the nuclear spin with the conduction electron spins. In particular, these data give no indication of magnetic fluctuations or local moments in the Ni layers. Thus, the normal and superconducting state magnetic and thermal properties measured to date indicate that YNi$_2$B$_2$C is a conventional moderately coupled electron-phonon, weakly exchange enhanced, 3D intermetallic d-band superconductor.

Extensive anisotropic magnetization measurements have been performed on single crystals of the RNi$_2$B$_2$C series with magnetic rare earths R = Gd (Canfield et al. 1995), Tb (B. K. Cho et al. 1995c), Dy (Dervenagas et al. 1995), Ho (Canfield et al. 1994, B. K. Cho et al. 1995d), Er (B. K. Cho et al. 1995b) and Tm (B. K. Cho et al. 1995e). The data for the Gd compound are nearly isotropic and exhibit AF ordering at $T_N = 20$ K. The remaining magnetic members of the series exhibit strongly anisotropic normal state behaviors, attributed mainly to crystalline electric field (CEF) effects, and AF ordering is observed for each compound. The influence of the CEF on the normal state magnetization of the Ho member has been quantitatively evaluated and fitted, and the CEF parameters have been determined (B. K. Cho et al. 1995d). From the magnetization measurements, the R^{+3} magnetic moments in the Tb, Dy, Ho and (below 150 K) Er compounds were found to have an easy-plane anisotropy, with the easy plane being the ab (RC) plane, whereas the Tm moments in the R = Tm compound were found to have an easy axis anisotropy with the easy axis along the c axis. These anisotropies are in agreement with point-charge CEF predictions (B. K. Cho et al. 1995c). In view of the existence of strong CEF effects on the magnetization of the magnetic RNi$_2$B$_2$C compounds, it is somewhat surprising that the T_N values are nearly linearly proportional to the de Gennes factor DG = $(g_J - 1)^2 J(J + 1)$ (B. K. Cho et al. 1995a) as shown in Fig. 13, which would be expected in the *absence* of CEF effects. On the other hand, the variation of T_c with DG factor does show strong deviations from de Gennes scaling, as shown in Fig. 14; these deviations likely arise mainly from CEF effects.

Magnetic neutron diffraction measurements have determined the magnetic structures of several members of the RNi$_2$B$_2$C series. The Dy (Dervenagas et al. 1995) and, below 5 K, the Ho (Goldman et al. 1994, Grigereit et al. 1994, Huang

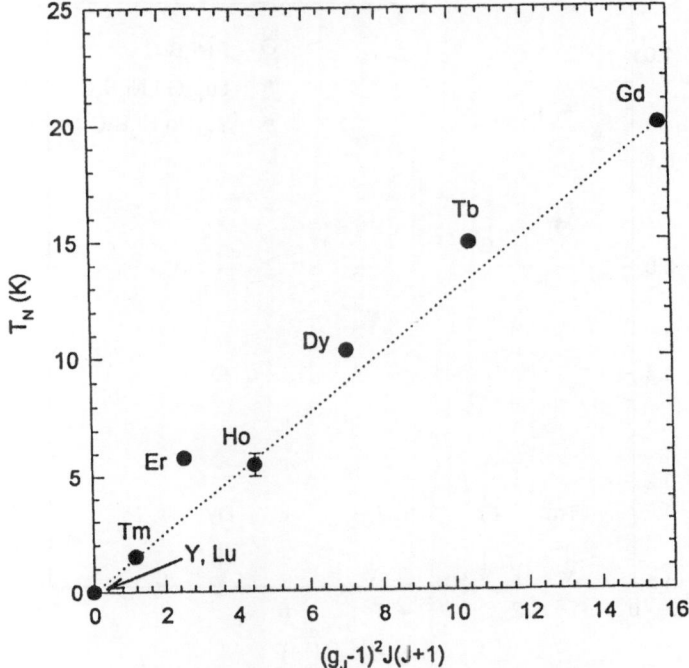

Fig. 13. Néel temperature T_N vs de Gennes factor $(g_J - 1)^2 J(J + 1)$ of the R^{+3} ions in $R\mathrm{Ni_2B_2C}$ single crystals

et al. 1995) members exhibit AF order which is commensurate with the lattice, with the R moments within a given RC plane aligned ferromagnetically in the plane and with the moments in adjacent planes aligned antiferromagnetically. On the other hand, the Er member (Zarestky et al. 1995, Sinha et al. 1995) and, between 5 and 6 K, the Ho member exhibit incommensurate AF order. This incommensurability may be associated with electronic features of the band structure (Rhee et al. 1995). The easy magnetization plane or axis in the magnetically ordered state of each compound was found to be the same as inferred from the above magnetization measurements.

The anisotropic temperature variations of the upper critical magnetic field $H_{c2}(T)$ for the Tm (B. K. Cho et al. 1995e), Er (B. K. Cho et al. 1995b) and Ho (Canfield et al. 1994) members of the $R\mathrm{Ni_2B_2C}$ series are shown in Fig. 15. The $H_{c2}(T)$ for the Tm compound, which has $T_N = 1.5\,\mathrm{K}$, shows broad maxima around 4 K which are most likely due to the increasing magnetization of the Tm sublattice in a given field with decreasing T. The cusp in $H_{c2}(T)$ for the Er compound at about 6 K for $H \parallel c$ is associated with AF ordering of the Er sublattice at this temperature. The presence of the cusp for $H \parallel c$ and only a plateau for $H \perp c$ may be attributable to the observation (B. K. Cho et al. 1995b) that T_N is independent of H for $H \parallel c$ whereas T_N decreases rapidly

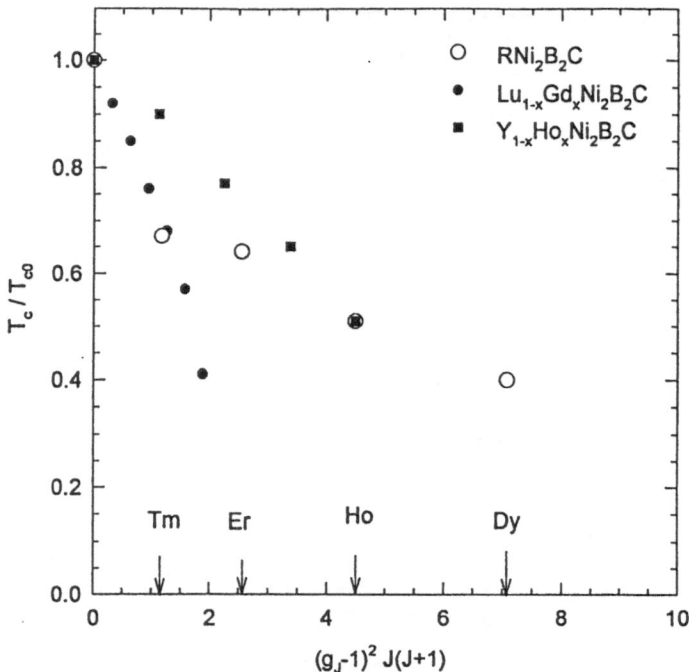

Fig. 14. Superconducting transition temperature T_c of RNi_2B_2C crystals, normalized to the transition temperature T_{c0} of the nonmagnetic Y or Lu members, vs de Gennes factor $(g_J - 1)^2 J(J + 1)$ of the R^{+3} ions

with H for $H \perp c$. The nearly re-entrant $H_{c2}(T)$ for the Ho compound at 5 K may be associated with the change in AF structure from incommensurate to commensurate with the lattice upon cooling below this T. The sign of the anisotropies in Fig. 15 are of the same sign as expected for electromagnetic effects on $H_{c2}(T)$: the smallest H_{c2} at each T is found for the field direction giving the largest magnetization of the R sublattice. In contrast with the anisotropic $H_{c2}(T)$ behaviors seen in Fig. 15, $H_{c2}(T)$ for the Dy member is found to be isotropic (B. K. Cho et al. 1995f), which in view of the strongly anisotropic susceptibility is not presently understood.

In summary, the RNi_2B_2C compounds exhibit interesting anisotropic super-conducting and normal state magnetic properties. Further work is necessary to determine whether the observed behaviors can be explained by existing theory, or whether new physics is required.

Acknowledgments

Ames Laboratory is operated for the U.S. Department of Energy by Iowa State University under Contract No. W-7405-Eng-82. The work at Ames was supported

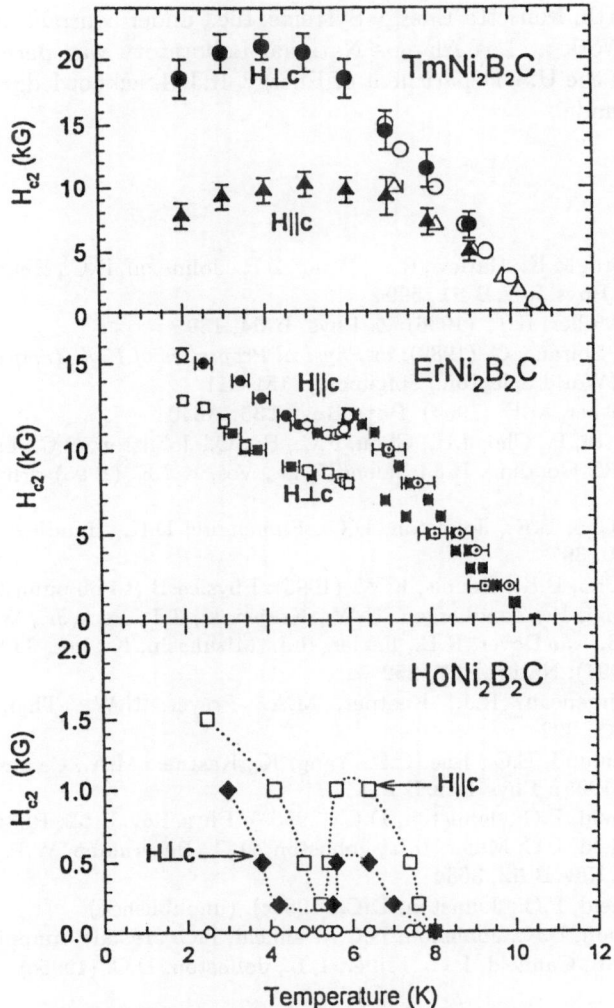

Fig. 15. Anisotropic upper critical magnetic field H_{c2} vs temperature of $TmNi_2B_2C$, $ErNi_2B_2C$ and $HoNi_2B_2C$ single crystals

by the Director for Energy Research, Office of Basic Energy Sciences. The work at the National Synchrotron Light Source, Brookhaven National Laboratory was supported by the U.S. Department of Energy, Division of Materials Sciences and Division of Chemical Sciences. Neutron diffraction work at Brookhaven was supported by the Division of Materials Sciences, USDOE under Contract No. DE-AC02-76Ch00016. Work performed at Oak Ridge National Laboratory was sponsored by the Division of Materials Sciences, U.S. Department of Energy. Oak Ridge National Laboratory is managed for the U.S. Department of

Energy by Martin Marietta Energy Systems, Inc., under contract no. DE-AC05-84OR21400. Work at Los Alamos National Laboratory was performed under the auspices of the U.S. Department of Energy. R.J.G. acknowledges support by NSERC of Canada.

References

Ami, T., Crawford, M.K., Harlow, R.L., Wang, Z.R., Johnston, D.C., Huang, Q., Erwin, R.W. (1995): Phys. Rev. B **51**, 5994

Bednorz, J.G., Müller, K.A. (1986): Z. Phys. B **64**, 189

Birgeneau, R.J., Shirane, G. (1989): in *Physical Properties of High Temperature Superconductors I* (World Scientific, Singapore), 151–211

Bonner, J.C., Fisher, M.E. (1964): Phys. Rev. **135**, A640

Borsa, F., Carretta, P., Cho, J.H., Chou, F.C., Hu, Q., Johnston, D.C., Lascialfari, A., Torgeson, D.R., Gooding, R.J., Salem, N.M., Vos, K.J.E. (1995): Phys. Rev. B (to be published)

Canfield, P.C., Cho, B.K., Johnston, D.C., Finnemore, D.K., Hundley, M.F. (1994): Physica C **230**, 397

Canfield, P.C., Cho, B.K., Dennis, K.W. (1995): Physica B (to be published)

Cava, R.J., Takagi, H., Zandbergen, H.W., Krajewski, J.J., Peck, Jr., W.F., Siegrist, T., Batlogg, B., van Dover, R.B., Felder, R.J., Mizuhashi, K., Lee, J.O., Eisaki, H., Uchida, S. (1994): Nature **367**, 252

Chen, C.Y., Birgeneau, R.J., Kastner, M.A., Preyer, N.W., Thio, T. (1991): Phys. Rev. B **43**, 392

Chen, C.Y., Branlund, E.C., Bae, C.S., Yang, K., Kastner, M.A., Cassanho, A., Birgeneau, R.J. (1995): Phys. Rev. B **51**, 3671

Cho, B.K., Canfield, P.C., Johnston, D.C. (1995a): Phys. Rev. B **52**, R3844

Cho, B.K., Canfield, P.C. Miller, L.L., Johnston, D.C., Beyermann, W.P., Yatskar, A. (1995b): Phys. Rev. B **52**, 3684

Cho, B.K., Canfield, P.C., Johnston, D.C. (1995c): (unpublished)

Cho, B.K., Harmon, B.N., Johnston, D.C., Canfield, P.C. (1995d): (unpublished)

Cho, B.K., Xu, M., Canfield, P.C., Miller, L.L., Johnston, D.C. (1995e): Phys. Rev. B **52**, 3676

Cho, B.K., Canfield, P.C., Johnston, D.C. (1995f): (unpublished)

Cho, J.H., Borsa, F., Johnston, D.C., Torgeson, D.R. (1992): Phys. Rev. B **46**, 3179

Cho, J.H., Chou, F.C., Johnston, D.C. (1993): Phys. Rev. Lett. **70**, 222

Chou, F.C., Borsa, F., Cho, J.H., Johnston, D.C., Lascialfari, A., Torgeson, D.R., Ziolo, J. (1993): Phys. Rev. Lett. **71**, 2323

Dervenagas, P., Zarestky, J., Stassis, C., Goldman, A.I., Canfield, P.C., Cho, B.K. (1995): Physica B **212**, 1

Eggert, S., Affleck, I., Takahashi, M. (1994): Phys. Rev. Lett. **73**, 332

Eggert, S. (1995): unpublished

Ginsberg, D.M. (1989): editor, *Physical Properties of High Temperature Superconductors*, Vols. I-IV (World Scientific, Singapore)

Goldman, A.I., Stassis, C., Canfield, P.C., Zarestky, J., Dervenagas, P., Cho, B.K., Johnston, D.C., Sternlieb, B. (1994): Phys. Rev. B **50**, 9668

Gooding, R.J., Salem, N.M., Mailhot, A. (1994): Phys. Rev. B **49**, 6067

Greven, M., Birgeneau, R.J., Endoh, Y., Kastner, M.A., Keimer, B., Matsuda, M., Shirane, G., Thurston, T.R. (1994): Phys. Rev. Lett. **72**, 1096

Greven, M., Birgeneau, R.J., Endoh, Y., Kastner, M.A., Matsuda, M., Shirane, G. (1995): Z. Phys. B **96**, 465

Grigereit, T.E., Lynn, J.W., Huang, Q., Santoro, A., Cava, R.J., Krajewski, J.J., Peck, Jr., W.F. (1994): Phys. Rev. Lett. **73**, 2756

Han, P.D., Chang, L., Payne, D.A. (1994): Physica C **228**, 129

Hasenfratz, P., Niedermayer, F. (1991): Phys. Lett. B **268**, 231

Hiroi, Z., Takano, M., Azuma, M., Takeda, Y. (1993): Nature **364**, 315

Huang, Q., Santoro, A., Grigereit, T.E., Lynn, J.W., Cava, R.J., Krajewski, J.J., Peck, Jr., W.F. (1995): Phys. Rev. B **51**, 3701

Johnston, D.C. (1991): J. Magn. Magn. Mater. **100**, 218

Johnston-Halperin, E., Fiedler, J., Farrell, D.E., Xu, M., Cho, B.K., Canfield, P.C., Finnemore, D.K., Johnston, D.C. (1995): Phys. Rev. B **51**, 12852

Keimer, B., Belk, N., Birgeneau, R.J., Cassanho, A., Chen, C.Y., Greven, M., Kastner, M.A., Aharony, A., Endoh, Y., Erwin, R.W., Shirane, G. (1992): Phys. Rev. B **46**, 14034

Keren, A., Le, L.P., Luke, G.M., Sternlieb, B.J., Wu, W.D., Uemura, Y.J., Tajima, S., Uchida, S. (1993): Phys. Rev. B **48**, 12926

Kipka, R., Müller-Buschbaum, Hk. (1977): Z. Naturforsh. Teil B **32**, 121

Lin, M.S., Shieh, J.H., You, Y.B., Hsu, Y.Y., Chen, J.W., Lin, S.H., Yao, Y.D., Chen, Y.Y., Ho, J.C., Ku, H.C. (1995): Physica C **249**, 403

Makivić, M.S., Ding, H.-Q. (1991): Phys. Rev. B **43**, 3562; Ding, H.-Q. (1992): Phys. Rev. Lett. **68**, 1927

Mattheiss, L.F. (1994): Phys. Rev. B **49**, 13279

Miller, L.L., Wang, X.L., Wang, S.X., Stassis, C., Johnston, D.C., Faber, Jr., J., Loong, C.-K. (1990): Phys. Rev. B **41**, 1921

Movshovich, R., Hundley, M.F., Thompson, J.D., Canfield, P.C., Cho, B.K., Chubukov, A.V. (1994): Physica C **227**, 381

Nagarajan, N., Mazumdar, C., Hossain, Z., Dhar, S.K., Gopalakrishnan, K.V., Gupta, L.C., Godart, C., Padalia, B.D., Vijayaraghavan, R. (1994): Phys. Rev. Lett. **72**, 274

Pickett, W.E., Singh, D.J. (1994): Phys. Rev. Lett. **72**, 3702

Rhee, J.Y., Wang, X., Harmon, B.N. (1995): Phys. Rev. B **51**, 15585

Shimakawa, Y., Jorgensen, J.D., Mitchell, J.F., Hunter, B.A., Shaked, H., Hinks, D.G., Hitterman, R.L., Hiroi, Z., Takano, M. (1994): Physica C **228**, 73

Siegrist, T., Zandbergen, H.W., Cava, R.J., Krajewski, J.J., Peck, Jr., W.F. (1994): Nature **367**, 254

Sinha, S.K., Lynn, J.W., Grigereit, T.E., Hossain, Z., Gupta, L.C., Nagarajan, R., Godart, C. (1995): Phys. Rev. B **51**, 681

Suh, B.J., Borsa, F., Miller, L.L., Corti, M., Johnston, D.C., Torgeson, D.R. (1995a): Phys. Rev. Lett. (in press)

Suh, B.J., Borsa, F., Torgeson, D.R., Cho, B.K., Canfield, P., Johnston, D.C. (1995b): (unpublished)

Teske, Chr.L., Muller-Büschbaum, Hk. (1969): Z. Anorg. Allg. Chem. **371**, 325

Vaknin, D., Sinha, S.K., Moncton, D.E., Johnston, D.C., Newsam, J.M., Safinya, C.R., King, Jr., H.E. (1987): Phys. Rev. Lett. **58**, 2802

Vaknin, D., Sinha, S.K., Stassis, C., Miller, L.L., Johnston, D.C. (1990): Phys. Rev. B **41**, 1926

Vanderah, T.A. (1992): editor, *Chemistry of Superconductor Materials*, (Noyes, Park Ridge, NJ, U.S.A.)

Wang, X.-L., Fernandez-Baca, J.A., Wang, Z.R., Vaknin, D., Johnston, D.C. (1995): Physica B (to be published)

Wang, Z.R., Wang, X.-L., Fernandez-Baca, J.A., Johnston, D.C., Vaknin, D. (1994): Science **264**, 402

Wang, Z.R., Johnston, D.C., Miller, L.L., Vaknin, D. (1995): Phys. Rev. B (in press)

Weller, M.T., Lines, D.R. (1989): J. Solid State Chem. **82**, 21

Xu, M., Cho, B.K., Canfield, P.C., Finnemore, D.K., Johnston, D.C. (1994): Physica C **235**, 2533

Xu, M., Canfield, P.C., Ostenson, J.E., Finnemore, D.K., Cho, B.K., Wang, Z.R., Johnston, D.C. (1995): Physica C **227**, 321

Zarestky, J., Stassis, C., Goldman, A.I., Canfield, P.C., Dervenagas, P., Cho, B.K., Johnston, D.C. (1995): Phys. Rev. B **51**, 678

HgBa$_2$Ca$_{n-1}$Cu$_n$O$_{2n+2+\delta}$ and Y$_2$Ba$_4$Cu$_{6+n}$O$_{14+n}$ Single Crystals: High Pressure Synthesis and Properties

J.Karpinski[1], H.Schwer[1], K.Conder[1], J.Löhle[1], R.Molinski[1,2], A.Morawski[2], Ch.Rossel[3], D.Zech[4], J.Hofer[4]

[1]Laboratorium für Festkörperphysik ETH 8093-Zürich,
[2]High Pressure Research Center PAS Warszawa,
[3]IBM Research Division Zürich,
[4]Physik Institut Universität Zürich

Abstract. An overview is presented of the phase diagram, crystal growth, physical and structural investigations of Y$_2$Ba$_4$Cu$_{6+n}$O$_{14+n}$ and HgBa$_2$Ca$_{n-1}$Cu$_n$O$_{2n+2+\delta}$ compounds synthesized at high pressure.

Keywords. Crystal growth, high pressure, Y$_2$Ba$_4$Cu$_{6+n}$O$_{14+n}$, HgBa$_2$Ca$_{n-1}$Cu$_n$O$_{2n+2+\delta}$.

1. Introduction

There are two main reasons for using high pressure for the synthesis of superconducting compounds:
1) Equilibrium partial pressure of a volatile component is high at the conditions of synthesis.
2) High hydrostatic pressure stabilizes the structure.
In the first case high pressure of an active gas component is required for a chemical equilibrium between a solid and a gas phase. Synthesis of several compound, YBa$_2$Cu$_4$O$_8$ (124), Y$_2$Ba$_4$Cu$_7$O$_{15-x}$ (247) and HgBa$_2$Ca$_{n-1}$Cu$_n$O$_{2n+2+\delta}$ (Hg-12(n-1)n) serve as examples. The required pressure of active gas component (O$_2$, HgO or Hg vapors) can reach values up to 3 kbar at the temperatures close to melting points, for these materials.
In the second case, the synthesis can be performed in a pure hydrostatic pressure of neutral medium. This can be a neutral gas or a quasi-hydrostatic solid-medium pressure in systems like belt, piston-cylinder or anvils. The value of pressure required for a stabilization of the high pressure structure is usually higher then 10 kbar and typically the solid-medium high pressure synthesis of superconductors have been performed at pressures 30-50 kbar. Using high hydrostatic pressure method, the one of the most interesting superconducting compounds has been synthesized: infinite layer - (Ca$_x$A$_{1-x}$)CuO$_2$ (where A-alkaline earth metal), which has the simplest superconducting structure containing only CuO$_2$ planes separated by Ca atoms [1]. Investigations of the origin of the superconductivity in this compound are very important for the understanding of the superconductivity phenomenon in all other cuprates. However, this method, which uses solid materials, usually a pyrophylite, as the pressure medium, does not allow to grow single crystals due to the absence of a free space and no gas atmosphere available for the growing crystals. On the other hand, single crystals are of crucial importance for the studies of both crystallographic and anisotropic physical properties.

Using the examples of $Y_2Ba_4Cu_{6+n}O_{14+n}$ and $HgBa_2Ca_{n-1}Cu_nO_{2n+2+\delta}$ families of compounds we will present data on the application of high pressure technique for the crystal growth of superconductors with volatile components.

Since the discovery of $YBa_2Cu_3O_{7-x}$ (123), the YBaCuO system has been studied intensively in a large range of oxygen pressure, $10^{-4} \leq Po_2 \leq 3000$ bar. In the course of these investigations two other superconducting phases 124 and 247 have been synthesized both as ceramic and single crystals [2,3,4].

In case of the $HgBa_2Ca_{n-1}Cu_nO_{2n+2+\delta}$ materials, whose synthesis at normal pressure causes many problems due to both the low decomposition temperature of HgO and the high O_2-gas and Hg-vapour pressures, the application of the solid-medium high-pressure technique has led to the almost single phase ceramic samples [5]. But only with the use of the high-pressure gas technique, single crystal growth was possible [6].

2. Phase diagram of YBaCuO system

The Y-Ba-Cu-O is a quaternary system, which makes its investigation rather complicated. In order to show phase-relationships more clearly, we will discuss section of the ternary Y_2O_3, BaO, CuO system along the 123-CuO tie line. Figure 1 shows a part of the ternary system at $Po_2 = 1$ bar. At 800°C all three superconducting compounds, 123, 247 and 124, are stable as single phases in equilibrium with the O_2 gas phase. However, in the presence of CuO, at the same P,T-conditions, 123 and 247 react with CuO to form 124. This indicates the importance of the composition of the sample for the phase stability. Therefore, we have investigated the P,T phase diagrams for 123, $123^1/_2(247)$ and 124, all of which lie on the tie line 123-CuO. Based on these P,T-diagrams, we were able to construct the T-x sections of the ternary system for various pressures ($1 \leq Po_2 \leq 3000$ bar). Another important section of the ternary phase diagram includes 123, 124 or 247 and the ternary eutectic. This pseudobinary diagram displays the conditions for the crystal growth.

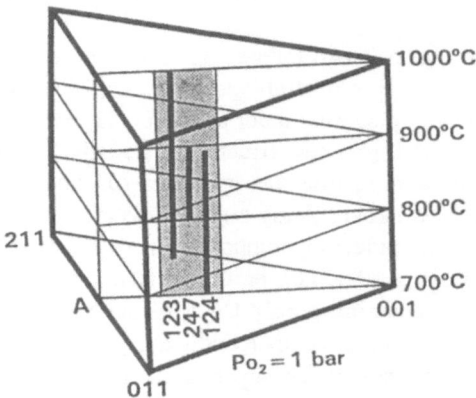

Fig.1. T-x section of the ternary Y_2BaCuO_5 (211)-$BaCuO_2$(011)-CuO(001) system at $Po_2 = 1$ bar. Parts of these sections are presented in Fig.3 for various Po_2.

2.1. P-T phase diagrams of the 123-CuO system

Results of annealings the 123 + CuO samples at various pressures and temperatures for two compositions 124 and $123^{1}/_{2}(247)$ are shown in Fig. 2. Experimental details have been published earlier [7]. The stability fields of 123, 247 and 124 phases are quite different and strongly dependent on the composition of the samples.

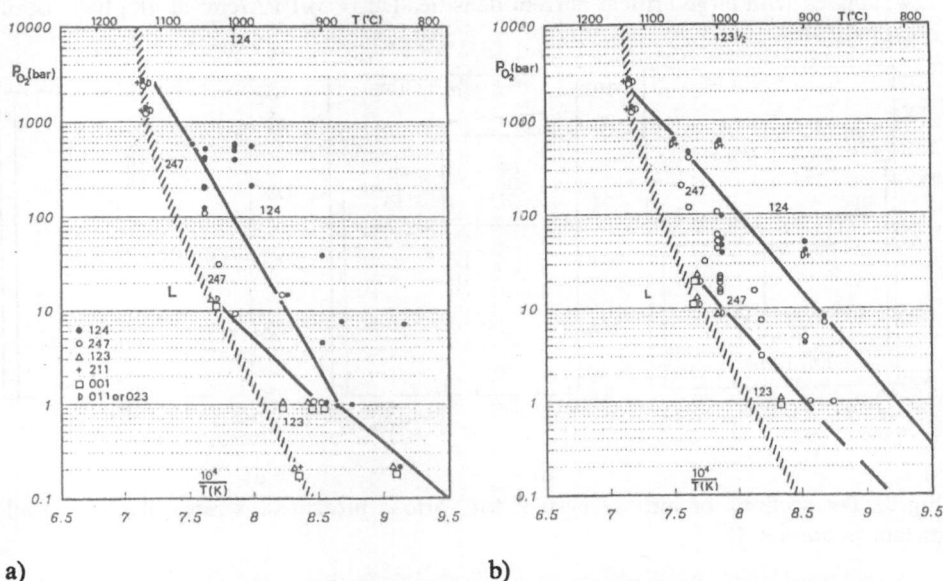

a)   b)

Fig.2. LogPo$_2$-1/T phase diagrams for the compositions Y:Ba:Cu; a)1:2:4, b)1:2:3$^1/_2$.

Composition 124: Our results show, that 124 decomposes at high temperature to 247 + CuO and 123 + CuO at Po$_2 \geq 1$ bar and Po$_2 \leq 1$ bar, respectively. The decomposition line of 124 intersects the peritectic decomposition line at Po$_2 \approx 3000$ bar, indicating that at this pressure it is possible to obtain 124 bulk by peritectic reaction from the melt. We have not observed a low temperature limitation of the stability range of 124. Therefore, we concluded that 124 is the only phase in $Y_2Ba_4Cu_{6+n}O_{14+n}$ system, which is thermodynamically stable at normal conditions.

Composition 247: The stability field of 247 in the phase diagram (Fig.2 b)) is limited at the high temperature side by the peritectic decomposition at higher pressures and the decomposition to 123 at lower pressures.

2.2. T-x sections of the ternary YBaCuO system

Phase equilibria will be presented using T-x section of the ternary YBaCuO system along the 123-CuO tie line for various pressures. The composition parameter, x corresponds to the ratio of CuO to all metals in at.%. Figure 1 shows a part of the ternary system with vertical T axes at Po$_2$=1 bar. In the Fig.3 three sections of this three-dimentional system are shown for various pressures. One can observe the variations of the phase stability ranges with Po$_2$. All three superconducting phases are stable at $Po_2 = 1$ bar in a limited temperature range (Fig.3a)). The stability field of 123 decreases with increasing Po_2 and completely disappears at $Po_2 = 100$ bar (Fig.3b)). The stability field of 247 is limited from both the high and low temperature side. Its decomposition temperature is always higher then that of 124. In the $Po_2 = 3000$ bar diagram (Fig.3c)), the 124 remains the only stable phase. Therefore, it is possible to crystalize bulk 124 from the melt at this pressure. Using this technique the textured 124 samples with large critical current densities (Jc = $3 \times 10^5 A/cm^2$ at 4K) have been obtained [9].

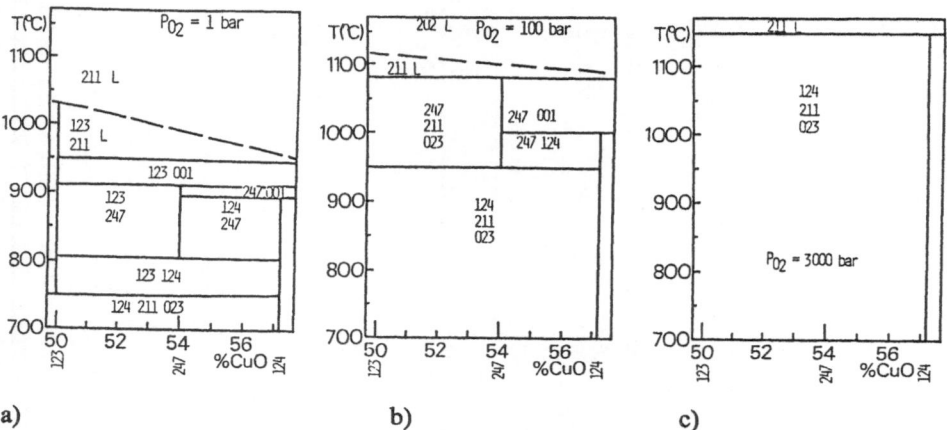

a)   b)   c)

Fig.3. T-x sections of ternary system for various pressures. x-ratio of CuO to all metals in atomic %.

2.3. P-T-x phase diagram and crystal growth

In order to determine the conditions for the crystal growth one has to consider the P-T-x phase diagram corresponding to the flux composition. Figure 4 shows T-x sections of the ternary system for two pressures (a) $Po_2 = 1$ bar, (b) $Po_2 = 1000$ bar.

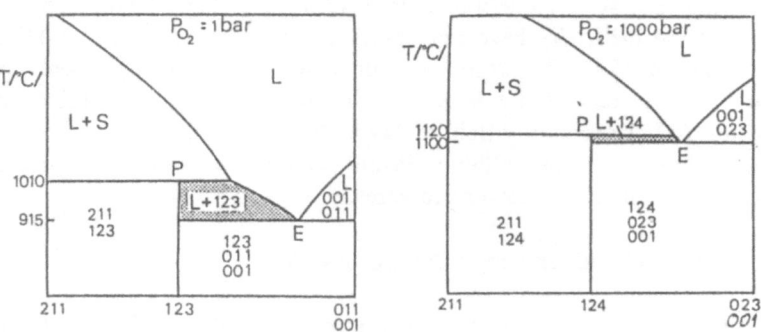

Fig.4. Simplified T-x phase diagram for the crystal growth for $Po_2 = 1$ bar and $Po_2 = 1000$ bar. Notice the increase of eutectic and peritectic melting temperatures with oxygen pressure.

The increase of the eutectic and peritectic melting temperatures with pressure are evident from this diagram. A lack of data for various flux compositions does not allow to give more details on these diagrams. Figure 5 shows P-T phase diagram for the total flux composition $YBa_3Cu_{7.5}O_x$ or $123:BaCuO_2:CuO$ in the ratio 1:1:3.5. Note the P-T conditions for the growth of 124, 247 and 123 crystals. Since the 247 crystal growth field is very narrow it is rather difficult to grow 247 crystals without the intergrowth of 123 or 124. The eutectic melting temperature of the flux increases with pressure from $\approx 900°C$ at $Po_2 = 1$ bar to $\approx 1100°C$ at $Po_2 = 1000$ bar.

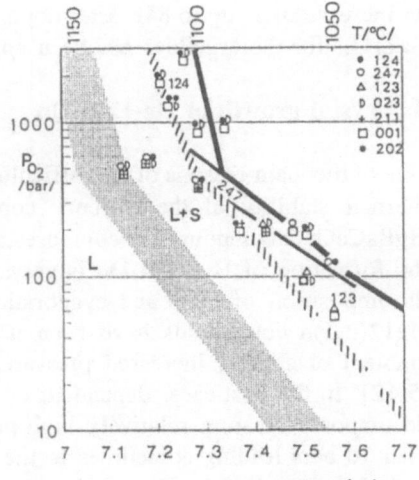

Fig. 5. Log Po_2-1/T phase diagram for total composition of the flux Y:Ba:Cu; 1:3:7.5. Solid line - eutectic melting.

Simultaniuosly, the peritectic melting temperature of the YBaCuO system increases from 1010°C to 1130°C. At $Po_2 = 1$ bar, the difference between peritectic decomposition and eutectic melting is 110°C, allowing the growth of 123 in this wide temperature range. At $Po_2 = 1000$ bar this difference is only 30°C. This narrow temperature range makes crystal growth very difficult. However, growth rate at high pressure and teperature is much faster than at normal pressure. Crystals of 1×2 mm^2 size were grown in 12 to 24 hours.

3. Crystal growth of the 124 and 247 from solution

The 124 and 247 compouds melt incongruently. Therefore, in order to decrease the melting temperature and avoid peritectic decomposition a use of a flux is necesary. In our experiments the following composition was used: $123:BaCuO_2:CuO$ in the ratio (1) 1:1:3.5 and (2) 1:4:8. The oxygen pressure has been varied for different experiments from 60 to 2800 bar. Three kinds of crucibles have been used for the crystal growth: Al_2O_3, ZrO_2 and Y_2O_3. The crystal growth runs typically consist of: (a) heating up to 1060-1120°C with rate 5°C/min, (b) dwelling for 1 to 48 h at maximum temperature, (c) cooling 1-5°C/h down to 1050°C, (d) cooling 1-5°C/min down to room temperature. Due to high reactivity between ZrO_2 and the YBaCuO, the dwell time has been reduced to 0.5h. The composition, number and size of the crystals depend on various experimental parameters, (Po_2, T_{max}, time, cooling rate, composition of the flux and crucible material). T_c of the 124 crystals varies depending on doping during crystal growth. The 124 crystals grown in Al_2O_3 crucible have $T_c = 72-73K$. Unfortunately, ZrO_2 reacts strongly with the YBaCuO. Nevertheless, fast crystal growth procedures with ZrO_2 crucibles allowed to grow 124 single crystals with $T_c \approx 80K$. The best results have been obtained with an Y_2O_3 crucible which allowed an increase of T_c up to 84K after doping with Ca. Recently new promising material, $BaZrO_3$, for the crucibles has been applied for the 123 crystal growth [10].

4. Crystal growth of Hg-12(n-1)n

One of the main reasons of the difficulties with the synthesis of Hg-12(n-1)n is the low thermal stability of the mercury compounds. HgO, used for the preparation of HgBaCaCuO compounds decomposes at ambient pressure at about 400°C. Kinetics of the formation of Hg-12(n-1)n is very slow at this temperature. In order to prevent decomposition of HgO and evaporation of Hg before the reaction takes place the Hg12(n-1)n compounds have been usually synthesized in quarz ampoules [11] at ambient or slightly increased pressure, or at high pressure in a belt type apparatus [5,12]. In the first case, dependent on a free volume of the ampule, a part of HgO decomposes creating relatively high pressure of Hg, HgO, and O_2 vapours of more than 10 bars leading sometimes to the explosion of the ampules. Due to difficulties with the control of the Hg partial pressure it is very difficult to obtain stoichiometric samples. Therefore, high pressure technique with a solid-medium, usually pyrophylite have been frequently applied. This technique has many limitations: (I) the sample size is limited up to ~ 0.5 cm^3, (II) the pressure and temperature distributions are not homogeneous and the partial oxygen pressure is difficult to control, resulting in an undefined preparation conditions, (III) single crystal growth is impossible, due to the absence of the gas atmosphere, (IV) there is a large possibility of introducing impurities from the pressure medium. These disadvantages are avoided in a gas pressure system. An Ar gas atmosphere with a controlled partial oxygen pressure leaves free space for the single-crystal growth. A temperature gradient in a multizone furnace is relatively easy to control. A maximum sample volume can be several cm^2. We have synthesized Hg12(n-1)n and infinite layer-CaCuO$_2$ single crystals and polycrystalline samples using such a system [6,13,14,15].

4.1. Thermodynamics of HgO

Although many papers appeared on the synthesis of the Hg superconductors, there is no information in the literature concerning thermodynamics of this system. As the HgO is a volatile chemical component of the system we have calculated partial pressures resulting from the decomposition of HgO only in closed crucibles (Fig.6). Dependent on the free volume of the crucible and the amount of the material, these pressures can reach several hundreds bars at the synthesis temperature. By a variation of the ratio between a free volume and the mass of the sample one can control partial pressures of the components. Fig.6 a) shows calculations for 1.5 cm^3 free volume in the crucible and 1g HgO, which corresponds to conditions of our experiment where Hg-1223 crystals have been grown. At the temperature of synthesis (1323-1343K) solid HgO and liquid Hg disapeared from the sample. In the cool zone of the crucible HgO condensates. The total vapor pressures reach value 400-500 bar. By increasing the amount of HgO up to 2g, liquid Hg remains in the sample and total pressure increases up to 600-800 bar. Application of starting O_2 pressure of 100 bar shifts the chemical equilibrium towards stabilization of HgO, which leads to chemical transport of HgO outside of the growth zone. In our crystal growth experiments, an increase of the partial pressures, by lowering of the crucible volume, leads to the synthesis of compounds with higher n.

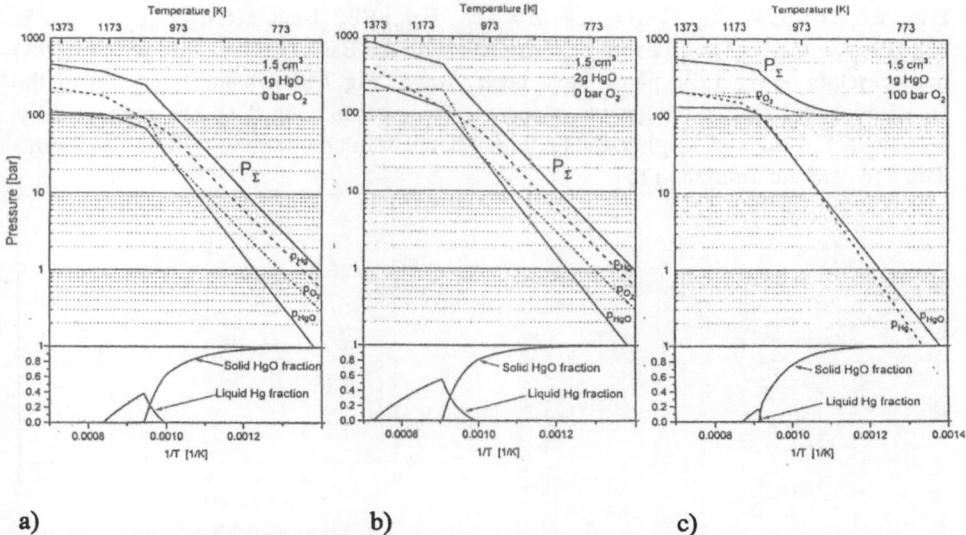

a)   b)   c)

Fig.6. Partial and total pressures created in a closed crucible of a free volume 1.5 cm³ by a decomposition of: a) 1g HgO; b) 2g HgO. c) shows results of the application of additional O_2 pressure 100 bar. In the lower part of the figure the fraction of solid HgO and liquid Hg is shown at certain temperature.

4.2. Crystal growth experiment

Hg-12(n-1)n compounds melt peritectically, but at ambient pressure they decompose before melting and the volatile components evaporate. Our goal was to prevent decomposition by an encapsulation of the sample with a high hydrostatic inert gas pressure until the peritectic temperature is reached. Due to the high density of Ar gas at this pressure the evaporation of Hg is strongly supressed. Additionally, application of a flux with a lower melting point allows growth of single crystals below the peritectic decomposition temperature. PbO or $BaCuO_2$-CuO mixtures have been used as a flux. The crystallization temperature was $1020° < T < 1070°C$. All experiments have been performed at Ar pressure of 10 kbar. Two crucible materials have been applied: alumina and yttria. Crystal growth in the alumina crucible leads to a considerable doping with Al and mixed (HgAl)BaCaCuO have been obtained. In yttria crucibles, we have crystallized HgPb-1223, HgPb-1234 and HgPb-1245 pure phase crystals as well as HgPb-1212 doped with Y. The detailes of the crystal growth procedure have been published previously [6,13,14,15]. The maximum T_c of HgPb-1234 single crystals is 130K. There is a relatively good agreement of T_c values for our single crystals and the literature data for ceramic samples for n =4 (130K) (Fig.8) and n=5 (125K). However, for n = 3 (130K) and n=1 (70K), the T_c's are lower, probably due to the substitution of Y for Ca. Unexpectedly, in the same batch, we found crystals of both, the 1245 and 1212 phases, but only 1212 contained large amount of Y. It seems, that the solubility of Y is larger in 1212 than in 1234 or 1245 crystals. The composition of our crystals grown from PbO flux determined by microprobe measurements were for n=2: $Hg_{0.5}Pb_{0.5}Ba_2Ca_{0.5}Y_{0.5}Cu_2O_{6-x}$, n=3: $Hg_{0.5}Pb_{0.5}Ba_2Ca_{1.9}Y_{0.1}Cu_3O_{8-x}$, n=4: $Hg_{0.8}Pb_{0.2}Ba_2Ca_3Cu_4O_{10+x}$, n=5: $Hg_{0.5}Pb_{0.5}Ba_2Ca_4Cu_5O_{12-x}$. From experiments with the $BaCuO_2$:CuO flux without PbO pure crystals of Hg-1223 phase have been grown. Fig.7 shows single crystal of the Hg-1234 compound with growth steps on the surface. Fig.8 shows magnetically measured T_c values of single crystals with various n in comparison with the literature data for ceramic samples [16].

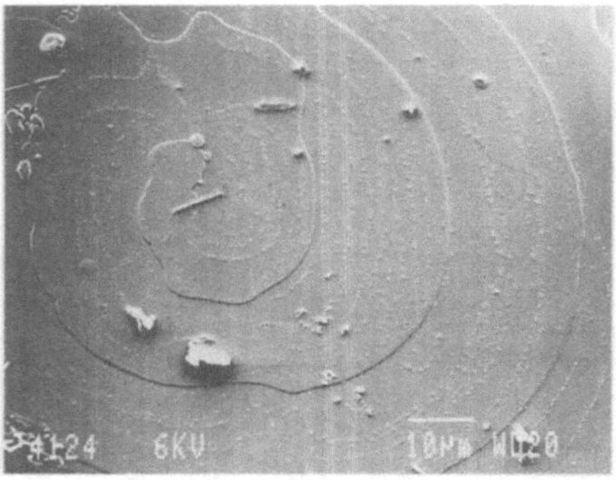

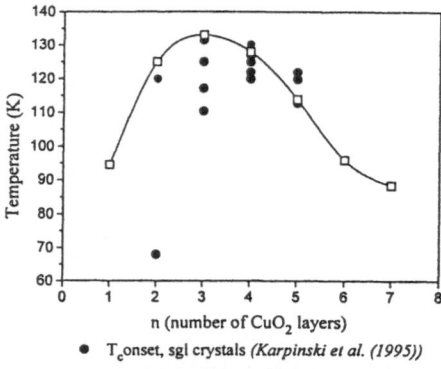

Fig.7. Single crystal of HgPb-1234 phase with spiral growth steps.

Fig.8. T_c values of Hg-12(n-1)n single crystals with various n in comparison with the literature data [16] for ceramic samples.

5. Transport measurements.

The in-plane resistance was measured on $Hg_{0.8}Pb_{0.2}Ca_2Ba_3Cu_4O_{10+x}$ single crystals by the van derPauw method [17]. The crystal was contacted at the edges in such a way that the current flows along the ab-plane and the magnetic field was applied parallel to the c-axis. Fig.9 shows the resistivity data of two single crystals with different transition temperatures. Sample A was a polished crystal with an onset temperature T_c=130K. Sample B was an as grown sample with a T_c of 125K. The shape of $\varrho(T)$ is not linear, unlike in other optimally doped superconductors, but has a downward curvature which is typical for underdoped cuprates. The linearity is reached at temperatures above 300 K. The resistive transition in a magnetic field up to 10 T have been used to determine the irreversibility line H*(T) (Fig.10). This line devides the H-T-phase diagram in two regions of irreversible (low H,T) and reversible magnetic behaviour. Above this line no lossless current transport is possible. In this region of the H-T-phase diagram the superconducting material is not used for technical applications. The upper limit criterion for the determination of the irreversible temperature T*(H) was $10^{-9}\Omega cm$. For the comparison the irreversiblity lines of $YBa_2Cu_4O_8$, $YBa_2Cu_3O_{7-x}$, and BSCCO single crystals are shown.

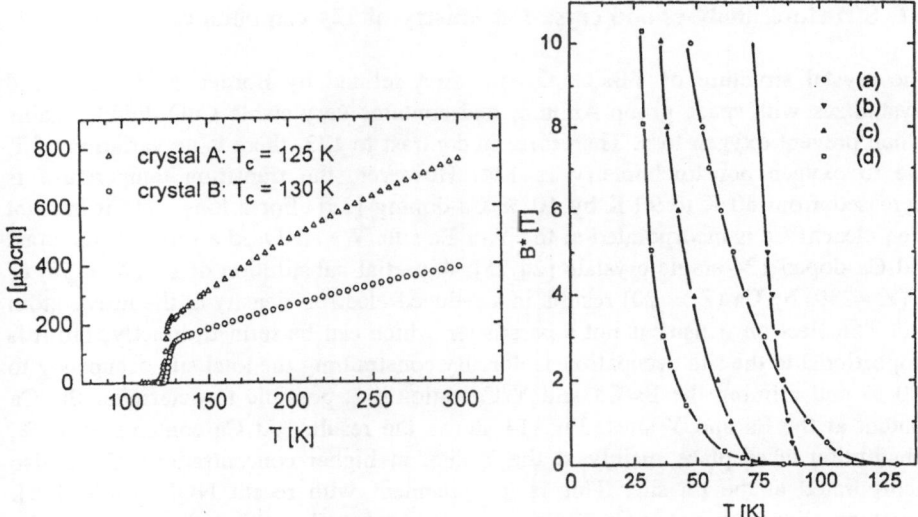

Fig.9. Resistivity versus temperature for two $Hg_{0.8}Pb_{0.2}Ba_2Ca_3Cu_4O_{10+x}$ single crystals with different T_c. Sample A: $T_c=130K$, B: $T_c=125K$.

Fig.10. Irreversibility lines for: HgPb-1234 single crystal (a), Y123(b), Y124(c) and $Bi_2Sr_2CaCuO_x$(d).

6. Structural investigations.

In order to understand changes of physical properties by a variation of growth conditions or chemical composition it is necessary to know the crystal structure on an atomistic scale. Very often superconducting properties are correlated with particular bondlengths in HT_c-cuprates: e.g. the transition temperature T_c is a function of the apical bondlength $Cu(2) - O(1)$ in YBaCuO compounds, connecting Cu of the superconducting CuO_2 plane with the bridging oxygen to the charge reservoir in the CuO chains [19,20].

Structure analysis is performed by X-ray diffraction of single crystals and refinement of structural parameters. Crystals for structure analysis have average dimensions of $0.2 \times 0.2 \times 0.02$ mm³ and are checked for intergrowth, twinning, and peak broadening with an X-ray precession camera, before mounting them on a Siemens P4 four-circle single crystal diffractometer. Peak profiles of reflections of the whole Ewald spere in the range $3° < 2\Theta < 70°$ are recorded with an ω-2Θ scan, using graphite monochromated MoK radiation ($\lambda = 0.71073$ Å). Between 2000 and 6000 intensity data were collected, corrected for Lorentz, polarisation, and absorption effects, and reduced to symmetry independent structure factors. Positional, thermal, and occupation parameters are obtained by full-matrix least-squares refinements with weighted structure factors, using the SHELXTL program [21]. The quality of a refinement is expressed by a disagreement factor R between observed and calculated structure factors. Bondlengths and distances are calculated from refined fractional atom coordinates and from unit cell dimensions.

6.1. Structure analyses and crystal chemistry of 124 compounds

The crystal structure of $YBa_2Cu_4O_8$ was first refined by Bordet et al. [22]. 124 crystallizes with space group Ammm, and contains very stable CuO double chains which prevent oxygen loss. Therefore, in contrast to 123, there is no variation of T_c due to oxygen nonstoichiometry in 124. However, the transition temperature is increased from 80 K to 90 K by 10 % Ca-doping [23]. For a long time it has not been clear if Ca is incorporated at the Y or Ba site. We analysed a series of undoped and Ca doped 124 single crystals [24,25]. A partial substitution of Ba (Z = 56) or Y (Z = 39) by Ca (Z = 20) results in a reduced electron density at the heavy atom site. The electron density is not a parameter which can be refined directly, but it is proportional to the site occupation factor. By constraining the total site occupancy to 100 % and refining the Ba/Ca and Y/Ca ratios it is possible to determine the Ca content at the Ba and Y sites. Fig. 11 shows the results: At Ca contents < 6 %, substitution takes place mainly at the Y site, at higher concentrations Ca is also incorporated at the Ba site. This is in agreement with recent NQR results [26]. Structural changes occur in Ca-124 crystals due to the substitution of trivalent Y^{3+} by divalent Ca^{2+} and due to ion size effects. The replacement of the small Y (r = 1.02 Å) by the larger Ca (r = 1.12 Å) increases the distance to the adjacent oxygen layers and diminishes the buckling of the CuO_2 superconducting planes. However, changes in the apical bond Cu(2) - O(1) were not observed upon Ca-doping which might explain the T_c increase.

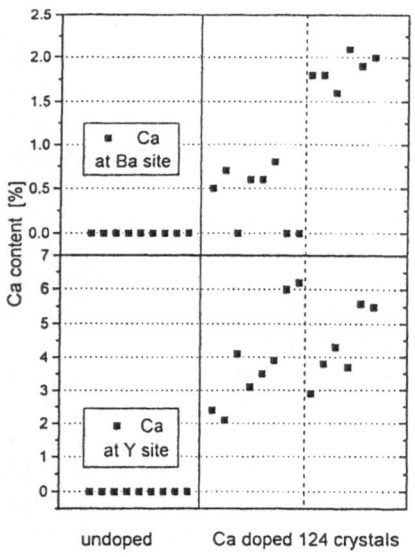

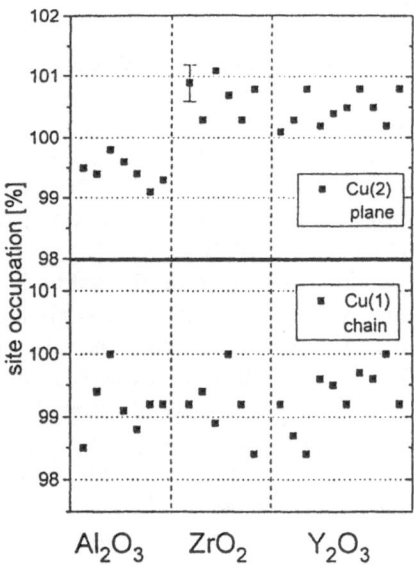

Fig. 11 Refined Ca contents at the Y and Ba site in (Ca-) 124 single crystals.

Fig. 12 Refined site occupancies of the Cu(1) chain and the Cu(2) plane site of 124 single crystals grown in different crucible materials

Differences in apical bondlengths between undoped and Ca-doped 124 crystals were observed in low temperature experiments in the range 100 K < T < 300 K [27]: At room temperature there is no difference, but with decreasing temperatures the contraction of Cu(2) - O(1) is much stronger in Ca-124. At T_c the apical bondlength is about 1 % shorter in Ca-124 compared to undoped crystals, which gives a possible explanation for the increase of the transition temperature by Ca-doping.

124 single crystals grown in Al_2O_3 crucibles have transition temperatures up to 73 K, whereas crystals grown in ZrO_2 and Y_2O_3 crucibles have a T_c of 78 and 79 K respectively. EDX and microprobe analyses have shown that the low-T_c crystals contain up to 0.15 atom-% Al. Due to the ionic size it is expected that Al (r = 0.53 Å) is incorporated at a Cu (r = 0.55 Å) site. It corresponds either to a 1 % replacement of Cu(1) in the double chains or to a 1 % substitution of Cu(2) in the CuO_2 superconducting planes. More than 20 structure analyses of 124 crystals grown in different crucible materials have been performed, and the occupancies of both the Cu(1) and the Cu(2) site have been refined [25,28]. Fig. 12 shows that the site occupation of the Cu(1) chain site did not show significant changes between crystals grown in different crucible materials. However, the occupancy of Cu(2) (plane site) is about 1 % lower in all crystals grown in alumina crucibles compared to those from ZrO_2 and Y_2O_3 crucibles. This result suggests that a 1 - 2 % substitution of Cu by Al in the superconducting CuO_2 layers is responsible for a decrease in T_c of 7 K.

6.2. Structure analyses and crystal chemistry of 247 compounds

The crystal structure of $Y_2Ba_4Cu_7O_{14+x}$ was first determined by Bordet et al. [3]. The structure contains in c-direction alternating blocks with CuO single chains (123-units) and CuO double chains (124 units) and crystallizes with space group Ammm. The transition temperature varies generally as a function of the oxygen content x, but samples with even the same oxygen concentration showed T_c differences of up to 40 K. We analysed a series of 247 crystals with 0 K < T_c < 93 K in order to find structural parameters which enhance T_c [29]. Single crystal structure refinements showed that structural changes only occur in the 123-units, whereas the 124 unit remains stable over the hole T_c range. With increasing T_c a reduction of the Cu(2) -O(1) apical bondlength is observed, like in 123 or in Ca-124 under cooling. Crystals with lower transition temperatures have a reduced electron density at the Cu(1) site in the single chain 123-unit (Fig. 13). This deficiency is explained by the introduction of defects at the Cu(1)

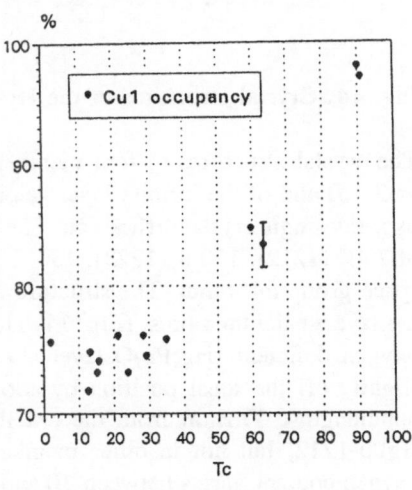

Fig. 13. Occupancy of Cu(1) in the single chain 123-unit of $Y_2Ba_4Cu_7O_{14+x}$ as a function of T_c.

site which may be Al, carbonate groups or vacancies. Probably, they destroy the local order, lead to a misorientation of single chains and reduce T_c. In contrast to 124, Ca doping in has no effect on the superconducting properties of 247. Surprisingly, Ca mainly replaces Ba(1) in the 123 unit and Y in the ratio 3 : 2, whereas Ba(2) in the 124 unit is not affected [24]. The partial substitution of the small Ca (r = 1.12 Å) for the large Ba (r = 1.42 Å) reduces the unit cell size of Ca-247.

6.3. Structure analyses and crystal chemistry of the HgPb-12(n-1)n family

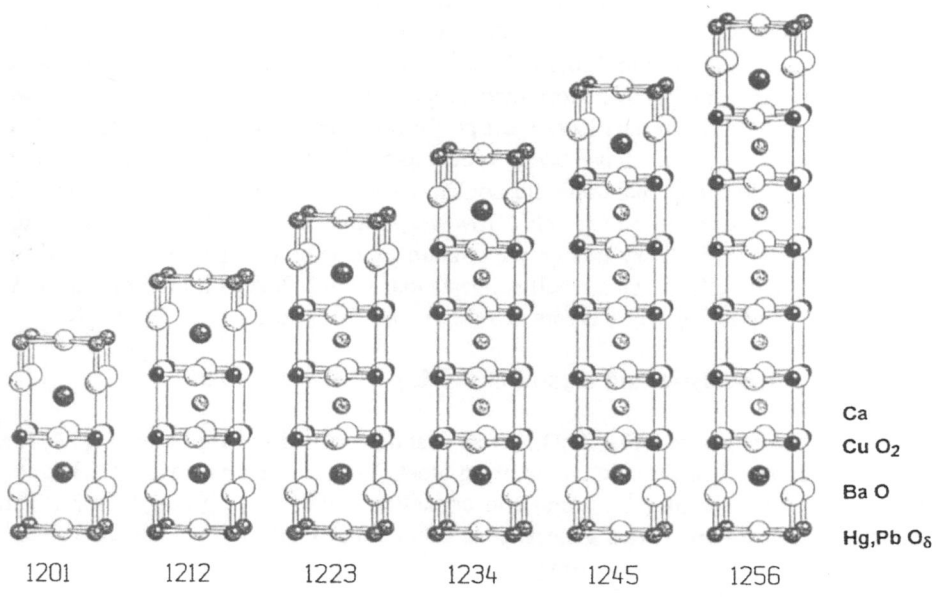

Fig. 14. Crystal structures of the Hg-12(n-1)n homologous series.

The crystal structures of four members of the homologous series HgPb-12(n-1)n (n = 2 - 5) and of the infinite layer calcium cuprate have been determined and refined by x-ray single crystal diffraction. The crystals have maximum transition temperatures of 70 K (1212), 130 K (1223), 130 K (1234), and 115 K (1245), and crystallize with space group P4/mmm. The structure consists of n layers copper oxide, separated by n-1 layers calcium atoms. (Fig. 14) These blocks are sandwiched by two BaO and one oxygen deficient (Hg,Pb)O layer. Lead is incorporated at the Hg site, but shifted slightly off the ideal position by about 0.35 Å, probably due to the shorter Pb-O bondlengths. Yttrium from the crucible is incorporated to 50 % at the Ca site in HgPb-1212, but not in other members of the HgPb-12(n-1)n family. The excess oxygen content varies between 10 and 35 %, and oxygen atoms are partially located at the site (0.5, 0.5, 0) and are partially shifted away from this ideal site by about 0.6 Å. Structure refinements of HgPb-1234 [30] and HgPb-1245 [31] show that the crystals contain up to 6 % stacking faults of material with one more or one less calcium cuprate layer than the main phase, which is also observed by TEM images. They introduce additional electron density in the Fourier maps at the position of the excess oxygen atom, and therefore cause an overestimation of the total oxygen content, if it is not corrected carefully. The excess oxygen content is very important for the superconducting properties of the HgPb-12(n-1)n materials. We developped a model for the determination of the amount of stacking faults and for the correction of the oxygen content which is valid for all members of the HgPb-12(n-1)n homologous series. The HgPb site has a reduced electron density, therefore an incorporation of defects like Cu, or carbonate seems likely. This site is quite similar like Cu(1) in 123 and 247, because defects accumulate at this position.

Structural parameters and bondlengths change in a systemtic way as a function of the number n of CuO layers in the structure. With increasing n, Ba moves towards the basal HgPb-O plane and the apical oxygen atom O(1) is shifted towards the superconducting CuO_2 layer. The changes in the $CaCuO_2$ units are less pronounced, but with increasing n they indicate the approach to the ideal $CaCuO_2$ structure.

7. Determination of superconducting parameters in $Y_2Ba_4Cu_8O_{16}$ and $HgBa_2Ca_3Cu_4O_{10+x}$

The determination of the penetration depth λ and the coherence length ξ is of considerable interest, since it provides crucial information concerning both the microscopic mechanism and the macroscopic phenomenology of the superconducting state. On a microscopic scale the penetration depth, interpreted in the clean limit in terms of the London formula $(\lambda^2=m^*/\mu_o e^2 n_s)$, yields a measure of the ratio of the effective mass m* and the superconducting carriers density n_s. The low-temperature behavior of λ is controlled by the symmetry of the pair wave function and is therefore of particular interest for testing theories predicting an unconventional symmetry of the order parameter. On a more macroscopic scale λ plays an important role in the magnetic interaction between vortices and as that characterizes the elastic properties of the flux line lattice. For instance, the reversible magnetization and the phenomenon of flux-lattice melting are closely connected with λ. The second important parameter is the coherence length ξ, which is the length scale on which pinning occurs. The third important parameter is the effective mass anisotropy γ, which is a measure of the effective coupling between adjacent blocks of CuO_2 planes. A reliable measurement of this fundamental parameters is due to the extreme values of these quantities still a challenging experimental task. Here we have used torque magnetometry which allows the simultaneous determination of all the basic superconducting parameters λ, ξ and γ. The power of this technique in the investigation of anisotropic magnetic properties of high-T_c superconductors has been widely demonstrated [32,33]. One of the important advantages of the torque magnetometer, as compared to commercial SQUID magnetometers is the ease in measuring continuously the angular dependence of the magnetic torque $\tau=m\times B$ produced by the sample magnetization m. Based on the continuous anisotropic Ginzburg-Landau model the reversible magnetic torque τ_{rev} of an anisotropic superconductor with volume V is given by:

$$\tau(\theta)_{rev} = \frac{\Phi_0 VB}{16\pi\mu_0\lambda_{ab}^2} \frac{\gamma^2-1}{\gamma} \frac{\sin(2\theta)}{\epsilon(\theta)} \ln\left(\frac{\gamma\eta B_{c2}^c}{B\epsilon(\theta)}\right)$$

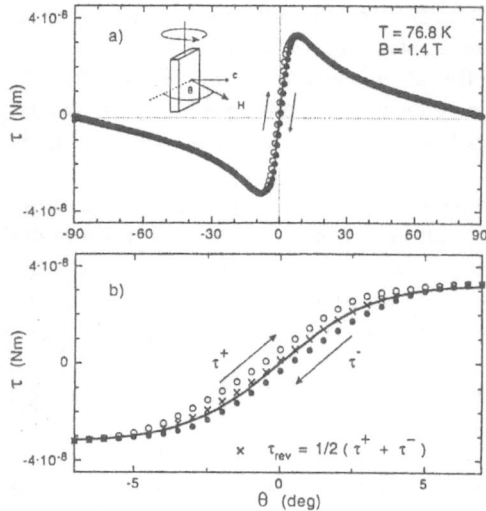

Fig. 15. a) Torque as a function of the orientation θ for a $Y_2Ba_4Cu_8O_{16}$ single crystal at T=76.8K and B=1.4T. The torque signal was monitored for both directions of rotation as indicated by the arrows. The solid line is the fit of the reversible torque data τ_{rev} using Eq. (1). The small hysteretic behavior close to the ab-plane is clearly seen in the enlarged scale b).

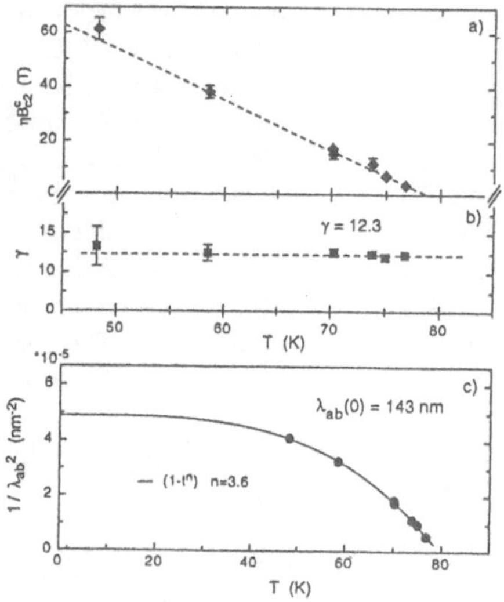

Fig. 16. Temperature dependence of (a) quantity ηB_{c2} (field parallel to the c-axis), (b) the effective mass anisotropy ratio γ, and (c) the penetration depth λ_{ab} as derived from the reversible torque (see text).

where θ is the angle between the applied magnetic field B and the ab-plane, Φ_0 is the flux-quantum, $\gamma=\sqrt{m_c^*/m_{ab}^*}$ is the effective-mass anisotropy, B_{c2}^c is the upper critical field in the c-direction, and λ_{ab} is an effective in-plane penetration depth. Here $\epsilon(\theta)^2=\cos^2\theta+\gamma^2\sin^2\theta$ is the angular scaling function. The numerical parameter η depends on the structure of the flux line lattice and is of the order of unity. As a first example we analyze the reversible torque measured in a high-quality $Y_2Ba_4Cu_8O_{16}$ sample (0.5x0.8x0.02mm) within the framework of the anisotropic London approach given in Eq. (1). In Fig. 15 a typical example of the angular dependence of the magnetic torque is shown. The data was taken at T=76.8K in an applied field B=1.4T and for both directions of rotation (τ^+,τ^-) from which the reversible torque $\tau_{rev}=1/2(\tau^++\tau^-)$ was further extracted. The solid line corresponds to a fit of Eq. (1) to the reversible torque, with λ, B_{c2}, and γ as the only adjustable parameters. The best fit was obtained by using the parameters $\gamma=12.3\pm0.1$, $B_{c2}^c(76.8K)=3.0\pm0.1T$, and $\lambda_{ab}(76.8K)=448\pm25nm$.

The same analysis has been performed on the torque data in the temperature regime $48K < T < 76K$. The results are shown in Fig. 16. The quantity ηB^c_{c2} (Fig. 16a), is linearly increasing with decreasing temperature with a slope of $B^c_{c2}/dT = -1.6T/K$ ($\eta \approx 1$). The zero temperature value $B^c_{c2}(0) \approx 87T$ is found through the relation $B^c_{c2}(0) \simeq -0.69T_c(dB^c_{c2}/dT)_{Tc}$ and corresponds to an in-plane coherence length of $\xi_{ab}(0) = \sqrt{\Phi_0/2\pi B^c_{c2}(0)} \simeq 1.92nm$, in good agreement with values as reported from magnetization measurements on polycrystalline samples [34]. The effective mass anisotropy parameter ($11.0 < \gamma < 12.5$) (Fig. 16b) is found the be almost temperature independent. The slightly higher anisotropy found in $Y_2Ba_4Cu_8O_{16}$ as compared to the $YBa_2Cu_3O_x$ system ($\gamma = 5-8$) may reflect the weaker coupling of neighbouring CuO_2 double layers separated by double CuO chains. Since our data are not restricted to the vicinity of T_c we can investigate the actual temperature dependence of λ_{ab} and extrapolate it to 0K. In Fig. 16c the temperature dependence of $1/\lambda_{ab}^2$ is shown, because it can be well fitted by a power-law of the form $1/\lambda_{ab}^2(t) = 1/\lambda_{ab}^2(0)(1-t^n)$. For a variable exponent n this empirical power-law provides a good approximation for the various theoretical models. For instance, the weak-coupling BCS temperature dependence may be approximated with an exponent $n \approx 2.3$, while the empirical two-fluid model corresponds to an exponent n=4. The solid line in Fig. 16c corresponds to a fit with n=3.6, a value which is close to the expectations for the two-fluid model, and also for an s-wave superconductor in the strong-coupling limit. The present value of $\lambda_{ab} = 143 \pm 15$ nm is smaller than earlier reported values ranging from 160nm to 200nm. [35] However, by using the independently measured $\kappa = 70$ from Ref. [36] and the above value of $\xi_{ab} = 1.92nm$, we get $\lambda_{ab} = \kappa\xi_{ab} = 135nm$, in reasonable agreement with our λ_{ab} obtained by torque-magnetometry. Moreover, a recent far infrared study on a 248 single crystal [37] revealed a large a/b anisotropy of the penetration depth ($\lambda_a = 200nm$, $\lambda_b = 80nm$). By taking the mean value of the penetration depths in the two planar directions we obtain an effective

Fig. 17. a) Angle dependent torque of a micro crystal ($m \approx 380ng$) of $HgBa_2Ca_3Cu_4O_{10}$micro crystallite at $T = 109.77K$ and for $B = 0.5T$. The solid line is a fit of the data to Eq. (1). b)Temperature dependence of the anisotropy parameter γ and ηB^c_{c2} of a $HgBa_2Ca_3Cu_4O_{10}$ microcrystal as determined by torque magnetometery.

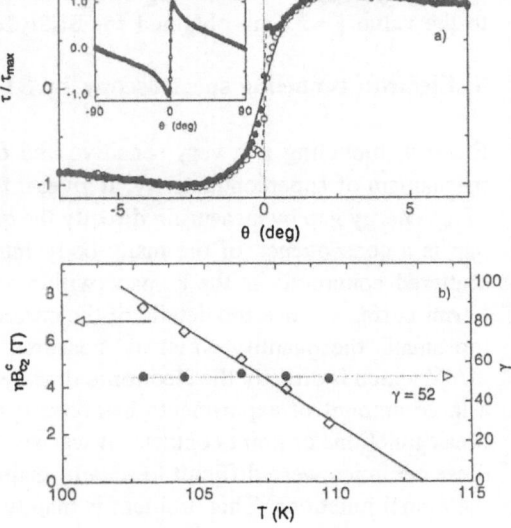

$\lambda_{ab}=\sqrt{\lambda_a\lambda_b}\simeq 135$nm which is again close to our own result $\lambda_{ab}=143$nm. As a further example we discuss the torque data measured on a $HgBa_2Ca_3Cu_4O_{10}$ crystal. The measurements have been performed with a novel microfabricated torque magnetometer developed in our group [38]. The device has a sensitivity of $\Delta\tau/B =$ $10^{-14}Am^2$ which exceeds that of the best commercial SQUID magnetometers by three orders of magnitude. For the torque measurements *one* micro crystal with dimensions of $60\times 80\times 10\mu m^3$ and a mass of only $m\simeq 380$ng was used. In Fig. 17a the angular dependence of the torque signal is shown for $T=109.77$K and for an applied field of $B=0.5$T. The torque was monitored for both directions of rotation, and was found to be almost perfectly reversible. The angle dependent torque was analyzed within the framework of the anisotropic London approach as discussed above. With the parameters $\gamma=51.35$ and $\eta B^2_{c2}=4.9$T Eq. (1) describes almost perfectly the experimental data over the whole range of θ (solid line). Only for $|\theta| <0.4°$, where the torque becomes irreversible clear deviations from the theoretical curve are observed, and this data has not been used for the fit. The quality of the fit was further tested for various strengths of the applied field and for various temperaturesin the regime $103K<T<110K$. The results are summarized in Fig. 17b. Due to the lack of an absolute calibration of the torque sensor, we consider here only the calibration independent parameters γ and ηB_{c2}^2. The anisotropy γ remains almost constant resulting in a mean value of $\gamma=52\pm 1$ for $103K<T<110K$. The large value of $\gamma=52\pm 1$ for Hg-1234 is much larger than previous estimates [39,40] ($\gamma=2$-6) as obtained from the ratio of physical quantities measured parallel and perpendicular to the CuO_2 layers, but is in good agreement to the value $\gamma=35$-70, as obtained from a 3D-2D dimentional crossover criterium [41]. The present value of $\gamma=52$ corroborates the quasi 2D character of the mercury based cuprates with enhanced vortex fluctuations. [42] The quantity ηB^c_{c2} increases linear with a slope of $dB_{c2}^2/dT=0.69\pm 0.05$T ($\eta\approx 1$), which gives an estimate of the in-plane coherence length $\xi_{ab}=2.5\pm 0.1$nm for Hg-1234. This value is slightly larger than $\xi_{ab}=1.5$-2.1nm previously reported for the Hg-1201 and Hg-1223 compound [39,42] but is similar to the value $\xi=2.5$nm obtained for $Bi_2Sr_2CaCu_2O_8$.

8. Electron tunneling spectroscopy by STM

Electron tunneling is a very sensitive and direct technique to elucidate the pairing mechanism of superconductivity. It gives information on the amplitude and symmetry of the energy gap by measuring directly the quasiparticle density of states. The energy gap is a consequence of the many-body interaction between quasiparticles that are scattered coherently in the k-space within a region of the order of kT_c , around the Fermi surface. Since the details of the current vs. voltage characteristics are usually too small, the quantity which is measured is the differential conductance $G(V)=$ dI/dV which is directly the electronic density of states. In the high T_c superconductors a large amount of experiments has been carried out using either planar junctions, break junctions or point contacts. It was soon realized that reliable tunneling data on these ceramics were difficult to obtain, mainly because of the low reproducibility of the tunnel junctions. This problem is mainly intrinsic to the nature of the parameters involved in high-T_c superconductors, in particular the short coherence length ξ_0, which reduces the superconducting order parameter at the surface as compared to the bulk. Another perturbing parameter is the extreme metastability of these materials, which reflects in a rapid degradation of the surface either by oxygen losses, disorder or changes in the stoichiometry.

A partial solution to these problems is to access an unperturbed surface by performing non-contact or true vacuum tunneling using a scanning tunneling microscope (STM) in a clean environnement. This can be achieved by cleaving single crystals in situ, what allows further to perform reliable spatially resolved tunneling spectroscopy.

We present here measurements performed with a low temperature STM [42] based on the beetle principle and working either in UHV conditions or in He gas. The single crystals which were investigated are $YBa_2Cu_4O_8$ and $HgBa_2Ca_3Cu_4O_{10+\delta}$.

$YBa_2Cu_4O_8$ is expected to be an ideal candidate for tunneling investigation since it has a fixed oxygen concentration and its structure, stabilized by the double CuO chains, does not produce twin plane defects. It is naturally underdoped, in the sense that its critical temperature can still be raised under pressure from 80 K up to 108 K [43]. By cleaving single crystals under UHV, the (001) cleavage plane exhibits terraces with step height of 1/2, 1,...n unit cells along the c-axis (c= 2.73 nm). In the regions where the pristine surface is smooth, atomic resolution can be achieved. A square atomic arrangement is usually observed with a lattice constant of about 0.27nm which corresponds to the O(2)-O(3) spacing in the CuO_2 layers or possibly to the Ba-O(1)

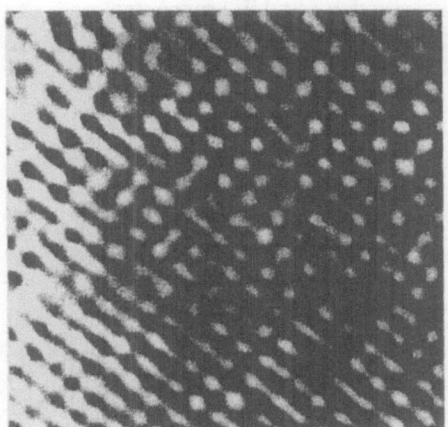 

Fig. 18. Atomic lattice with periodicity of 0.27nm on the (001) surface of YBCO-124. Image size : 4.8 nm x 4.8 nm

Fig. 19. STM image at 77 K of a cleaved surface of YBCO-124 showing in alternance CuO_2 layers (smooth) and CuO chains (rough). Image size 600 nm x 600 nm. ($I_t =$ 100 pA, $V_t = 780$ mV)

length of the adjacent layer (Fig.18). Observation of a square lattice with spacing of 0.38nm corresponding to the lattice constant a and b was also reported on uncleaved samples [44]. In some specific areas of the cleaved surfaces an alternance of smooth and rough terraces with characteristic tooth-shaped edges are observed as shown in Fig.19. From the height of the successive subunit cell steps (0.4 and 1 nm) we can attribute the smooth surfaces to the CuO_2 planes and the rough ones to the CuO chains [45,46]. Tunneling conductance curves measured on the smooth surfaces, obtained by cleavage at low temperature (< 25 K), present a well defined gap structure with a gap value of $\Delta \cong 22$ meV corresponding to a ratio $2\Delta/kT_c \cong 6.5$. This contrasts with the v-shaped curves measured on the rough areas, showing weaker gap features, indicating the distinct character of the CuO chains, possibly with proximity-induced superconductivity [46].

In the case of $Hg_{0.8}Pb_{0.2}Ba_2Ca_3Cu_4O_{10+\delta}$ crystals ($T_{c\ onset} = 127$ K) the tunnnel spectroscopy were done on several cleaved and uncleaved surfaces [47]. The cleaved surfaces display terraces with steps corresponding to one (c = 1.9 nm) or several unit cells. In several cases, true non contact tunneling could be achieved (as compared to point contact tunneling) with well characterized and reproducible conductance curves (Fig. 20). An interesting property of the various surfaces is the linear scaling between the gap value Δ at 4.2 K and an effective surface critical temperature T_c^* (Fig.21) derived from the temperature dependence of the zero bias conductance G_0. The values for Δ were determined by fitting the normalized conductance curves using a BCS density of states, smeared by quasi-particle lifetime effects. The depression of T_c^*, as compared to the bulk value T_c which is measured by SQUID magnetometer, can be rather large, showing the strong perturbation of the surface properties in this materials.

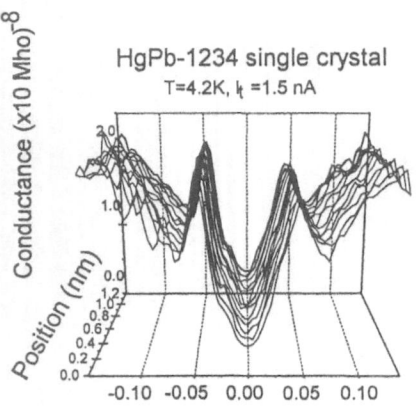

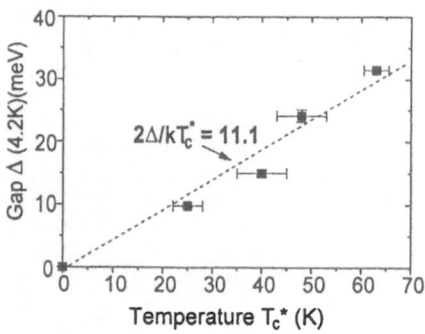

Fig. 20. Series of conductance curves measured at 4.2K along a line on a HgPb-1234 single crystal. ($I_t = 1.5$ nA).

Fig. 21. Scaling between the gap value $\Delta(4.2K)$ and the effective transition temperature onset T_c^* measured at the surface of different HgPb-1234 single crystals. The dashed line is a least-square fit to the data. Its slope leads to a rather high ratio $2\Delta/kT_c = 11.1$.

The detailed shape of the conductance curves, i.e., of the quasiparticle density of states, is still partly controversial. In particular the subgap filling and the asymmetry which deviates from a BCS behavior are still subject to various interpretations.

References

1. M.Takano, et al. Physica C 159 (1989) 375.
2. J.Karpinski, E.Kaldis, E.Jilek, S.Rusiecki, B.Bucher, Nature 336 (1988) 830.
3. P. Bordet, C. Chaillout, J. Chenavas, J.L. Hodeau, M. Marezio, J. Karpinski, E. Kaldis, Nature 334, (1988), 596.
4. J.Karpinski, E.Kaldis, S.Rusiecki, E.Jilek, P.Fischer, P.Bordet, C.Chaillout, J.Chenavas, J.L.Hodeau, M.Marezio,etal. J.Less-Commmon Metals, 164(1990)3-19.
5. J.Capponi et al. Physica C 235-240 (1994) 146-149.
6. J.Karpinski, H.Schwer, I.Mangelschots, K.Conder, A.Morawski, T.Lada, A.Paszewin, Nature.371(1994) 661.
7. J.Karpinski, K.Conder, C.Krüger, H.Schwer, E.Kaldis, in Materials and Crystallographic Aspects of HT$_c$-Superconductivity, NATO ASI Series Kluver 1994 Ed.E.Kaldis 555-583.
8. H.Murakami et al. Jap.J.App.Phys. 29 (1990) 2720.
9. J.Karpinski, I.Mangelschots, P.Wägli, Ch.Krüger, E.Kaldis, Proceedings of 7th Int.Workshop on Critical Currents in Superconductors 94 Alpbach Austria 455.
10. A.Erb, Physica C 245 (1995) 245-251.
11. A.Schilling, M.Cantoni, J.D.Guo, H.R.Ott, Nature 363(1993)56.
12. M.Marezio et al., Physica B 197 (1994) 570.
13. J.Karpinski, H.Schwer, I.Mangelschots, K.Conder, A.Morawski, T.Lada, A.Paszewin, Physica C 234 (1994) 10-18.
14. J.Karpinski, K.Conder, H.Schwer, J.Löhle, A.Morawski, T.Lada, A.Paszewin, Journal of Superconductivity, Vol.8, No.4,(1995) 515-518.
15. A.Morawski, T.Lada, A.Paszewin, J.Karpinski, K.Conder, A.Wisniewski, Physica C 235-240 (1994) 921-922.
16. B.Scott et al., Physica C 230 (1994) 239.
17. J.Löhle, J.Karpinski, A.Morawski, P.Wachter, submitted.
19. J.D.Jorgensen, B.W.Veal, A.P.Paulikas, L.J.Nowicki, G.W.Crabtree, H.Claus, W.K Kwok, Phys. Rev. B 41, (1990), 1863.
20. R.J.Cava, A.W.Hewat, E.A.Hewat, B.Batlogg, M.Marezio, K.M.Rabe, J.J. Krajewski, W.F. Peck Jr., L.W. Rupp Jr., Physica C 165, (1990), 419.
21. G.M. Sheldrick, SHELXTL, Siemens Analytical X-ray Instruments (1990).
22. P.Bordet, J.L.Hodeau, R.Argoud, J.Muller, M.Marezio, J.C.Martinez, J.J. Prejean, J. Karpinski, E. Kaldis, S. Rusiecki, B.Bucher, Physica C 162, (1989), 524.
23. T. Miyatake, S. Gotoh, N. Koshizuga, S. Tanaka, Nature 341, (1989), 41.
24. H. Schwer, E. Kaldis, J.Karpinski, C.Rossel, J.Solid State Chem.111,(1994), 96.
25. H.Schwer, J.Karpinski, E.Kaldis, C.Rossel, Z.Krist. Supp.Issue 8(1994), 667.
26. M. Mali et al. submitted Phys. Rev. B.
27. H. Schwer, J. Karpinski, E. Kaldis, Physica C 235 - 240, (1994), 801.
28. H. Schwer, J. Karpinski, E. Kaldis, C. Rossel, in preparation
29. H. Schwer, E. Kaldis, J. Karpinski, C. Rossel, Physica C 211, (1993), 165.

30. H. Schwer, J. Karpinski, K. Conder, L. Lesne, C. Rossel, A. Morawski, T. Lada, A. Paszewin, Physica C 243, (1995), 10.

31. H. Schwer, J. Karpinski, L. Lesne, C. Rossel, A. Morawski, T. Lada, A. Paszewin, Physica C in print.

31. D.E.Farrell, C.W.Williams, S.A.Wolf, N.P.Bansal, and V.G.Kogan, Phys.Rev. Lett. 61,2805 (1988).

32. B. Janossy, R. Hergt, L. Fruchter, Physica C 170, 22 (1990).

33. W. C. Lee and D. M. Ginsberg, Phys. Rev. B 45, 7402 (1992).

34. G.Triscone, A. F.Khoder, C.Opagiste, J.Y.Genoud, T.Graf, E.Janod, T. Tsukamoto, M. Couach, A. Junod, and J. Muller, Physica C 224, 263   (1994).

35. J. Sok, M. Xu, W. Chen, B. J. Suh, J. Gohng, D. K. Finnemore, M. J. Kramer, L.   A. Schwartzkopf, and B. Dabrowski, Phys. Rev. B 51, 6035 (1995).

36. D.N.Basov, R.Liang, D.A.Bonn, W.N.Hardy, B.Dabrowski, M.Quijada, D.B. Tanner, J.P.Rice, D.M.Ginsberg, and T.Timusk, Phys. Rev. Lett. 74,   598 (1995).

37. D.Zech, J.Hofer, C.Rossel, P.Bauer, H.Keller, and J.Karpinski, to be published.

38. R. Puzniak, R. Usami, K. Isawa, H. Yamauchi, Phys. Rev. B 52, 3756 (1995).

39. Y.C.Kim, J.R.Thompson, J.G.Ossandon, D.K.Christen, M.Paranthaman, Phys. Rev. B 51, 11767 (1995).

40. R.Puzniak, R.Usami, K.Isawa, H.Yamauchi, Phys.Rev.B 52, 3756 (1995).

41. Y.C.Kim, J.R.Thompson, D.K.Christen, Y.R.Sun, M.Paranthaman, E.D.Specht, Phys.Rev. B 52, 4438 (1995).

42. R.R Schulz and C. Rossel, Rev. Sci. Instrum. 65, 1918 (1994)

43. E.vanEenige, R.Griessen, R.Wijngaarden, J.Karpinski, E.Kaldis, S.Rusiecki, High Pressure Research 7 (1991) 58.

44. H.P.Lang, J.P.Ramseyer, D.Brodbeck, T.Frey, J.Karpinski, E.Kaldis T.Wolf, Ultramicroscopy 42-44, 715 (1992).

45. R.R. Schulz, PhD Thesis, Universität des Saarlandes, 1994

46. C. Rossel, P. Bauer, R.R. Schulz and J. Karpinski, to be published

47. C. Rossel, P. Bauer, J. Karpinski, A. Schilling, and A. Morawski, to appear in Phys. Rev. B

Tl–Based High Temperature Superconducting Cuprates: Structure and Properties

Martha Greenblatt

Department of Chemistry, Rutgers, The State University of New Jersey,
New Brunswick, NJ 08855–0939, USA

Abstract. The structure and properties of the Tl–based high temperature superconducting cuprates are reviewed. The single Tl–O layered phases $Tl(Sr/Ba)_2Ca_2Cu_3O_9$ stabilized by substitution of Pb or Bi in the Tl site are being developed for technological applications, because of their high T_c, excellent magnetic flux pinning and minimal weak–link behavior. Some of the factors that affect the high temperature superconducting properties, including substitutions, electronic structure, oxygen nonstoichiometry and defects are discussed in relation to the structure.

1 Introduction

All of the Tl–based high compounds can be described by the general formula, $Tl_mA_2Ca_{n-1}Cu_nO_{2n+m+2\pm\delta}$, where $m = 1$ or 2; $n = 1 - 5$; A = Ba,Sr. For convenience, the names of these compounds are abbreviated as 1223 and 2223 for $Tl_2Ba_2Ca_2Cu_3O_{10}$ and $TlSr_2Ca_2Cu_3O_9$, respectively, for example, where each number denotes the number of Tl, Ba/Ca and Cu respectively. The compounds with $m = 1$ and $m = 2$ are usually referred to as single and double Tl–O layered compounds. Until the recent discovery of the Hg–based high temperature superconductors (HTSC), the Tl–2223 material exhibited the highest superconducting transition temperature (T_c) at 125 K.

All of the Tl–based phases are quite unstable and are difficult to synthesize; the single layered compounds ($m = 1$) have been prepared with A = Ba or Sr or solid solutions thereof, while single phase compounds, with A = Sr only, for the $m = 2$ phases have not been stabilized for any value of n thus far. The synthesis of many of these phases has been reviewed recently and will not be treated here [1].

There have been many structural investigations of the various Tl–cuprates by X-ray and neutron diffraction methods of both single crystalline and polycrystalline forms [2]. It is well established that the Tl–based compounds are closely related structurally to the rare–earth–based phases. Figure 1 illustrates schematically the crystal structures of these compounds; the structure of $YBa_2Cu_3O_7$ (denoted as 123) is included for comparison.

Similar to the other family of cuprates the Tl–based materials are amenable to many chemical substitutions, which dramatically change the physical properties. Substitutions improve the stability of the compounds; a stable phase is critical for reproducible and reliable measurements of the physical properties

104

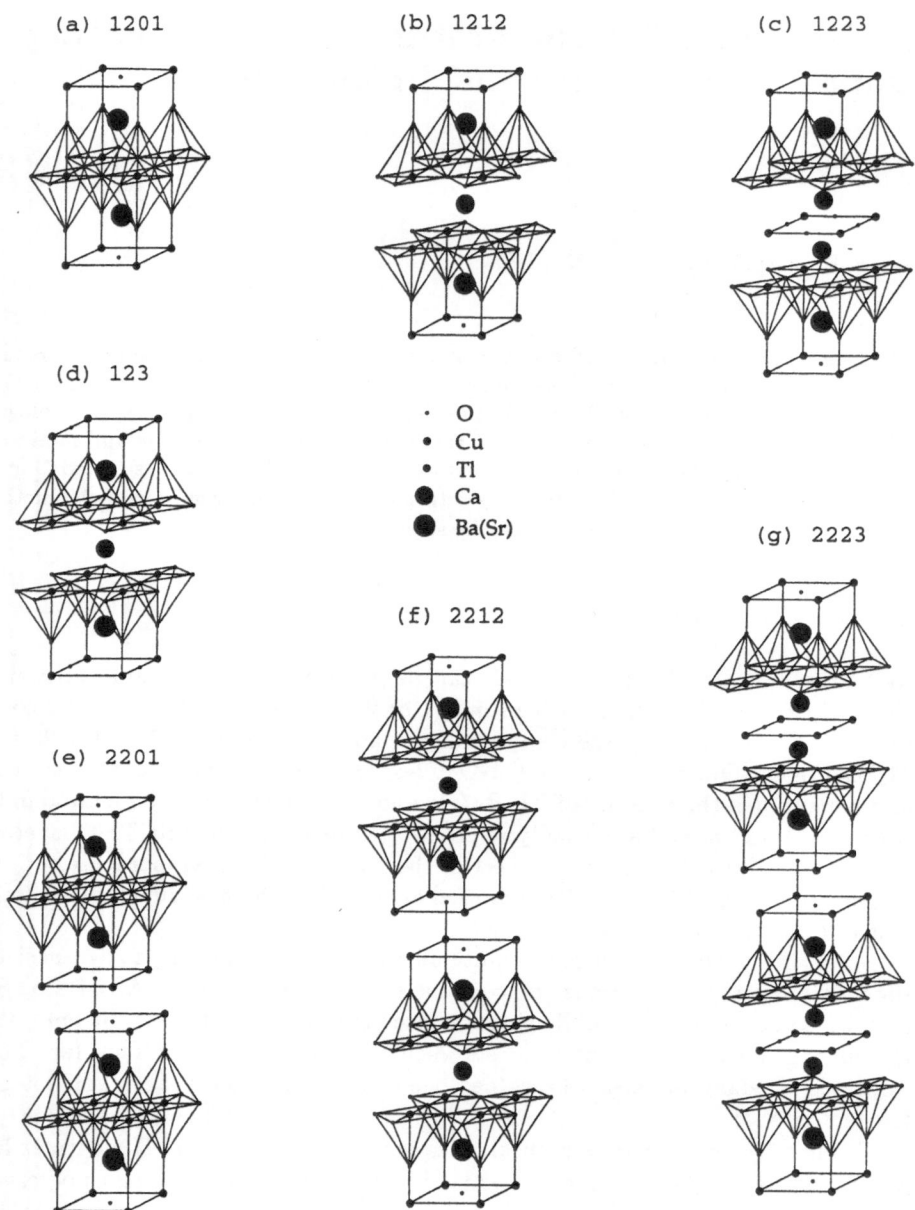

Fig. 1. The schematic representation of the structures of the mono- and bi–layered thalium cuprates, $Tl_m A_2 Ca_{n-1} Cu_n O_{2n+m+2\pm\delta}$; the structure of $YBa_2 Cu_3 O_7$ (123) is given for comparison.

and understanding the mechanism of high T_c superconductivity. Moreover, substitutions often enhance many of the technologically relevant superconducting parameters including T_c and the critical current, J_c. There have been literally hundreds of Tl–based compounds and their substituted analogs reported in the literature. Polycrystalline–bulk, single–crystal and thick and thin film forms have been investigated respectively. A summary of the most interesting and important phases reported in the recent past are given in Table 1.

Table 1. $TlBa_2Ca_{n-1}Cu_nO_{2n+3+\delta}$ and Substituted Phases

Compound	T_c (K)	n	Ref.and Comm.
$TlBa_2CuO_{5\pm\delta}$	50	1	3
$TlBa_2CuO_{5-x}F_x$	37–45	1	4; $0.1 < x \leq 0.6$
$TlBaSrCuO_{5\pm\delta}$	43	1	5
$Tl_{1-x}Cd_xBaLaCuO_5$	38–48	1	6
$TlBa_2CaCu_2O_{7\pm\delta}$	80–110	2	7
$TlBa_2Pr^{IV}Cu_2O_{7-\delta}$	*	2	8; *): $\rho_{RT} = 9.6$ kΩ
$TlBa_2Ca_{1-x}Y_xCu_2O_7$	90	2	9
$TlBa_2Ca_2Cu_3O_{9\pm\delta}$	117–126	3	10
$TlBa_2Ca_3Cu_4O_{11}$	114	4	11
$TlBaSrCaCu_2O_{7-\delta}$	94	2	12
$TlBaSrCa_2Cu_3O_{9-\delta}$	103/116	3	13
$TlBa_{1.5}Sr_{0.5}Ca_2Cu_3O_{9-\delta}$	124	3	14
$TlSr_2CuO_{5\pm\delta}$	NSC	1	15
$TlSrLaCuO_5$	46	1	16
$TlSr_{2-x}La_xCuO_5$	40	1	17; ($x = 1$)
$TlSr_{2-x}Pr_xCuO_5$	0–40	1	18
$TlSr_{1-x}Ba_xLaCuO_5$	42–37	1	19
$TlSr_2CaCu_2O_{7\pm\delta}$	100	2	20
$TlSr_2Ca_2Cu_3O_{9\pm\delta}$	106	3	21; (90% pure)
$TlSr_2CaCu_2O_{7-\delta}$	55	2	22; (Single crystal)
$TlSr_2Ca_{1-x}Ln_xCu_2O_7$	60-90	2	23
$TlSr_{2.6}Nd_{0.4}Cu_2O_y$	80	2	17
$(Tl,Pb)Sr_2CuO_{5\pm\delta}$	60	1	24
$(Tl,Pb)Sr_2CaCu_2O_{7\pm\delta}$	90	2	25
$(Tl,Pb)Sr_2Ca_2Cu_3O_{9\pm\delta}$	124	3	26
$Tl_{0.5}Pb_{0.5}Ca_{0.8}Y_{0.2}Cu_2O_y$	110	2	27
$Tl_{0.5}Pb_{0.5}Sr_{1.6}Ca_{2.4}Cu_3O_9$	24		28
$(Tl,Bi)Sr_2CuO_{5\pm\delta}$	45	1	29
$(Tl,Bi)Sr_2CaCu_2O_{7\pm\delta}$	95	2	30
$(Tl,Bi)Sr_2Ca_2Cu_3O_{9\pm\delta}$	115	3	31
$Tl_2Ba_2CuO_6$	0-90	1	32
$Tl_2Ba_2CaCu_2O_{8-\delta}$	110	2	33
$Tl_2Ba_2Ca_2Cu_3O_{10-\delta}$	128	3	34
$Tl_2Ba_2Ca_3Cu_4O_{12-\delta}$	116	4	35

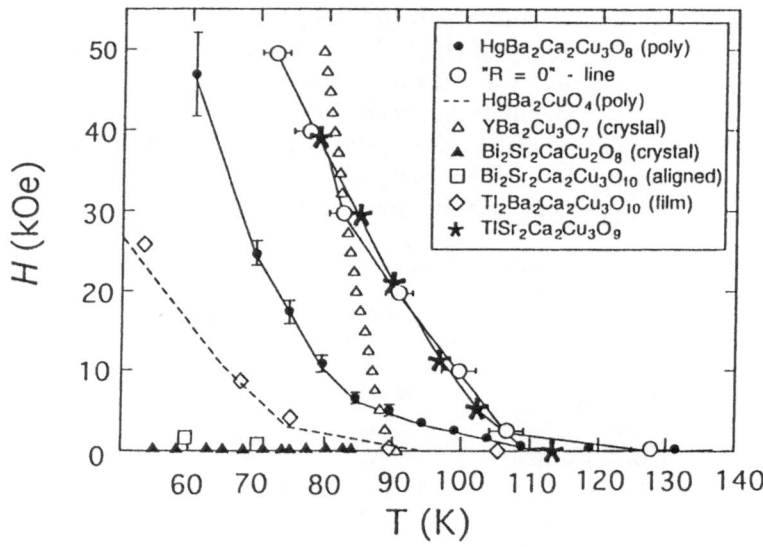

Fig. 2. Irreversibility boundaries H_{irr} as a function of temperature of various superconducting cuprates (after ref. 41)

Jorgensen (see his paper in this volume) has summarized the optimal HTSC parameters: high T_c ($¿$ 77 K) which increases with n, up to $n = 3$ (the number of Cu–O layers per unit cell), longer Cu–O_{apical} distances, metallic blocking layers, and fewer defects; a high J_c, in contrast, requires shorter blocking layers (*i.e.*, shorter Cu–O_{apical} distance), and pinning centers (*i.e.*, defects), but also improves with metallic blocking layers as does T_c. For applications of HTSC (*e.g.*, generators, motors, magnets), a $J_c \approx 10^4 - 10^6$ A/cm^2 at $\approx$ 77 K in an applied magnetic field, $H \approx$ 1 T is required. Ideally polycrystalline samples with good intergranular connectivity and excellent flux pinning are desirable. The 123 material has $J_c \approx 10^4 - 10^5$ A/cm^2 at 77 K and $H \approx$ 1 T, and a $T_c \approx$ 92 K, however its intergranular connectivity is poor [36].

The HTSC materials exhibit a remarkable broadening of the resistive transition in a magnetic field, which increases with increasing H and T. This phenomenon has been attributed to thermally assisted flux creep [37], super conducting fluctuations [38] and the vortex glass phase [39]. Because this magnetic–field induced transition prevents the use of HTSC's in high magnetic fields and at high temperature, it is necessary to understand the broadening mechanism in order to improve the HTSC materials for practical applications. The broadening of the transition generally increases with anisotropy, which originates from an anisotropic order parameter with small coherence lengths. Recently, it has been reported that there is a strong correlation between the resistive broadening and the Cu–O layer spacing and the irreversibility magnetic field decreases with increasing Cu–O interlayer spacing [40]. One important feature that sets the upper H/T limit of the HTSC applications, is the detection of the irreversibility line in the (H, T) phase diagram, which separates the superconducting region in two separate regimes: a reversible, dissipative regime where vortices are free to move and an irreversible non–dissipative region of pinned magnetic flux. The irreversible behavior, as the magnetic–field induced resistive transition, is dependent on the coupling between the Cu–O layers. This coupling is characterized by the distance between these layers and is significantly different in the various superconducting families: the coupling is strong in the 123 and monolayer Tl–phases, weaker in the Tl–2223 and Bi–2223 compounds and intermediate in the Hg–1223 family. The H/T phase diagram is illustrated in Fig.2 for the above–mentioned phases [41].

Recently, there has been a great deal of interest in the single layer Tl–O compounds. Although the T_c of Tl–1223, for example, is somewhat lower than that of the Tl–2223, (Table 1), the interlayer coupling between the Cu–O planes is stronger when there is only a single Tl–O separating the Cu–O planes. This causes a stronger pinning of flux in the Tl–1223 than in the Tl–2223 phase. In addition, the Tl–1223 phases exhibit excellent intergranular connectivity [42]. Because of the high T_c (¿ 110 K), the excellent intrinsic magnetic flux pinning and minimal weak link behavior, the Tl–1223 cuprates are candidates for development as superconducting tapes and wires for operation at 77 K. In this paper some of the properties of single Tl–O layer compounds will be reviewed.

2 Factor Affecting the HTSC Properties of the Tl–Based Cuprates

2.1 Structural and Electronic Properties

In the stoichiometric, 2201, 2212, 2223 and 2234 phases the Cu valence is $+2$ (*i.e.*, there are no holes or electrons) and therefore for the existence of superconductivity, there must be either Tl deficiency, substitution of lower valent ions on the Tl^{3+} site (*e.g.*, Ca^{2+}), oxygen nonstoichiometry, or possibly all of the above processes occurring simultaneously, leading to an optimal hole concentration in the Cu–O layers. An internal redox process on the Tl–O layers corresponding to $Tl^{3+} + Cu^{2+} \rightarrow Tl^{3-\delta} + Cu^{2+\delta}$ can also lead to hole formation in the Cu–O layers and electronic interaction between the Tl–O and Cu–O layers [43]. Indeed electronic band structure calculations by the tight binding method show that the Tl $6s$ level crosses the Fermi surface (FE) of the Cu $3d$ and O $2p$ band in the double Tl–O layer cuprates [44]. Such redox reaction could render the Tl–O layers metallic.

In contrast, in the stoichiometric 1201, 1212, 1223 and 1234 Tl–cuprates, the Cu oxidation states are $+3$, $+2.5$, $+2.33$ and $+2.25$ respectively. Thus the Cu valence, or the hole concentration is too high in most cases, and must be reduced. These phases are unstable and those with lower n values more so. In addition the $n = 1$ phase is semiconducting without Pb or Bi substitution on the Tl site [24,29]. The band structure calculations, which show that the Tl $6s$ level is high above the F_E, while the Bi and Pb levels could cross the Fermi level (the Bi level could conceivably even fall below it) are consistent with these observations [45].

2.2 Flux Pinning Properties

In the Introduction it was indicated that the Tl–1223 phases show good magnetic flux pinning properties, mainly because the distance between the supeconducting Cu–O layers is small, which leads to good electronic coupling in the c direction. In Table 2 a summary of T_c and the Cu–O interlayer distances (d_{Cu-Cu}) suggest that the best materials might be $Tl_{0.5}Pb_{0.5}Sr_2Ca_2Cu_3O_9$–based phases; this material has a high $T_c \approx 124$ K with a $d_{Cu-Cu} \approx 3$ Åsmaller than that of $Tl_2Ba_2Ca_2Cu_3O_{10}$. Substitution of Ca for the Sr site could further improve the flux pinning properties. Similarly, analogous Bi–substituted $Tl_{0.5}Bi_{0.5}Sr_2Ca_2Cu_3O_{9-\delta}$–based phases would be worth investigating for their flux pinning characteristics in search of the optimal HTSC materials.

Table 2. Selected Superconducting Parameters in Various Tl–Cuprates

Compound	T_c (K)	a (Å)	c (Å)	d_{Cu-Cu} (Å)	Ref.
$Tl_2Ba_2Ca_2Cu_3O_{10}$	128	3.8503(6)	35.88(3)	11.50	34
$TlBa_2Ca_2Cu_3O_{9-\delta}$	117	3.853(1)	15.913(4)	9.55	46
$TlSr_2Ca_2Cu_3O_{9-\delta}$	106	3.814	15.302	8.76	21
$Tl_{0.5}Pb_{0.5}Sr_2Ca_2Cu_3O_9$	124	3.814	15.267	8.74	28
$(Tl_{0.58}Pb_{0.38}Sr_{1.9}Ca_{2.1}Cu_3O_9)$					

2.3 Oxygen Stoichiometry and Flux Pinning

Several structural studies of the Tl–cuprates indicate unambiguously that the Tl–O layers in both the single and double layer phases are highly defective. Both the oxygens and the Tl's are disordered in the plane with respect to their ideal crystallographic site (Fig.3a,b).

It has been well established that T_c is extremely sensitive to the oxygen stoichiometry in all of the layered cuprates, but particularly in the Tl–2 phases; extremely small changes in oxygen content (¡0.1 mol%) can lead to large variation of T_c [32]. T_c is not affected by oxygen content as much in the single Tl–O layered compounds. For example, when Tl–2223 is treated in Ar, O_2 or H_2 atmosphere, the Tc is 128, 124 and 116 respectively; for a similarly treated Tl–1223, T_c varies from 98, to 95, to 115 K respectively. However it is noteworthy that while the irreversibility line (IL) is unaffected by oxygen nonstoichiometry, in the monolayered Tl–O cuprates, it is dramatically affected by it in the bilayered phases. Wahl et al. based on their beautiful study of single crystals of $TlBa_{0.8}Sr_{1.2}Ca_2Cu_3O_{9-\delta}$ explain these observed differences between mono and

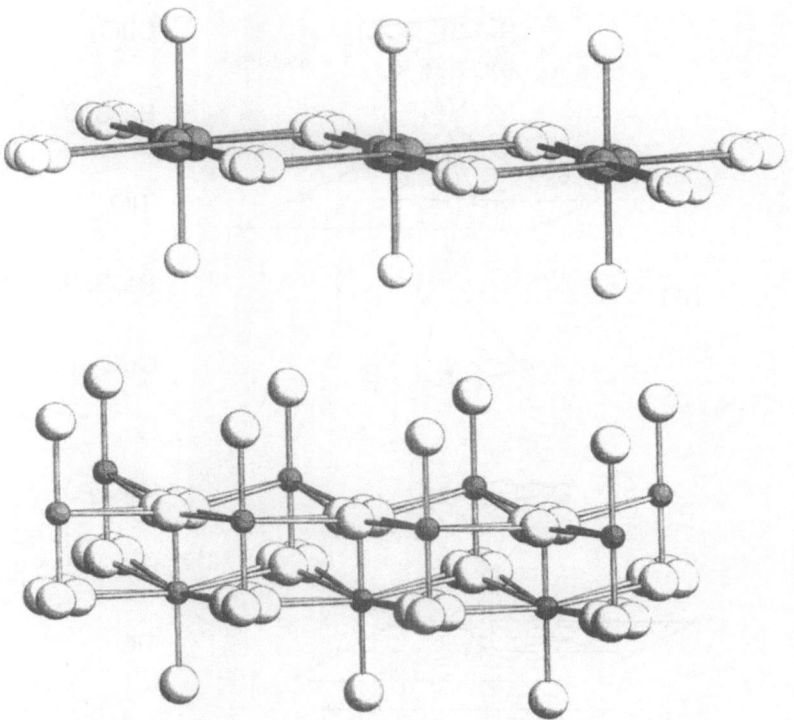

Fig. 3. (a) A schematic diagram of the single Tl–O layers in the
Tl(Ba/Sr)$_2$Ca$_{n-1}$Cu$_n$O$_{2n+3}$ showing the Tl–O oxygen bonds and the disordering of
both the thaliums and oxygens in the ab plane.
(b) A schematic diagram of the double Tl–O layers in the Tl$_2$Ba$_2$Ca$_{n-1}$Cu$_n$O$_{2n+4}$
showing the Tl–O oxygen bonds and the disordering of oxygens in the ab plane.

bilayered Tl–cuprates by the differences in the oxygen environment in the (Tl–
O)$_\infty$ layers and the modification of the electronic structure *via* the hole–reservoir
effect [47]. Figure 4a shows the stacking of layers in the monolayer compound
based on neutron–diffraction studies [48]. It was established that the (Tl–O)$_\infty$
layer consists of distorted TlO$_4$ tetrahedra, and oxygens are missing only from
the Tl–O oxygen plane sites (open circles), and not from the apical sites (hatched
circles). Consequently, the chain of "Cu–O–Tl–O–Cu" along the c direction and
the coupling are not affected by oxygen non–stoichiometry.

In contrast, structural studies of the bilayer Tl–cuprates established a distorted octahedral coordination for Tl [49]. Again, oxygen nonstoichiometry takes
place in the Tl–O planes (open circles, Fig.4b), however, electronic communica-
tions and consequently the coupling along the c-axis in the "Cu–O–Tl–O–Tl–
O–Cu" chain direction are now affected by missing oxygens in the plane, even

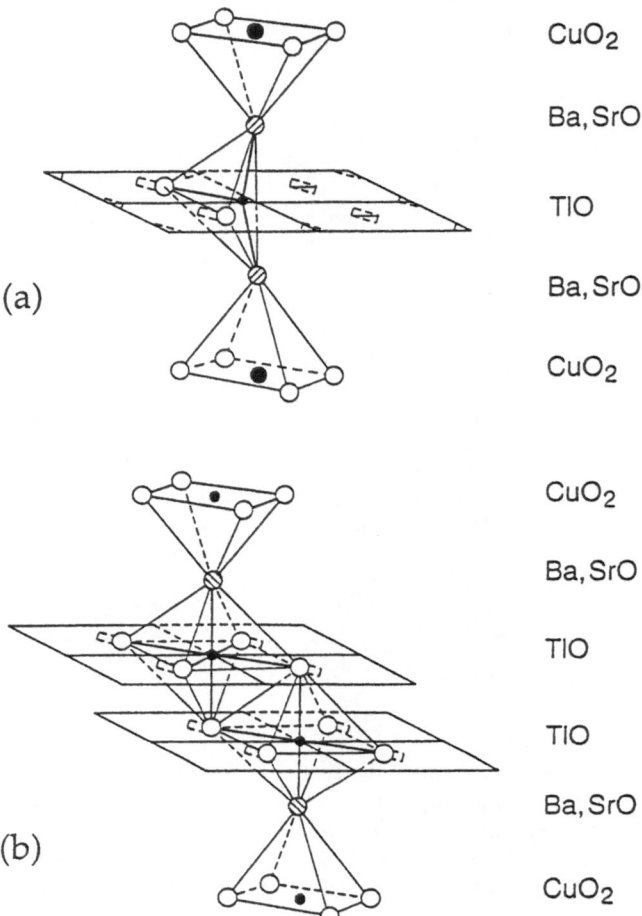

CuO₂

Ba, SrO

TlO

Ba, SrO

CuO₂

CuO₂

Ba, SrO

TlO

TlO

Ba, SrO

CuO₂

Fig. 4. (a) The tetrahedral environment in the thalium mono–layer cuprates. (b) The octahedral coordination of the thalium in the bilayer cuprates. The split positions in the Tl–O layer are drawn as dotted lines (after Ref.47).

though the apical oxygens of the CuO$_5$ pyramids are fully occupied.

In addition, the annealing conditions, which affect the oxygen content of the Tl–O layers are expected to have a greater influence on the electronic structure and the coupling along the c–axis in the bilayered Tl–cuprate, due to the hole reservoir effect [43]. In the monolayer Tl–phases, there is a large concentration of Cu^{III} already in the stoichiometric compound, and there is no evidence of charge reservoir behavior of the (Tl–O)$_\infty$ layers in these phases [50]. That the IL is not affected significantly by oxygen nonstoichiometry in the monolayered Tl–cuprates is an important advantage for technological applications.

2.4 The Effect of Hg Substitution in $Tl_{1-x}Hg_xBa_2Ca_2Cu_3O_{8+\delta}$

The recent discovery of superconductivity in mercury–based cuprate oxides generated new interest in HTSC [51]. These materials are the Hg analogs of the monolayered $TlBa_2Ca_{n-1}Cu_nO_{2n+3}$–phases and form an isostructural homologous series with Hg replacing Tl. However, the Hg–HTSC's are extremely unstable and difficult to prepare in pure form. Moreover, although the T_c of these materials is the highest (134 K for Hg–1223) of all of the known superconductors so far, the IL is intermediate between those of the Tl–1223 (or 123) and Tl–2223 phases (Fig.2). Recently it has been shown that a continuous solid solution exists in $Tl_{1-x}Hg_xBa_2Ca_2Cu_3O_{8\pm\delta}$ [52]. The T_c has been optimized close to 134 K; the Tl substituted sample has enhanced stability, intergranular connectivity, a substantially larger superconducting volume fraction, excellent magnetic flux pinning, and high J_c. It is likely that the HTSC parameters can be further optimized in this system, for example, substitution of significant amounts of Sr for Ba should improve the c–directional coupling between the Cu–O layers and consequently the flux pinning.

References

1. S.Li, L.E.H.McMills, and K.V.Ramanujachary, in: *Studies of High Temperature Superconductors*, ed.by A.V.Narlikar (Nova Sci.Publ., Commack, NY 1990) p.143; M.Greenblatt, L.E.H.McMills, S.Li, K.V.Ramanujachary, M.H.Pan, and Z.Zhang, *Inorg.Synth.* **30**, 201 (1995).

2. A.W.Hewat, in: *Materials and Crytallographic Aspects of HTC Superconductivity*, ed.by E.Kaldis, NATO ASI Series (Kluwer Acad.Publ., Netherlands 1994) p.17; articles in: *Thalium Based High-Temperature Superconductors*, ed.by A.M.Herman and J.V.Yakhmi (Marcel Dekker, New York 1993); C.N.R.Rao and A.K.Ganguli, *Chem.Soc.Rev.* **24**, 1 (1995).

3. S.S.P.Parkin, V.Y.Lee, A.I.Nazzal, R.Savoy, R.Beyers, and S.J.LaPlaca, *Phys.Rev.Lett.* **61**, 750 (1988).

4. M.A.Subramanian, *Mater.Res.Bull.* **29**, 119 (1994).

5. I.K.Gopalakrishnan, J.V.Yakhmi, and R.M.Iyer, *Physica* C **175**, 183 (1991).

6. M.A.Subramanian and A.K.Ganguli, *Mater.Res.Bull.* **26**, 91 (1991).

7. B.Morosin, D.S.Ginley, P.F.Hlava, M.J.Carr, R.J.Baughman, J.E.Schriber, E.L.Venturini, and J.F.Kwak, *Physica* C **152**, 413 (1988).

8. C.C.Lai, B.S. Chiou, Y.Y.Chen, J.C.Ho, and H.C.Ku, *Physica* C **202**, 104 (1992).

9. R.Vijayaraghavan, J.Gopalakrishnan, and C.N.R.Rao, *J.Mater.Chem.* **2**, 327 (1992).

10. B.Morosin, D.S.Ginley, J.E.Schirber, and E.L.Venturini, *Physica* C **156**, 587 (1988).

11. D.M.Ogborne and M.T.Weller, *Physica* C **230**, 153 (1994).

12. I.K.Gopalakrishnan, J.V.Yakhmi, and R.M.Iyer, *Physica* C **172**, 450 (1991).

13. C.Martin, M.Huve, M.Hervieu, A.Maignan, C.Michel, and B.Raveau, *Physica* C **201**, 362 (1992).

14. A.Maignan, C.Martin, M.Huve, C.Michel, M.Hervieu, and B.Raveau, *Chem.Mater.* **5**, 571 (1993).

15. A.K.Ganguli and M.A.Subramanian, *J.Solid State Chem.* **93**, 250 (1991).

16. M.A.Subramanian, *Mater.Res.Bull.* **25**, 191 (1990).

17. A.K.Ganguli, V.Manivannan, A.K.Sood, and C.N.R.Rao, *Appl.Phys.Lett.* **55**, 2664 (1989).

18. D.Bourgault, C.Martin, C.Michel, M.Hervieu, J.Provost, and B.Raveau, *J.Solid State Chem.* **78**, 326 (1989).

19. M.A.Subramanian and M.-H.Whangbo, *J.Solid State Chem.* **96**, 461 (1992).

20. W.L.Lechter, M.S.Osofsky, R.J.Soulen Jr., V.M.LeTourneau, E.Skelton, S.B.Qadri, W.T.Elam, H.A.Hoff, R.A.Hein, L.Humphreys, C.Skowronek, A.K.Singh, J.V.Gilfrich, L.E.Toth, and S.A.Wolf, *Solid State Commun.* **68**, 519 (1988).

21. P.E.D.Morgan, T.Doi, and V.R.Housley, *Physica* C **213**, 438 (1993).

22. C.Martin, A.Maignan, Ph.Labbe, J.Chardon, J.Hejtmanek, and B.Raveau, *Chem.Mater.* **7**, 1415 (1995).

23. R.Vijayaraghavan, A.K.Ganguli, N.Y.Vasanthacharya, M.K.Rajumon, G.U.Kulkarni, G.Sankar, D.D.Sarmat, A.K.Sood, N.Chandrabhas, and C.N.R.Rao, *Supercond.Sci.Techn.* **2**, 195 (1989).

24. M.-H.Pan and M.Greenblatt, *Physica* C **176**, 80 (1991).

25. A.Maignan, C.Martin, V.Hardy, and Ch.Simon, *Physica* C **228**, 323 (1994).

26. M.A.Subramanian, C.C.Torardi, J.Gopalakrishnan, P.L.Gai, J.C.Calabrese, T.R.Askew, R.B.Flippen, and A.W.Sleight, *Science* **242**, 249 (1988).

27. R.S.Liu, P.P.Edwards, Y.T.Huang, S.F.Wu, and P.T.Wu, *J.Solid State Chem.* **88**, 334 (1990).

28. R.S.Liu, S.F.Hu, D.A.Jefferson, and P.P.Edwards, *Physica* C **198**, 318 (1992).

29. M.-H.Pan and M.Greenblatt, *Physica* C **184**, 235 (1991).

30. S.Li and M.Greenblatt, *Physica* C **157**, 365 (1989).

31. H.Takei, H.Kugai, Y.Torii, and K.Tada, *Physica* C **210**, 109 (1993).

32. K.V.Ramanujachary, S.Li, and M.Greenblatt, *Physica* C **165**, 377 (1990).

33. M.Paranthaman, M.Foldeaki, D.Balzar, H.Ledbetter, A.J.Nelson, and A.M.Hermann, *Supercond.Sci.Techn.* **7**, 227 (1994).

34. C.C.Torardi, M.A.Subramanian, J.C.Calabrese, J.Gopalakrishnan, K.J.Morissey, T.R.Askew, R.B.Flippen, U.Chowdery, and A.W.Sleight, *Science* **240**, 631 (1988); A.Maignan, C.Martin, V.Hardy, Ch.Simon, M.Hervieu, and B.Raveau, *Physica* C **219**, 407 (1994).

35. D.M.Ogborne and M.T.Weller, *Physica* C **223**, 283 (1994).

36. T.Doi, M.Okada, A.Soeta, T.Yuasa, K.Aihara, T.Kamo, and S.-P.Matsuda, *Physica* C **183**, 67 (1991).

37. M.Tinkham, *Phys.Rev.Lett.* **61**, 1658 (1988); T.T.M.Palstra, B.Batlog, R.B.van Dover, L.F.Schneemeyer, and J.V.Waszczak, *Phys.Rev.B* **41**, 6621 (1990); T.Matsushita, Fujiyoshi, K.Toko, and K.Yamafuji, *Appl.Phys.Lett.* **56**, 2039 (1990).

38. R.Ikeda, T.Ohmi, and T.Tsuneto, *J.Phys.Soc.Jpn.* **60**, 1051 (1991).

39. D.S.Fisher and M.P.A.Fisher, *Phys.Rev.B* **43**, 130 (1991).

40. T.Nabatame, J.Sato, Y.Saito, K.Aihara, T.Kamo, and S.Matsuda, *Physica* C **193**, 390 (1992).

41. A.Schilling, O.Jeandupeux, J.D.Guo, and H.R.Ott, *Physica* C **216**, 6 (1993).

42. D.S.Ginley, J.F.Kwak, E.L.Venturini, B.Morosin, and R.J.Baughman, *Physica* C **160**, 42 (1989); K.Jagannadham and J.Narayan, *Mater.Sci.Eng.* B **8**, 201 (1991).

43. T.Suzuki, M.Nagoshi, Y.Fukuda, Y.Syono, M.Kikuchi, N.Kobayashi, and M.Tachiki, *Phys.Rev.*B **40**, 5184 (1989); N.Merrien, L.Coudrier, C.Martin, A.Maignan, F.Studer, and A.M.Flank, *Phys.Rev.*B **49**, 9906 (1994); H.Romberg, N.Nucker, M.Alexander, J.Fink, D.Hahn, T.Zetter, H.H.Ott, and K.F.Renk, *Phys.Rev.*B **41**, 2609 (1990).

44. D.Jung, M.-H.Whangbo, N.Herron, and C.C.Torardi, *Physica* C **160**, 38 (1989).

45. D.Jung, M.-H.Whangbo, N.Herron, and C.C.Torardi, *Physica* C **160**, 381 (1989).

46. M.A.Subramanian, *J.Solid State Chem.* **77**, 192 (1988).

47. A.Wahl, A.Maignan, V.Hardy, J.Provost, Ch.Simon, and B.Raveau, *Physica* C **244**, 341 (1995).

48. C.Michel, E.Suard, V.Caignaert, C.Martin, A.Maignan, M.Herviu, and B.Raveau, *Physica* C **178**, 29 (1991).

49. D.M.Ogborne, M.T.Weller, and P.C.Lanchester, *Physica* C **200**, 167 (1994); D.M.Ogborne, M.T.Weller, and P.C.Lanchester, *Physica* C **200**, 207 (1994).

50. F.Studer, C.Michel, and B.Raveau, *Mater.Sci.Forum* **137–139**, 187 (1993).

51. S.N.Putilin, E.V.Antipov, O.Chmaissem, and M.Marezio, *Nature* **362**, 226 (1993).

52. J.Z.Liu, I.C.Chang, M.D.Lan, P.Klavins, and R.N.Shelton, *Physica* C **246**, 203 (1995).

Hydrogen in High–T_c Superconductors

Henryk Drulis[1] and Jan Klamut[1,2]

[1] *W. Trzebiatowski* Institute of Low Temperature and Structure Research,
Pol.Acad.Sci., P.O.Box 937, 50-950 Wrocław, Poland;
[2] Int.Lab.of High Magnetic Fields and Low Temperatures, Gajowicka 95,
53-529 Wrocław, Poland

Abstract. The role of hydrogen atom doping on the physical properties of high–T_c superconductors is presented. The main features are illustrated by presenting some chosen studies of the thermopower, resistivity, *ac* losses and Mössbauer effect in the hydrogenated SmBaCuO, YBaCuO and La(Sr)CuO systems. Also a brief survey of the current knowledge in this field is given.

Keywords: oxide superconductor, hydrogen

1 Introduction

Almost immediately after the discovery of superconductivity in La-Sr-Cu-O by Bednorz and Müller in 1986 [1] and superconductivity in Y-Ba-Cu-O by the team from the Universities of Alabama and Houston [2], intensive studies have been launched to investigate the influence of hydrogen on physical properties of this new class of superconductors.

The first papers on this subject appeared in 1987 and came from Brookhaven National Laboratory [3]. These works concerned Y-Ba-Cu-O superconductors and, as one might suppose, were motivated by the hope that T_c could be increased by hydrogenation. Very soon other researchers [4–10] carried out their own research programs in this field. This time the investigation had more ambitious goals. It was determination to get an answer to the following questions:

1. Can we, and how can we modify superconducting properties by charging superconducting material with hydrogen, *i.e.* how would T_c , critical current densities and other properties be changed?
2. How would the hydrogen atoms change crystallographic and magnetic properties of the materials?
3. Would the hydrogen atom be an electron donor or an electron acceptor in superconducting compounds?
4. Would the hydrogen atom change the carrier concentration by introducing an electron to the conduction band of superconductors?
5. Is the hydrogenation an opposite process to oxygen depletion in Y-Ba-Cu-O systems?

Most research activity on the hydrogenated high T_c materials has been done in the Y-Ba-Cu-O system. In spite of many contradictions and varied results [4,11–17] which appeared in the literature in the last eight years, the situation is more or less clear now, and a few general conclusions have been formulated:

1. Keeping the appropriate temperature and pressure conditions during hydrogenation in a sealed volume ($T < 180^\circ C; P < 200$ mmHg), the oxygen content remains unchanged — it means that there is no oxygen removal from the material *via* water molecules.

2. The hydrogen atoms are acting like electron donors.

3. There are three distinct hydrogen content ranges in superconducting ceramics:

   (a) small doping when the hydrogen content $x < 0.2$, and where:
      i. hydrogen forms a solid solution in the superconducting material,
      ii. cuprates remain superconducting,
      iii. the T_c onset is almost constant or slightly increases with hydrogen concentration increase;

   (b) middle doping range $0.5 < x < 1.0$, where:
      i. the protons induce global magnetic ordering for $x > 0.5$ and probably local ordering for smaller x,
      ii. hole neutralization appears to be the cause of magnetic ordering restoration,
      iii. hydrogen–induced magnetic ordering is similar to that observed upon oxygen removal,
      iv. a gradual decrease of superconducting volume fraction takes place during hydrogen uptake;

   (c) hydrogen concentration $x > 1$, where:
      an unknown "hydride" phase is formed, no longer superconducting.
      We call this phase "hydride" though we are aware that it is not a hydride like metal hydrides from which the hydrogen can be removed. For metal hydrides x up to 4 has been obtained without destroying the basic cation sublattice.

Some of these conclusions are illustrated in Figs.1–2 where ac susceptibility measurements for hydrogenated Y123 and Gd123 are shown [6]. In order to understand the influence of hydrogen doping, it is important to ascertain the hydrogen lattice site and to which atoms it is bonded. Thanks to many NMR [12,18–20], NQR [21–22] and μSR [23–29] investigations in H_xYBaCuO we know that the protons are probably localized at interstitial sites called B2 and bonded preferentially to an oxygen in the BaO planes, as is shown in Fig.3. Other H^+–O bond sites, such as the B1 or Lin sites are also observed but are less populated, depending on the oxygen concentration and presence of dopants. What is important, in all cases hydrogen is bonded to the oxygen atoms located out of the superconducting Cu–O planes. Fig.4 shows the possible hydrogen atom positions in La_2CuO_4 unit cell.

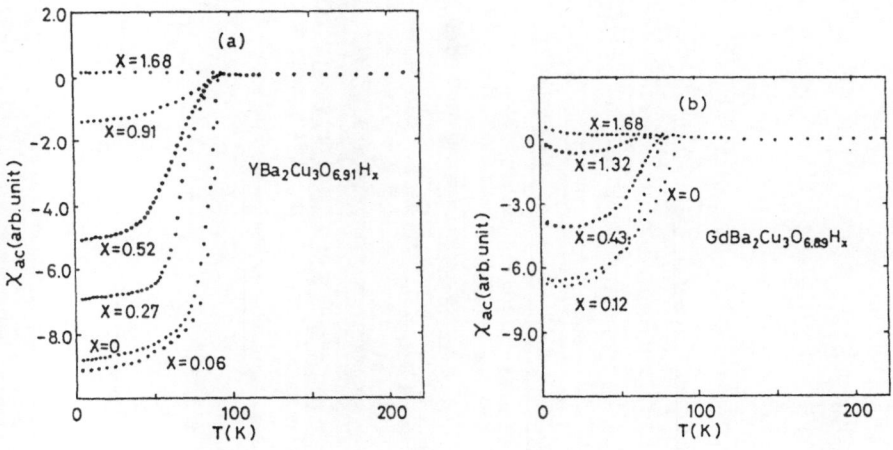

Figure 1: *ac* susceptibility as a function of temperature for (a) $YBa_2Cu_3O_{6.91}H_x$ and $GdBa_2Cu_3O_{6.89}H_x$ [6].

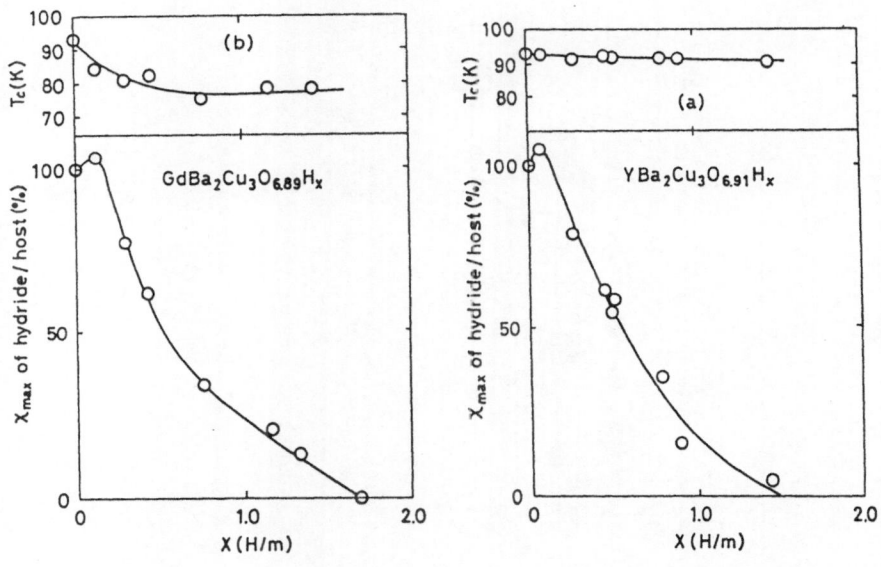

Figure 2: Superconducting transition temperature T_c and relative superconducting volume fraction as a function of hydrogen concentration x for (a) $YBa_2Cu_3O_{6.91}H_x$ and $GdBa_2Cu_3O_{6.89}H_x$ [6].

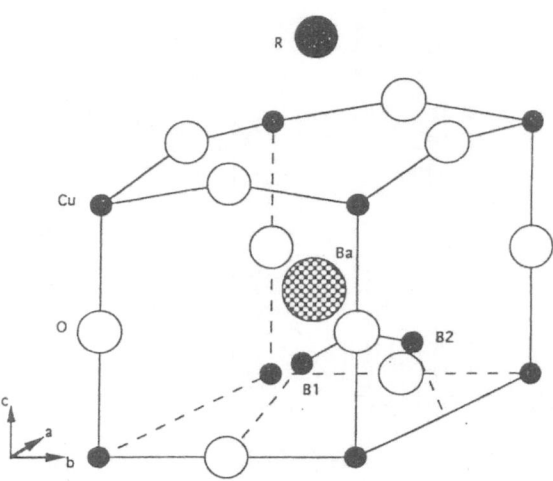

Figure 3: Probable H^+ and muon–stopping sites location depicted in the lower half of the unit cell of $ReBa_2Cu_3O_{7-\delta}$; the B2 site is the most probable one. The B1 site lies approximately in the $c - b$ plane, whereas the B2 site is located near the $c - a$ plane [23–29].

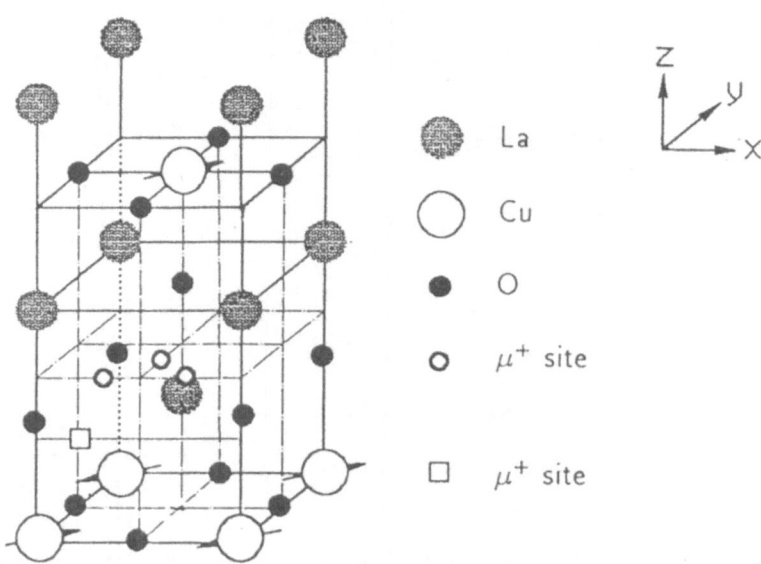

Figure 4: The results of stable muon site for La_2CuO_4 [28] (open small circles) and [45] (squares).

2 Thermopower and Resistivity Measurements in the $H_xSmBaCuO$ System

As it has been mentioned before, both hydrogenation and deoxygenation can be considered as an electron doping processes. However, the position of hydrogen in the structure as well as its charge and radius are quite different from those of oxygen vacancies. Oxygen occupies the appropriate lattice positions, whereas hydrogen ions enter the material interstitially. In spite of this, many results, for example ^{89}Y NMR measurements showed that the addition of hydrogen causes similar effects to that of oxygen removal [12]. To have a more detailed insight into that problem, the thermoelectric power and resistivity measurements of $H_xSmBaCuO_{7-y}$ have been performed in the composition ranges of $x = 0$–0.085 and $y = 0$–0.25 [16]. The results are summarized in Fig.5. Both hydrogenation and deoxygenation change the temperature dependence of the thermopower in the same way. As can be seen, both $S(T)$ and dS/dT (at 275 K) increase with electron doping (hole consuming). However, it is accompanied by very different resistivity variations (Figs.5c,5d). A negligible change of $\rho(T)$ by deoxygenation is in contrast to the dramatic change of resistivity by hydrogenation. In the latter case the $\rho(T)$ behavior is changed from metallic to semiconducting.

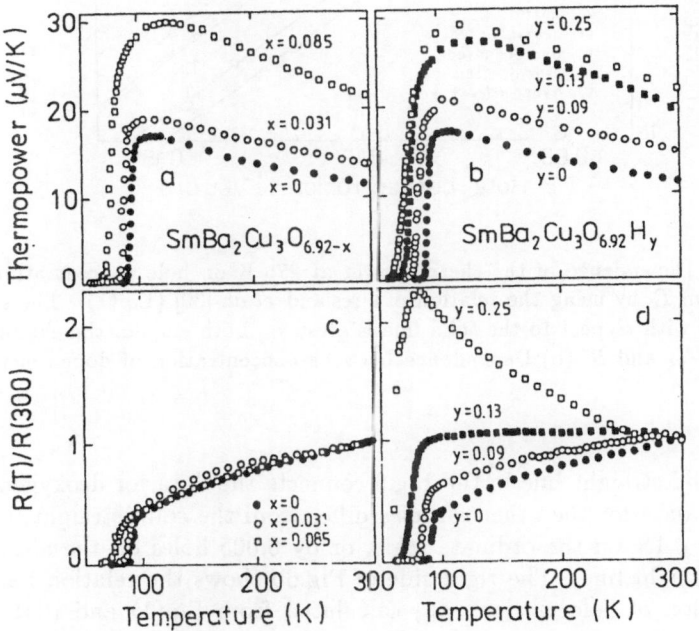

Fig. 5. Temperature dependence of the thermoelectric power ((a) and (b)) and resistivity ((c) and (d)) of ceramic specimens of $SmBa_2Cu_3O_{6.92-x}H_y$, deoxygenated ((a) and (c)) or hydrogenated ((b) and (d)) [16]

Analysis of the thermopower data has been done using the relation between transition temperature T_c and hole carrier density proposed by Presland [30].

$$(T_c^{max} - T_c)/T_c^{max} = 82.6(p_c - 0.16)^2 \,, \tag{1}$$

where p_c is the hole concentration per Cu atom in the CuO plane.

We apply (1) to find the relationship between the effects of hydrogenation and deoxygenation. We have used $T_c^{max} = 92$ K, the highest T_c ever obtained in the system examined here. Our $S(275)$ data plotted in $(\log S, p_c)$ coordinates are shown in Fig.6a.

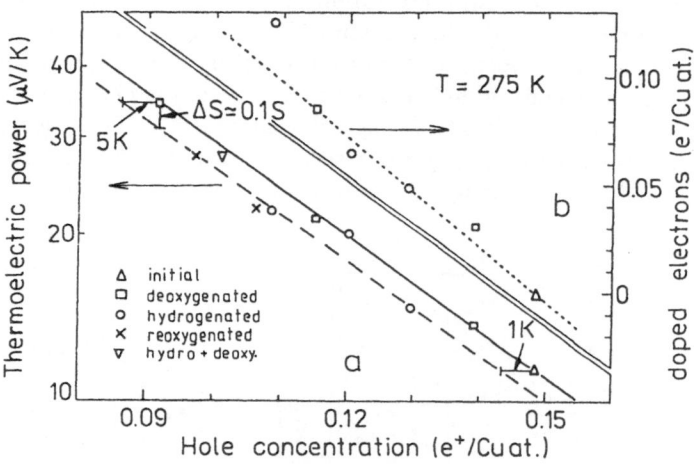

Fig. 6. (a) Dependence of the thermopower at 275 K on hole concentration, p_c, calculated from T_c by using the relation of Presland *et al.* [30] (Eq.(1)). The shift of the broken line with respect to the solid line is given on both ends of the examined range in terms of T_c and S. (b) Dependence between concentration of doped electrons and p_c [16].

The solid straight line in the Fig.6 connects the data for deoxygenated systems. The data for the other samples differ from the solid straight line by not more than $0.1S$ on the ordinate scale, or by 0.005 holes on the abscissa scale (broken straight line). The right side of Fig.6b shows the relation between the concentration of hole carriers, p_c, calculated from Eq.(1) and that of doped electrons, p_d, determined from stoichiometry concentrations (*i.e.* hydrogen concentration per formula unit per CuO$_2$ plane). It was assumed that every oxygen removed or every 2H atoms added reduce the carrier density by nominally one hole per Cu atom (from CuO$_2$ plane). For slight deoxygenation or hydrogenation of $SmBa_2Cu_3O_{6.92-x}H_x$ p_c depends linearly on p_d. However, of each 5 doped electrons (by any method) only about 2 fill the holes in the CuO_2 plane. Regardless of the source of doping electrons to $SmBa_2Cu_3O_{6.92-x}H_y$, its dS/dT data systematically increase with electron doping to more and more negative values, as shown in Fig.7.

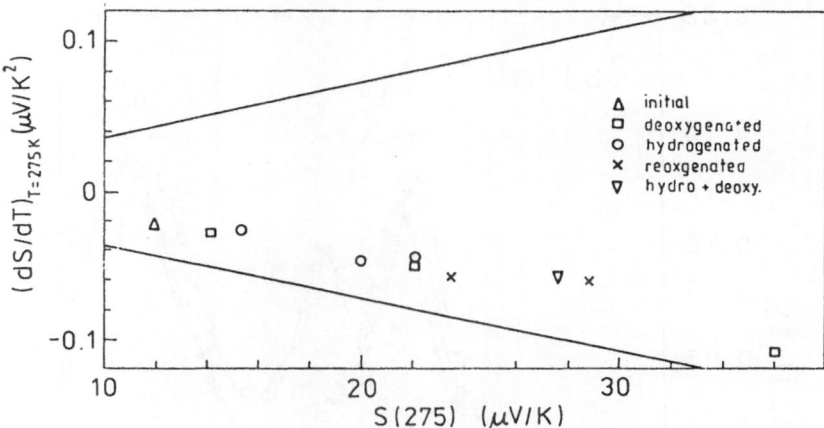

Fig. 7. Dependence of temperature derivative of thermopower on thermopower at 275 K for samples of $SmBa_2Cu_3O_{6.92-x}H_y$. Upper solid line gives the dependence for wide–band (Mott diffusion thermopower) thermopower behavior and lower solid line is the predicted one for a narrow–band conductor in the high–temperature limit [16].

The lower solid line in Fig.7 shows the S *versus* dS/dT dependence expected for a narrow–band hole conductor in the high–temperature limit; there is expected $S \sim 1/T$ dependence [31]. The upper line presents the case of $S \sim T$, which is characteristic of wide band conductors according to the well–known Mott diffusive thermopower [32]. From literature data on thermopower studies it is now known that the — generally speaking — thermopower can be decomposed into two parts: one linear in T and another identified as a narrow band. The last figure shows that with increasing electron doping the term $\sim T$ (narrow band) becomes increasingly dominant.

The strong influence of hydrogen on the resistivity behavior indicates that hydrogen absorption significantly changes the intergrain contribution to the global resistivity.

3 *ac* Losses in the Hydrogenated YBaCuO Systems

The influence of hydrogenation on the inter– and intragrain critical current density in Y123 system has been studied by *ac* losses [33]. The *ac* permeability measurement is one of the best methods to distinguish both types of critical currents. The dependence of the loss component μ'' of the *ac* permeability on the external field amplitude b_o for samples of Y123 with different hydrogen contents x, is shown in Fig.8.

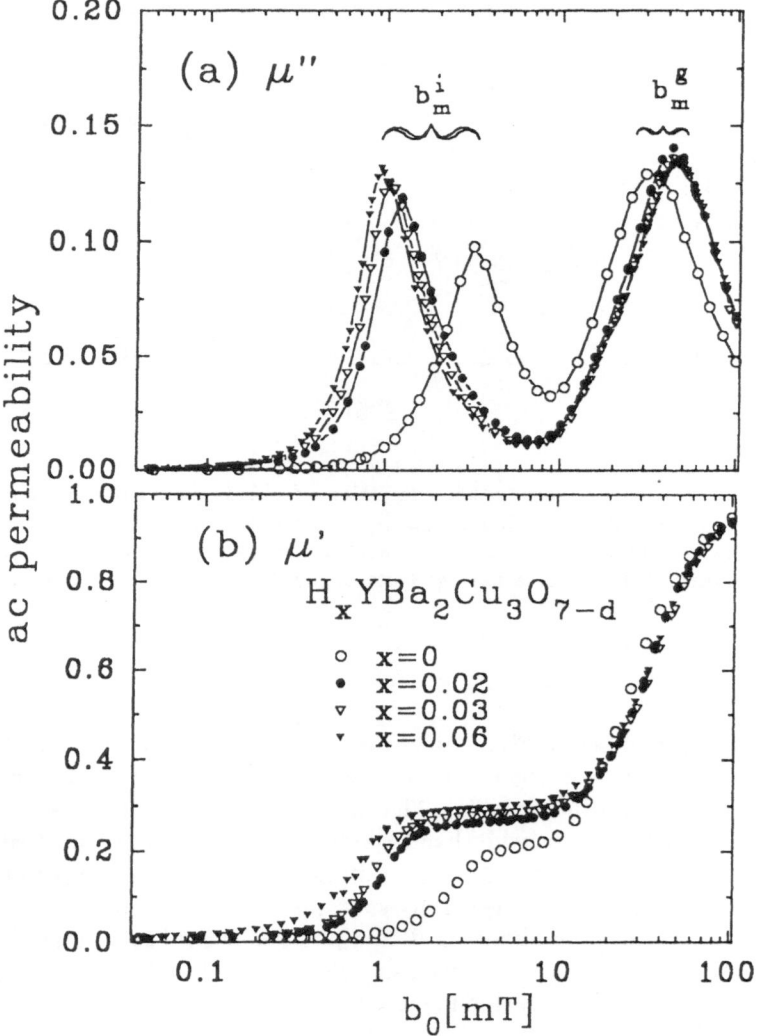

Fig. 8. Complex *ac* permeability as a function of the external magnetic field amplitude b_o for different hydrogen contents x — a) loss component, μ'' and b) dispersive component μ' [33]

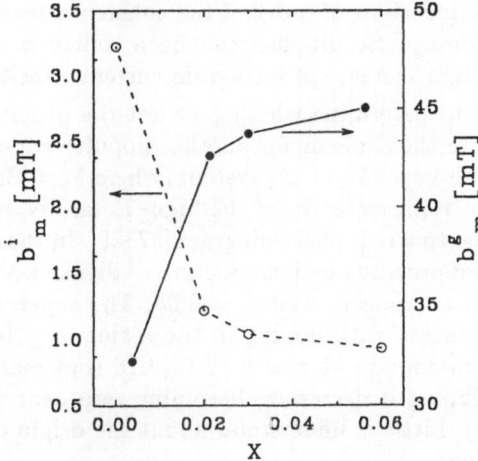

Fig. 9. Change of the intergrain (open circles) and intragrain (filled circles) full penetration fields with the hydrogen concentration x in the sample [33]

The $\mu''(b_o)$ plots are characteristic for granular high T_c superconductors where two distinct peaks are commonly observed. The first peak is related to the field value at which the magnetic flux just reaches the center of the sample (intergranular losses), whereas the maxima appearing at higher b_o are related to losses within the grains (grains are already fully penetrated by the magnetic flux in the form of Abrikosov type vortices for $b_o > B_{c1}$ — the first critical field for the grains). The hydrogenation of the sample changes both the position of the maxima of $\mu''(b_o)$ as well as their heights. The first peak (connected with the intergrain losses) shifts to lower value of b_o whereas the second one (intragrain losses) moves towards higher b_o. The values of $b_o(\mu''_{max})$, as a function of hydrogen index x are shown in the Fig.9. The intergrain peak is shifted to lower amplitudes by a factor of 3 whereas the peak related to the intragranular losses appears at higher b_o but the increase is only by a factor 1.4.

The shift of the intergrain maxima to lower amplitude by hydrogenation might be associated with a pulverization of the sample. Some, rich in hydrogen, hydride phase probably appears on the surface of the grains [34]. The phase is less conducting or even insulating, and decreases the Josephson coupling between the grains and thus drastically lowers the intergranular critical current density. Any attempts to identify this phase have been unsuccessful, up to now.

The increase of the intragrain critical current density after loading the sample with hydrogen probably indicates that hydrogen creates additional pinning centers inside a grain. From many investigations, first of all from μSR spectroscopy [26–28], it appeared that — through the hole neutralization — hydrogenation can restore antiferromagnetic interactions within a small area in a superconducting medium with dimensions of order of the coherence length [27,35–36]. These centers or antiferromagnetic droplets can be a source of new pinning centers responsible for a slight increase of intragrain current density.

It is of interest to determine whether or not the observations made for hydrogenated Y123 are the same as for another popular superconducting system, $La_{2-x}M_xCuO_4$. The $La_{2-x}M_xCuO_4$ system (where M = Ba or Sr), being structurally the simplest representative of the high-T_c family, has attracted a great deal of attention due to a rich phase diagram [37–39]. In these compounds charge carrier (hole) doping proceeds by introduction of divalent M ions on the La sites and superconductivity appears near $x = 0.06$. The superconducting transition temperature T_c increases with doping in the region $x \leq 0.16$ (underdoped region), reaches the maximum at $x = 0.15$ to 0.16 (optimal hole concentration) and with further doping it decreases, becoming zero near $x = 0.30$ (overdoped region) (see Fig.10). Little is understood about the origin of the T_c decrease in the overdoped region.

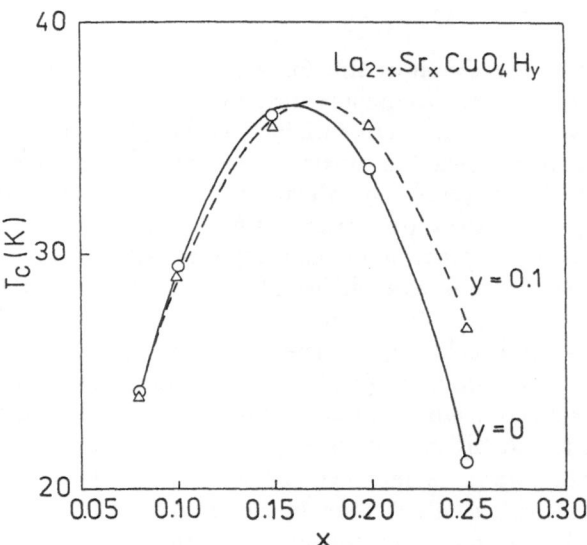

Fig. 10. T_c *vs.* composition for $La_{2-x}Sr_xCuO_4H_y$ system with $y = 0$ (o) and $y = 0.1$ (Δ). The lines are only guides to the eye [36].

The aforementioned facts suggest experiments aimed at changing the properties of LaSrCuO by means of electron doping in the reverse direction, *i.e.* towards lower hole concentration, with x set at a given level. This might be done by hydrogen H^+ doping.

4 Mössbauer Effect in the Hydrogenated $H_zLa_{2-x}Sr_xCu(Fe)O_4$ System

To get more insight into the properties of the studied compound at the microscopic level, advantage of such a powerful tool as Mössbauer spectroscopy has been taken. The lightly ^{57}Fe–doped $La_{2-x}Sr_x(Cu)FeO_4$ ($y = 0.005$) samples have been investigated [40]. Two Sr concentrations $x = 0.11$ and 0.20 have been chosen, i.e., belonging to the underdoped and overdoped regimes. The influence of hydrogenation on superconducting properties of $La_{2-x}Sr_xCu(Fe)O_4$ appeared to be similar to that reported in [36,41]. Hydrogenation slightly increased the critical temperature of the $x = 0.20$ series, whereas in the underdoped regime hydrogen caused a slight decrease of the critical temperature T_c^{md} whereas the onset temperature T_c^{on} remains unchanged (Fig.11).

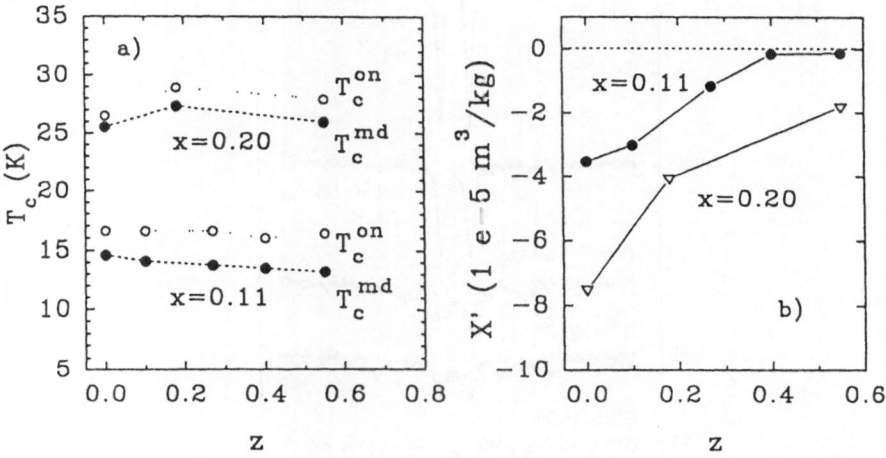

Fig. 11. Critical temperatures of superconducting transitions T_c^{md} and T_c^{on} (a), and real part of magnetic ac susceptibility χ' measured at 5 K (b) plotted vs hydrogen content for $x = 0.11$ and 0.20 samples [40]

It is possible to understand the independence of T_c of the hydrogen concentration in underdoped samples by assuming that the carrier concentration is reduced by hydrogen inhomogeneously, forming alternating regions with superconducting and insulating properties, as it has been mentioned earlier. In this case hydrogen atoms neutralize holes around the places where they are localized. It causes a local suppression of superconductivity creating nonsuperconducting areas. Evidently, in the areas where superconductivity survives, the hole concentration would be the same as before hydrogenation. That is why the T_c remains unchanged. A decrease of the superconducting volume fraction is the only result of hydrogenation. It is indeed experimentally observed (see Fig.11b).

Room temperature Mössbauer spectra of all the samples studied were similar to the spectrum presented in Fig.12, and were fitted assuming a unique doublet, *i.e.*, iron in one crystallographic position only. Isomer shift δIS values for the $x = 0.11$ series extracted from RT spectra are presented in Fig.13, *vs.* hydrogen index z. The increase of isomer shift δIS upon increase of z is observed. In nonhydrogented samples a decrease of Sr concentration x leads to an increase of δIS. Hence the $\delta IS(z)$ dependence observed here could be explained as being due to decrease of charge–carrier concentration.

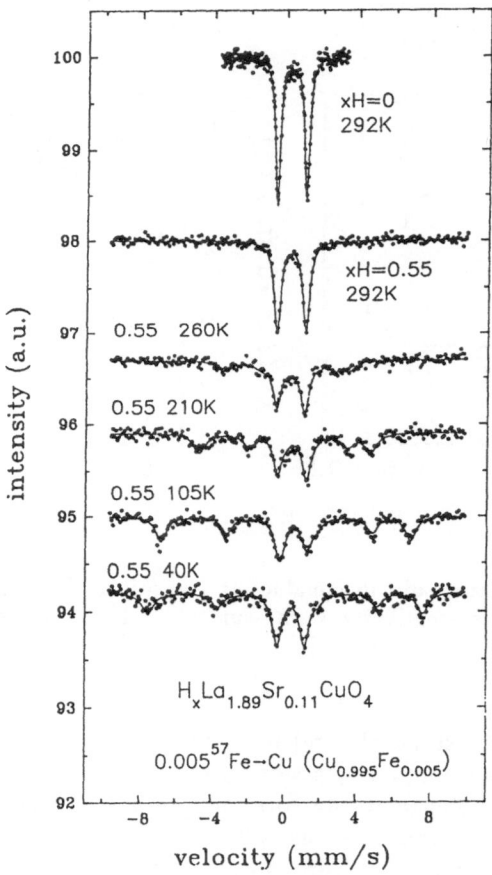

Fig. 12. Mössbauer absorption spectra of $H_y La_{1.89} Sr_{0.11} Cu_{0.995} Fe_{0.005} O_4$ taken at temperatures from $\approx$ 292 K to 40 K for $y = 0.55$ [40]

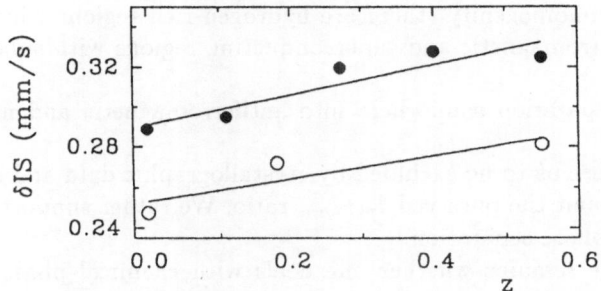

Fig. 13. Isomer shift obtained from room temperature Mössbauer spectra of $H_y La_{2-x} Sr_x Cu_{0.995} Fe_{0.005} O_4$ samples with $x = 0.11$ (closed symbols) and $x = 0.20$ (open symbols) plotted *vs* hydrogen index y [40]

Upon lowering the temperature below $Ts = 280$ K, Mössbauer spectra of highly hydrogenated $La_{2-x} Sr_x Cu(Fe)O_4$ samples (*i.e.* $x = 0.11$, $z = 0.4$ and 0.55, and $x = 0.2$, $z = 0.55$) split into magnetic and nonmagnetic parts (see Fig. 12). The relative intensities of magnetic (m) and nonmagnetic (nm) subspectra I_m and I_{nm} depicted in Fig. 14 do not change significantly below 200 K.

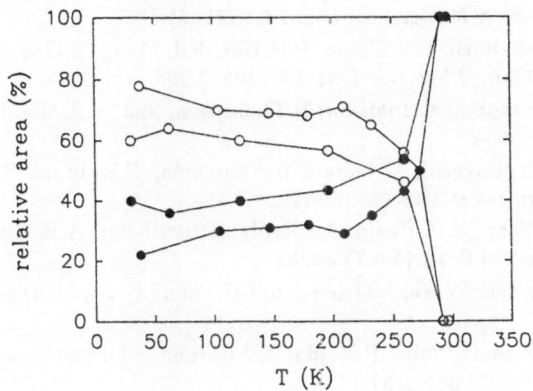

Fig. 14. Relative area of magnetic (open symbols) and nonmagnetic (closed symbols) parts of Mössbauer spectra of hydrogenated $H_y La_{1.89} Sr_{0.11} Cu_{0.995} Fe_{0.005} O_4$ for two hydrogen indexes $y = 0.40$ (circles) and $y = 0.55$ (squares) [40]

The occurrence of both magnetic and nonmagnetic parts in the MS spectra could again be explained as being due to:

a) sample inhomogenity (there are hydrogen rich regions which are insulating and antiferromagnetic and superconducting regions with small hydrogen content).

b) phase separation somewhere into antiferromagnetic and nonmagnetic domains.

The case (a) seems to be excluded by crystallographic data and especially if we take into account the observed I_m/I_{nm} ratio. We rather support the second explanation *i.e.* phase separation.

A major issue remains whether one deals with chemical phase separation effects or the so-called electronic phase separation.

The simultaneous observations of magnetism and superconductivity occur for proton implantation [42,43], hydrogen absorption, oxygen depletion [44], *i.e.* for the phenomena which are independent of the chemical substitutions. Therefore, coexistence of magnetism and superconductivity within the sample is very probable. This coexistence occurs on the atomic scale but not necessarily at the same sites.

A comparison of effects due to H doping, O depletion in superconducting cuprates points towards a probable magnetic origin of CuO–based superconductivity.

References

1. J.G.Bednorz and K.A.Müller, *Z.Phys.*B **64** (1986) 189.

2. M.K.Wu, J.R.Ashuburn, C.J.Torng, P.H.Hor, R.L.Meng, L.Gao, Z.J.Huang, Y.Q.Wang, and C.W.Chu, *Phys.Rev.Lett.* **58** (1987) 908.

3. J.J.Reilly, M.Suenaga, J.R.Johnson, P.Thompson, and A.R.Moodenbaugh, *Phys.Rev.*B **36** (1987) 5694.

4. T.Kamiyama, S.Tomiyoshi, M.Omari, H.Yamauchi, T.Kajitani, T.Matsunaga, and H.Yamamoto, *Physica* B **148** (1987) 491.

5. C.Y.Yang, X.Q.Yang, S.M.Heald, J.J.Reilly, T.Skotheim, A.R.Moodenbaugh, and M.Suenaga, *Phys.Rev.*B **36** (1987) 8798.

6. H.Fujii, H.Kawanaka, W.Ye, S.Orimo, and H.Fukuba, *Jap.J.Appl.Phys.* **27** (1988) L525.

7. M.Nicolas, J.N.Daou, I.Vedel, P.Vajda, J.P.Burger, J.Lesucur, and L.Dumoulin, *Solid State Commun.* **66** (1988) 1157.

8. Ch.Niedermayer, H.Glucker, R.Simon, A.Golnik, M.Rauer, E.Recknagel, A.Weidinger, J.I.Budnick, W.Paulus, and R.Schollhorn, *Phys.Rev.*B **40** (1989) 11 386.

9. S.D.Goren, C.Korn, V.Volterra, M.Schaefer, H.Riesemeier, E.Rossler, H.Stenschke, H.M.Vieth, and K.Luders, *Solid State Commun.* **70** (1989) 279.

10. H.Drulis, J.Klamut, and Z.Bukowski, *Mod.Phys.Lett.*B **4** (1990) 289.

11. V.V.Sinitsyn, I.O.Bashkin, E.G.Poniatovskii, V.M.Prokopenko, R.A.Dilanyan, V.Sh.Shekhtman, M.A.Nevedomskaya, I.N.Kremanskaya, N.S.Sidorov, R.K.Nikolaev, and Zh.D.Sokolovskaya, *Fiz.Tverd.Tela* **31** (1989) 54 (in Russian).

12. S.D.Goren, C.Korn, V.Volterra, H.Riesemeier, E.Rossler, M.Schaefer, H.M.Vieth, and K.Luders, *Phys.Rev.*B **42** (1990) 7949.

13. I.Felner, B.Brosh, S.D.Goren, C.Korn, and V.Volterra, *Phys.Rev.*B **43** (1991) 10 368.

14. N.M.Suleimanov, H.Drulis, A.D.Shengelaya, and G.Chadzynski, *Sverkhprov.: Fiz.Khim.Tekhn.* **4** (1991) 1925 (in Russian).

15. A.W.Hewat, *Physica* B **180&181** (1992) 369.

16. Z.Henkie, T.Cichorek, H.Drulis, and J.Klamut, *Physica* C **214** (1993) 138.

17. G.Dortmann, J.Erxmeyer, S.Blasser, J.Steiger, T.Paatsch, A.Weidinger, H.Karl, and B.Stritzker, *Phys.Rev.*B **49** (1994) 600.

18. T.Kebukawa, Y.Kumaki, K.Fujiwara, Y.Tatsumi, H.Taimatsu, and H.Kaneko, *Jpn.J.Appl.Phys.* **33** (1994) L507.

18. N.M.Suleimanov, H.Drulis, G.W.Chadzynski, A.D.Shengelaya, E.F.Kukovitskii, R.G.Mustafin, and J.Janczak, *Pis'ma v ZhÉTF* **51** (1990) 371 (in Russian).

19. N.M.Suleimanov, A.D.Shengelaya, R.G.Mustafin, E.F.Kukavitskii, P.W.Klamut, G.W.Chadzynski, H.Drulis, and J.Janczak, *Physica* C **185–189** (1991) 759.

20. S.D.Goren, C.Korn, V.Volterra, H.Riesemeier, E.Rossler, H.M.Vieth, and K.Luders, *Phys.Rev.*B **46** (1992) 14 142.

21. S.D.Goren, C.Korn, E.Rossler, H.M.Vieth, K.Luders, Z.Gavra, J.R.Johnson, and J.J.Reilly, *J.Alloys Compd.* **210** (1994) 9.

22. H.Lutgemeier, S.Schmenn, and Yu.Baikov, *Solid State Commun.* **94** (1995) 283.

23. W.K.Dawson, K.Tibbs, S.P.Weathersby, C.Boekema, and K.C.Chan, *J.Appl.Phys.* **64** (1988) 5809.

24. C.H.Halim, W.K.Dawson, W.A.Baldwin, and C.Boekema, *Physica* B **163** (1990) 453.

25. M.Weber, P.Birrer, F.N.Gygax, B.Hitti, E.Lippelt, H.Maletta, and A.Schenck, *Hyperf.Interact.* **63** (1990) 207.

26. N.Nishida and H.Hiyatake, *Hyperf.Interact.* **63** (1990) 183.

27. H Glucker, Ch.Niedermyer, G.Nowitzke, E.Recknagel, J.Weidinger, and J.I.Budnick, *Europhys.Lett.* **15** (1991) 355.

28. R.Saito, H.Kamimura, and K.Nagamine, *Physica* C **185–189** (1991) 1217.

29. S.Sulaiman, S.Srinivas, N.Sahoo, F.Hagelberg, T.P.Das, E.Torikai, and K.Nagamine, *Phys.Rev.*B **49** (1994) 9879.

30. M.R.Presland, J.L.Tallon, R.G.Buckley, R.S.Liu, and N.E.Flower, *Physica* C **176** (1991) 95.

31. V.V.Moschalkov, *Physica* B **163** (1990) 59.

32. N.F.Mott and E.A.Davis, *Electronic Processes in Non-crystalline Materials*, 2nd ed. (Clarendon, Oxford 1979).

33. M.Ciszek, J.Klamut, A.J.Zaleski, H.Drulis, P.W.Klamut, and J.Olejniczak, *J.Appl.Phys.* **76** (1994) 4.

34. J.M.Coey, X.Z.Wang, K.Donnelly, and J.F.Lawler, *J.Less-Comm.Met.* **151** (1989) 195.

35. H.Drulis, J.Klamut, A.Zygmunt, N.M.Suleimanov, and E.F.Kukovitskii, *Solid State Commun.* **84** (1992) 1069.

36. A.D.Shengelaya, H.Drulis, J.Klamut, A.Zygmunt, and N.M.Suleimanov, *Physica* C **226** (1994) 147.

37. G.M.Luke *et al.*, *Phys.Rev.*B **42** (1990) 7981.

38. H.Takagi, T.Ido, S.Ishibashi, M.Uota, S.Uchida, and Y.Tokura, *Phys.Rev.*B **40** 2254 (1989).

39. H.Romberg, M.Alexander, N.Nucker, P.Adelman, and J.Fink, *Phys.Rev.*B **42** 8768 (1990).

40. J.Olejniczak, A.J.Zaleski, A.D.Shengelaya, and J.Klamut, *Phys.Rev.*B **51** 8641 (1995).

41. T.Ekino, K.Matsukuma, T.Takabatake, and H.Fujii, *Physica* B **165&166** 1530 (1990).

42. T.Kato, K.Aihara, J.Kuniya, T.Kamo, and S.P.Matsuda, *Jpn.J.Appl.Phys.* **27** L564 (1988).

43. G.Wang, G.Pang, C.Luo, S.Yang, Y.Li, Z.Ji, and Z.Sun, *Phys.Lett.* **130** 405 (1988).

44. H.Alloul, P.Mendels, G.Collin, and P.Monod, *Phys.Rev.Lett.* **61** 746 (1988).

45. B.Hitti *et al.*, *Hyperf.Interact.* **63** 287 (1990).

The Superconductor-Insulator Transition in the LaSrCuO System[*]

Marta Z. Cieplak[1,2], K. Karpińska[1], and A. Malinowski[1,3]

[1] Institute of Physics, Polish Academy of Sciences, 02 668 Warsaw, Poland
[2] Department of Physics and Astronomy, Rutgers University, Piscataway, NJ 08855, USA
[3] Institute of Physics, Warsaw University, Branch at Białystok, 15 354 Białystok, Poland

Abstract. This paper describes recent experiments which probe the nature of the magnetic–field tuned superconductor-insulator (SI) transition in the LaSrCuO system. A summary is given of various experiments on the SI transition in other high-T_c systems and in conventional superconducting thin films. Similarities and differences between the behavior of LaSrCuO and the behavior of conventional films are discussed.

Keywords. Metal-insulator transitions, High-T_c cuprate films

1 Introduction

The phase diagrams of the materials that exhibit high-T_c superconductivity are rich in variety and information. Superconductivity develops upon doping of carriers into magnetically ordered, insulating "parent" compounds. The nature of the transition from the insulating to the superconducting state is only partially understood. Spectroscopic experiments [1] and theoretical considerations [2] suggest that the insulators are of the charge-transfer type. Transport experiments indicate hopping as the mechanism of conduction in the insulating state [3]. With the increasing doping of carriers, which are introduced by changes in chemical composition, the conductivity increases and almost simultaneously superconductivity appears at low temperatures. The transition from the insulating to the superconducting state is very sharp. This has led to predictions that the underlying normal state in the absence of superconductivity is a three-dimensional (3D) localized state [4], or that a peculiar charge-spin separation leads to an extended state confined to 2 dimensions (2D) [5]. These theories are difficult to test experimentally because of the high superconducting transition

[*]Work done in collaboration with T. Skośkiewicz, W. Plesiewicz, M. Berkowski (Institute of Physics, Polish Academy of Sciences) and S. Guha, S. Gershman, P. Lindenfeld (Physics Department, Rutgers University). Supported by the Polish Committee for Scientific Research, KBN, under grants 2 P03B 05608 and 2 P03B 06708, by the National Science Foundation under grant DMR 95-01504, and by the Commission of the European Communities under contract CIPA-CT93-0032.

temperature and high upper critical field (H_{c2}). Both the transition temperature, and H_{c2} may be considerabely lower in specimens in the vicinity of the SI transition thus creating an opportunity to study the nature of transition, and to elucidate the properties of the underlying normal state.

The sharp SI transition in high-T_c systems resembles SI transitions induced by disorder in conventional superconducting thin films. It is known that, in addition to disorder, a magnetic field may induce SI transitions in conventional 2D films [6]. Considerable efforts have been made recently to study both the disorder-induced, and the field-induced transitions [7-9]. At the center of attention is the problem of the relative importance of fluctuations in two components of the superconducting order parameter, that is, the phase and the amplitude. Fluctuations of both components may lead to the destruction of the long-range order in the superconducting state and may induce transitions to the insulating phase. The suggestion has been made that the field-induced transition which is driven by phase fluctuations should result in a so-called "bosonic insulating phase" in which Cooper pairs remain bound while undergoing localization [10]. On the other hand, the amplitude-driven transition should result in localization of single carriers. Because of the inherent quasi-2D nature of the high-T_c materials it has been suspected that high-T_c systems may undergo similar field-induced transitions. Initial experiments, performed on single specimens [11,12] were interpreted in terms of an SI transition driven by phase fluctuations. As we will show in the following sections, the study of the LaSrCuO system indicates the presence of quite different SI transition which occurs in the vicinity of H_{c2} and is probably driven by fluctuations of the amplitude of the superconducting order parameter.

The plan of this paper is as follows. In section 2 we review the experiments on conventional superconductivity in 2D, and discuss the initial experiments on high-T_c compounds. In section 3 our study of the LaSrCuO system is described. Our experiment provides an interesting glimpse of the low-temperature behavior of the upper critical field. We find that cooling the samples down to the mK region results in sharpening of the superconducting transition, an effect which has not been previously seen in underdoped specimens. This allows the observation that H_{c2} increases without saturation as the temperature is lowered. This divergent behavior is inconsistent with theories of conventional superconductivity.It has been observed previously in overdoped specimens of some high-T_c compounds [13,14].

2 Superconductor-Insulator Transition in 2D

By definition, in the insulating state the conductivity approaches zero as the temperature goes to zero, while in the superconducting state it diverges in this limit. Tuning from one state to the other results in an abrupt change of the conductivity by many orders of magnitude, provided that the experiment is performed at sufficiently low temperatures. For an assessment of the properties of the SI transition it is thus necessary to go to very low temperatures, a requirement which has not always been fulfilled in the experiments. Tuning may be achieved in various ways, such as by a change of external parameters like the magnetic field, or by changes in the properties of the specimens that lead to localization. One of these changes is intentionally introduced potential disorder (Anderson localization), which has been studied intensively in conventional superconductors. In bulk (3D) materials weak disorder has only a small effect on superconductivity unless it is magnetic in nature, while strong disorder lowers the superconducting transition temperature and eventually leads to the destruction of superconductivity. In 2D superconductors localization is induced by arbitrarily small disorder [15,16].

In the context of high-T_c systems it is important to remember that electron localization occurs also in the absence of disorder but in the presence of Coulomb repulsion (Mott localization). High-T_c superconductors with strong electron-electron correlations are presumably in-between these two limiting cases of localization, and the SI transition in high-T_c systems may differ from the disorder-induced transition in conventional superconducting films.

2.1 Tuning of the SI transition in zero magnetic field

A common way of inducing an SI transition in thin films of conventional superconductors is to introduce disorder by changing the thickness of the films. In the left panel of Fig. 1 we reproduce a set of curves which display the dependence of the resistance on the temperature for Bi films deposited on Ge, with the thickness ranging from 4.36 Å to 74.27 Å [17]. An abrupt swiching from superconducting to insulating behavior occurs in the vicinity of $R_c = 6450\Omega/\square$. Other examples of disordered films include amorphous materials, like Mo-Ge in which disorder is tuned by a change in the composition [18], or α-InO$_x$ amorphous composite films in which disorder depends on the oxygen pressure during deposition [19]. The value of R_c seems to be nearly the same in various materials, ranging from $3.2k\Omega/\square$ for Mo-C [20] to $10\ k\Omega/\square$ for α-InO$_x$ [21]. With the exception of granular films, all other thin films display transport behavior similar to that shown in Fig. 1. In granular superconductors metallic grains of substantial size are immersed in an insulating matrix. In this case the T-dependence of the resistance at the superconducting transition aquires a characteristic "two-step" feature: there is a well- defined decrease in the resistance at the temperature at which the grains become superconducting, followed by a longer resistance "tail" which extends to a lower temperature at which the whole matrix becomes superconducting [22].

In contrast to granular films amorphous materials are much more homogeneously disordered, and the superconducting transitions are smooth without any signs of the two-step behavior. The transitions are usually broader compared to bulk samples as a result of thermally-induced unbinding of the vortex-antivortex pairs which gives rise to fluctuations of the phase of the superconducting order parameter and to a nonvanishing resistance. The temperature at which true long-range order is destroyed by the phase fluctuations, the Kosterlitz-Thouless (KT) transition temperature, T_{KT}, is substantially lower than the temperature T_{co} at which the amplitude of the order parameter van-

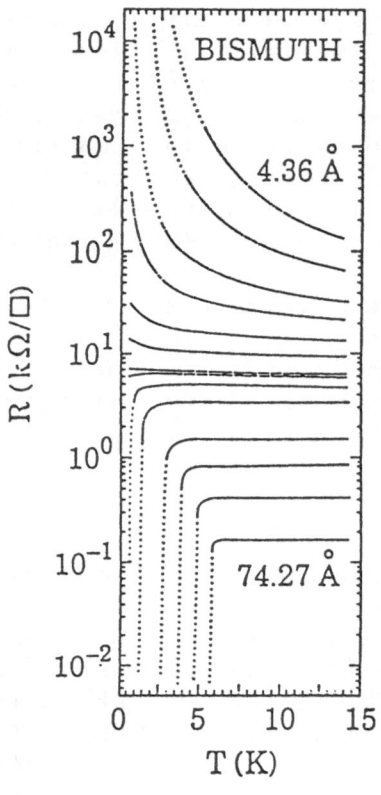

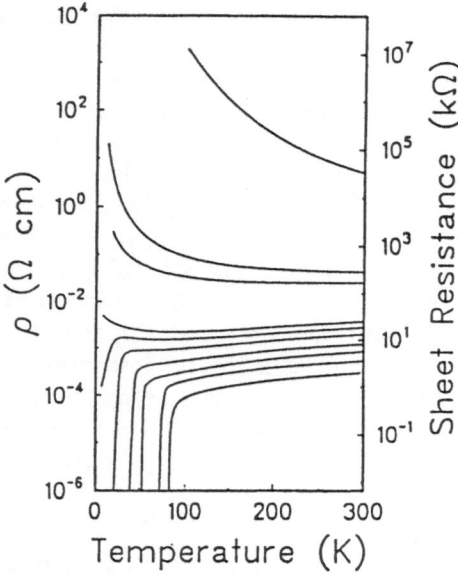

Fig. 1 Temperature dependence of the resistance of *(left panel)* Bi films deposited on Ge [17] and *(right panel)* $Bi_2Sr_2Y_xCa_{1-x}Cu_2O_8$ single crystals with various Y content (increasing upwards) [25].

ishes. The KT transition may be observed provided the specimen is truly 2D with respect to the vortex-antivortex interaction, i.e. when the film thickness d is smaller then the penetration depth, and provided the vortex interaction is not cut off by sample inhomogeneities or pinning. These conditions are fulfilled in ultrathin films of conventional superconductors [23]. In the case of high-T_c materials the structural anisotropy leads to quasi-2D character for each CuO_2 plane, and the KT transition is observed in bulk crystals [24].

An example of an SI transition in a high-T_c system is shown in the right panel of Fig. 1. Here tuning of the SI transition is achieved by a change of the yttrium content in $Bi_2Sr_2Y_xCa_{1-x}Cu_2O_8$ single crystals [25]. At first glance the similarity to the behavior of conventional thin films is striking. The switching between superconducting and insulating behavior seems to occur at a value of the resistance per square which is of the same order of magnitude as for conventional films. Similar tuning of the SI transition occurs when the composition is altered in most of the high-T_c systems. In LaSrCuO both a change of the strontium content and a change of the oxygen content induces an SI transition [26]. We note, however, very important differences between high-T_c systems and conventional superconducting films. Changes in composition of high-T_c materials affect the degree of disorder, but disorder does not drive the SI transition. It is the change of the carrier density, and the related change of the electronic structure which are the most important factors. It is thus essential to monitor the zero-T limit of the conductivity in high-T_c systems in order to understand the nature of the SI transition. Unfortunately the experiment shown in the right panel of Fig. 1 does not investigate low enough temperatures to reach unambiguous conclusions about the zero-T behavior of the conductivity.

2.2 Bosonic description of the SI transition

The KT transition is an example of a transition tuned by phase fluctuations, with the amplitude of the superconducting order parameter remaining finite. Fisher [10] argues that this type of transition may be correctly described by a model of charge-$2e$ bosons, representing Cooper pairs moving in a random potential. The phase diagram resulting from this model is shown in Fig. 2. The three different axes represent magnetic field (B), temperature (T), and disorder (Δ). Three different phase transitions are possible in this boson system. At $B = 0$, in the $T-\Delta$ plane, the system undergoes a KT transition at $T = T_c$. The amplitude of the superconducting order parameter remains constant at this temperature, and decreases to zero at the higher temperature $T = T_{co}$. As the disorder Δ increases, the point Δ_c is approached. It is the single metallic point at $T = 0$. For a disorder $\Delta < \Delta_c$ the system is superconducting, and for $\Delta > \Delta_c$ the system is insulating. At $\Delta = 0$, in the $B-T$ plane, a melting transition at $B = B_M$ occurs from a vortex lattice to a vortex fluid, followed by a transition to the normal state at H_{c2}. Finally, at $T = 0$, in the $B - \Delta$ plane, the ground state in zero magnetic field is a vortex glass in which the Cooper pairs form a superconducting

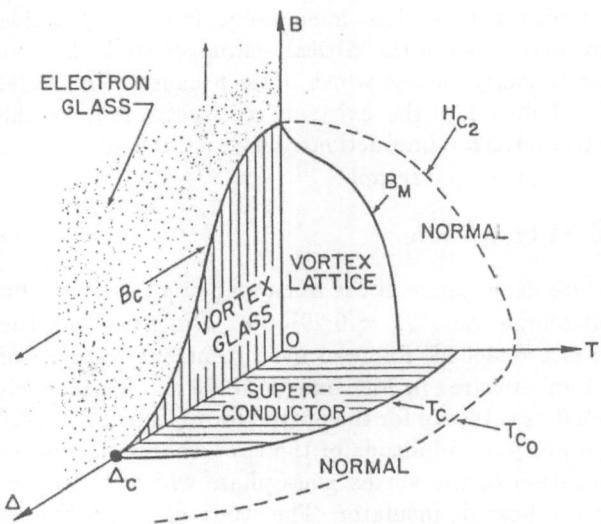

Fig. 2 The phase diagram of a 2D superconducting system (Ref.[10]).

condensate, and the vortices (bosons) are localized by disorder. On increasing the magnetic field the density of vortices increases, they become delocalized, and eventually condense into a macroscopic quantum state. This "bosonic insulator" is described as a phase "dual" to the superconducting phase, in the sense that the Cooper pairs are localized and the vortices undergo Bose condensation.

The theory predicts a set of scaling relations at the SI transition. At the zero-field transition, when the disorder Δ approaches criticality, the superconducting coherence length ξ diverges as $(\Delta - \Delta_c)^{-\nu}$. The KT transition temperature should vanish as $T_c \sim (\Delta - \Delta_c)^{z\nu}$, where z is the dynamical exponent which describes the dependence of T_c on ξ. The critical magnetic field at the field-tuned transition should vary as $B_c \sim (\Delta - \Delta_c)^{2\nu}$. Thus in the vicinity of Δ_c we have

$$B_c \sim T_c^{2/z}. \tag{1}$$

For the zero-field transition the resistance is predicted to scale as

$$R = (h/4e^2)\tilde{R}_\Delta[c_0(\Delta - \Delta_c)/T^{1/z\nu}], \tag{2}$$

whereas for a field-tuned transition the scaling takes the form

$$R = (h/4e^2)\tilde{R}_B[c_1(B - B_c)/T^{1/z_B\nu_B}]. \tag{3}$$

In (2) and (3) c_0 and c_1 are nonuniversal constants, and $\tilde{R}_\Delta$, $\tilde{R}_B$ are dimensionless scaling functions with the property that $\tilde{R}(0)$ is a universal number.

Experimentally, the existence of a universal value of the resistance is still an open question. As mentioned before, the values of R_c at the zero-field transition are of the same order but nevertheless not quite universal. It has been pointed out by Cha *et al.* [27] and by Hebard [7] that there are several reasons why experiments may not test theoretical predictions adequately. It is also possible that some experiments are not performed in the critical regime, or that different real systems are in various universality classes which differ because of material properties, such as the type of disorder, the existence of frustration, or the strength and range of the electron-electron interactions. The case of field-induced transitions is discussed in the following paragraph.

2.3 Magnetic–field–tuned SI transition

Fig. 3 shows the temperature dependence of the resistance of an α-InO$_x$ film with zero-field KT transition temperature $T_c = 0.29$K [6]. On increasing the magnetic field the transition is broadened, followed by a reentrant increase of the resistance at the lowest temperatures in high magnetic fields. This quasi-reentrant behavior is interpreted as evidence for the partial formation of a super-conducting condensate with a nonzero amplitude of the superconducting order parameter, followed by a transition to the vortex glass phase which undergoes a field tuned SI transition to the bosonic insulator. The study of a set of specimens with various T_cs shows that the scaling relation (1) is fulfilled with the dynamical exponent z close to 1, in excellent agreement with the theoretical predictions [10]. For each specimen the scaling of the resistance (3) could also

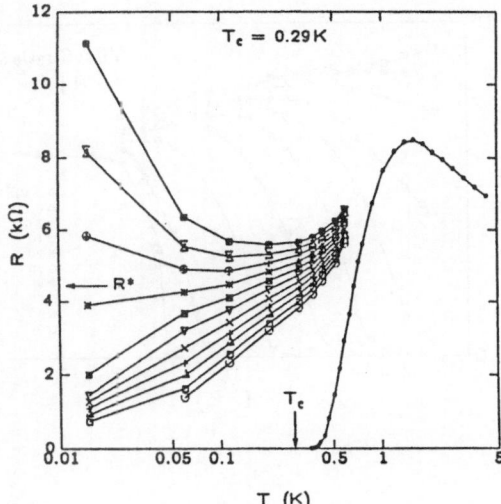

Fig. 3 Temperature dependence of the resistance of an α-InO$_x$ film in zero magnetic field (black dots) and in magnetic fields increasing from 4 kG (open circles) to 6 kG (open squares) in 0.2 kG intervals (Ref.[6]).

be evaluated, with the product of the scaling exponents $z_B\nu_B$ close to 1.3.

Values of the critical exponents of $z_B = 1.0$ and $\nu_B = 1.36$ were estimated independently from recent measurements of resistance and dynamical resistance ($\delta V/\delta I$) for a series of amorphous MoGe films deposited on Si with a Ge buffer layer [8]. While scaling seems to work perfectly for all samples in the series, the value of the critical resistance was found to be sample dependent, and therefore not universal. It was suggested that the lack of universality results from fermionic excitations which introduce a finite contribution to the resistance. This contribution is presumed to have small impact in the case of highly disordered samples in which unpaired electrons are localized and significant impact in the weakly localized limit.

There were two attempts to observe field-tuned SI transitions in high-T_c systems. In one experiment [11] an oxygen-defficient film of YBa$_2$Cu$_3$O$_{6.38}$ (YBCO) was cooled down to 50 mK, and in the second experiment [12] the measurements, down to 1.7 K, were performed on an electron-doped Nd$_{1-x}$Ce$_x$CuO$_4$ (NCCO) superconducting film. The influence of the magnetic field on $R(T)$, shown in Figs. 4 and 5, resembles results obtained for the indium-oxide films. In both cases it has been possible to test the predicted scaling behavior of the resistance (3), and the value of the product of the scaling exponents $z_B\nu_B$ obtained from the analysis of the data was close to those found for α-InO$_x$. However, the state of the technology used to grow the films was somewhat unsatisfactory, and in each experiment only one sample was investigated. This precluded testing of the scaling relation (1). We note also that the temperatures at which the experiment on NCCO was performed are much too high to consider them as being close to $T = 0$. Furthermore, it has been reported that the films used in this study display a negative magnetoresistance which is about 50% isotropic, so that there is substantial spin scattering in the film [28]. Spin scattering is explicitly excluded

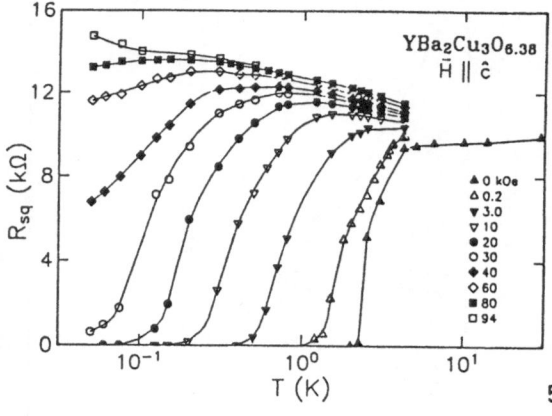

Fig. 4 Temperature dependence of the sheet resistance of $YBa_2Cu_3O_{6.38}$ for a series of magnetic fields perpendicular to the CuO_2 planes [11].

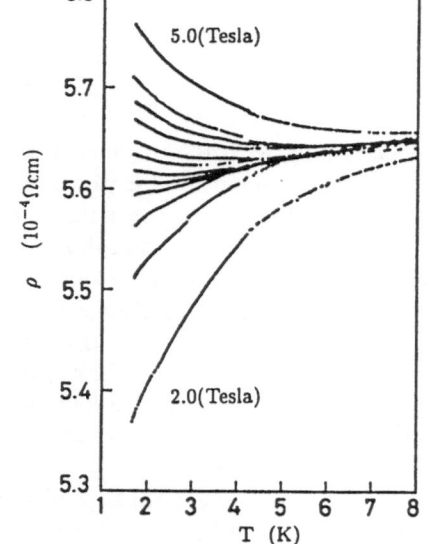

Fig. 5 Temperature dependence of the resistivity of $Nd_{1-x}Ce_xCuO_4$ films in the presence of perpendicular magnetic fields from 2.0 T (bottom) to 5.0 T (top) [12].

from the theory which describes the bosonic transition since pair breaking resulting from the spin scattering should break the Cooper pairs before they become localized [27].

Finally, we note that there is a marked difference between the behavior of $R(T)$ observed in these high-T_c films, and the behavior observed in α-InO_x: high-T_c films do not show the minimum in the resistance followed by reentrance, which in α-InO_x films indicates a partial formation of the superconducting condensate. It is thus not obvious that the SI transition is tuned by phase fluctuations only. It is possible that amplitude fluctuations play some role as well.

The relative importance of the amplitude fluctuations compared to the phase fluctuations has been studied recently in the case of conventional superconducting ultrathin films of PbBi evaporated on Ge underlayer [9]. In Fig. 6(a) we show the $R(T)$ dependence in the presence of a magnetic field for one of the films investigated in that study. It is seen that the large broadening of the transition, without any obvious minimum in the $R(T)$ curve, is qualitatively similar to the results for high-T_c films. Hsu *et al.* [9] also studied the tunneling conductance on the same films, as shown in Fig. 6(b). The increase of the magnetic field suppresses the peaks in the conductance and increases the zero bias conductance. Estimates of the superconducting carrier density in the absence of the field indicate that the average number of Cooper pairs in a coherence volume is comparable to 1. It is thus very likely that the Cooper pair density fluctuates from coherence volume to coherence volume, leading to fluctuations in the amplitude of the superconducting order parameter. The magnetic field greatly enhances the amplitude fluctuations leading to an SI transition. The physical origin of the enhancement may be related to the fact that the field diminishes the size of the coherence volume, and induces pair-breaking effects. Some theoretical attempts to take these effects into account[29] have so far failed to describe the experimental results adequately. However, it is clear that the phase-driven SI transition is not necessarily a universal feature of 2D superconducting systems. Interestingly, Hsu *et al.* [9] mention that the resistance of PbBi/Ge films follows the scaling relation (3) despite the fact that the transition is clearly tuned by fluctuations of the amplitude and not of the phase.

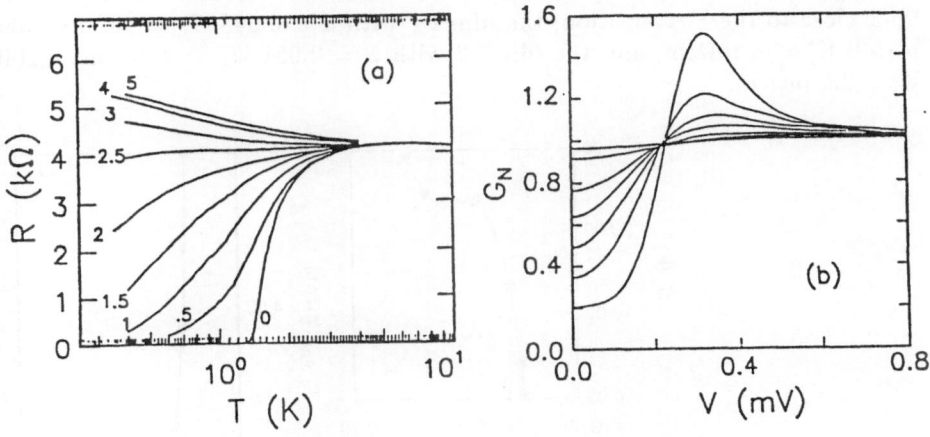

Fig. 6 Temperature dependence of the resistance (a), and normalized junction conductance at $T = 360$ mK (b) in the presence of a magnetic field for the PbBi/Ge film. In (b) the fields are (from the bottom up) at $V = 0$: 0, 0.5, 1, 1.5, 2, 3, and 4 T [9].

3 SI transition in the LaSrCuO system

Now we turn our attention to our recent experimental studies of the SI transition in $La_{1-x}Sr_xCuO_4$ (LSCO). Studying this system offers several advantages over studies of other high-T_c materials. Unlike YBCO, LSCO contains only one copper-oxide plane per unit cell, so that there is no contribution from the one-dimensional chain structure which complicates the understanding of the transport properties in YBCO. In addition, LSCO is relatively well studied compared to other systems. An SI transition is known to occur as a result of a change of the strontium content in the vicinity of $x = 0.5$ [30]. The main effect of a change in x is a change of the carrier concentration, which is known to be roughly proportional to x for metallic samples with small Sr-content [31]. Technological improvements make it possible to obtain both single crystals and c-axis aligned epitaxial films with well-characterized properties [32,33]. Sets of films with various Sr-content obtained by magnetron sputtering have been used to study the evolution with x of the normal-state resistivity [34] and the Hall effect [35].

The films described in this study were grown by pulsed-laser deposition on $SrLaAlO_4$ substrates, which have a smaller lattice mismatch to LSCO than other substrates that have been used [36]. This method allows the preparation of reproducible sets of specimens with small changes in x in the vicinity of the SI transition. The zero-resistance transition temperatures and ab-plane resistivities, ρ_{ab}, at $T = 40$ K for one such set are shown in Fig. 7. Superconducting transition temperatures of the specimens with $x < 0.12$ closely follow those for the bulk ceramic targets from which the films are made. Samples with values of x close to the optimal value for superconductivity have been shown normally to be deficient in oxygen [37], leading to a lowering of T_c. Here we discuss the study of two films close to the SI transition, the film F1 with $x = 0.048$, $T_c = 450$ mK and $\rho_{ab}(40$ K$) = 8$ mΩcm, and the film F2 with $x = 0.051$, $T_c = 4$ K and $\rho_{ab}(40$ K$) = 3.5$ mΩcm.

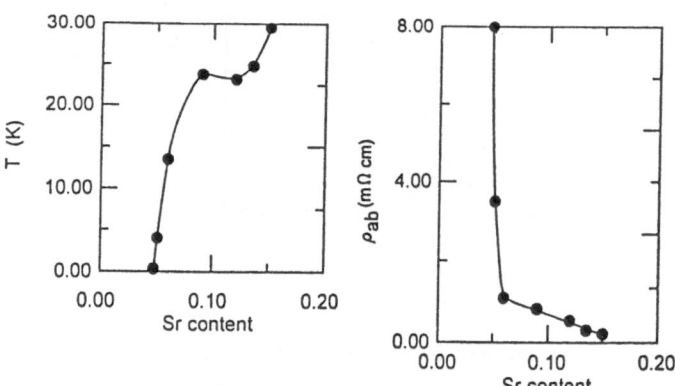

Fig. 7 The zero-resistance transition temperatures *(left)* and ab-plane resitivities at 40 K *(right)* for a set of $La_{1-x}Sr_xCuO_4$ films with various values of x.

3.1 Field tuned SI transition

Fig. 8 shows logarithmic plots of the resistance transitions in a magnetic field perpendicular to the ab-plane for film F1. We note first that if one restricts the temperatures to $T > 1$ K, the data look qualitatively similar to the results obtained by Tanda *et al.* [12], that is one observes a change in the curvature of $R(T)$ from superconducting-like to insulating-like for a magnetic field close to about 1 T. However, further cooling of the film reveals that after an initial increase of the resistance there is a maximum followed by a decrease of R down to the lowest temperatures. The reentrance observed in α-InO$_x$, which we consider to be a hallmark of the phase-fluctuations tuned SI transition, is absent here. Instead, we see a maximum in $R(T)$ which shifts towards lower temperatures as the magnetic field grows, until finally it is reduced below the lowest measurable temperature at a field of 2.8 T.

The absence of the reentrance casts doubt on the likelihood that the transition is related to phase fluctuations. The doubts are further enhanced when we attempt to test the scaling prediction (3). The usual way to pick a critical field is to analyze a slope of $R(T)$ dependence in the limit of lowest experimentally accesible temperature. The critical field is then defined as the field for which the slope of $R(T)$ crosses zero [6-8]. It is seen in Fig. 8 that the "critical field" defined in this way depends on the lowest T limit reached in the experiment. It follows that there is no critical field observed in this experiment, in the temperature range studied, and therefore we cannot observe the scaling characteristic for the SI transition tuned by the phase fluctuations.

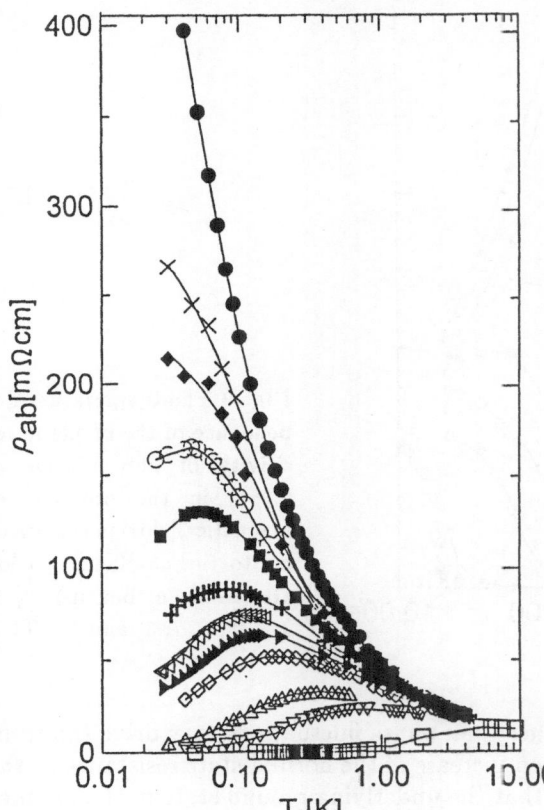

Fig.8 The temperature dependence of the *ab*-plane resitivity of film F.1 ($x = 0.048$) in the presence of magnetic fields perpendicular to the *ab*-plane. The fields are (from bottom up): 0, 1.3, 1.7, 2, 2.2, 2.3, 2.4, 2.6, 2.7, 2.8, 2.9, and 3 T.

The nature of the SI transition becomes clear when we look at the resistive behavior of film F2 (Fig. 9). T_c for this film is an order of magnitude larger then T_c for film F1, so that quenching of the superconductivity requires higher magnetic fields. Nevertheless, the character of the $R(T)$ curves is the same, that is, we see a broadening of the transition for small magnetic fields, followed by the appearance of a well-defined maximum which shifts to lower temperatures with increasing magnetic field. The maximum is much more pronounced then in the case of film F1. We identify this maximum as the field-affected "onset" of superconductivity in the vicinity of which the amplitude of the superconducting order parameter becomes finite. The appearance of the maximum stems from the fact that the normal-state resistance increases steeply with the lowering of the temperature. The high magnetic field uncovers this dependence, which is normally hidden by the transition to the superconducting state. The field-induced SI transition occurs near the H_{c2}, and so it is most likely tuned by the amplitude rather then the phase fluctuations.

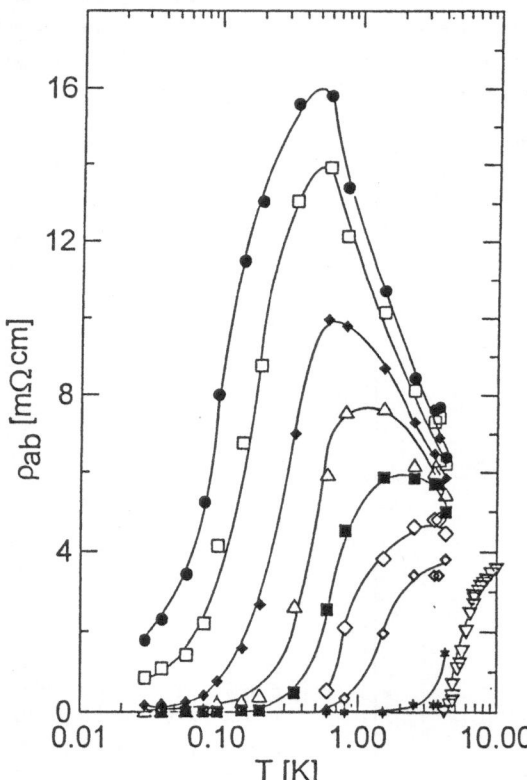

Fig. 9 The temperature dependence of the ab-plane resitivity of film F2 ($x = 0.051$) in the presence of magnetic fields perpendicular to the ab-plane. Fields are (from bottom up): 0, 1, 3, 4, 5, 6, 7, 8, and 8.5 T.

Why is it that amplitude rather than phase fluctuations may drive the transition? We note first that the steep increase of the normal-state resistance as the temperature is lowered suggests that the underlying ground state in the absence of superconductivity is likely to be insulating in the $T = 0$ limit. We find that the temperature dependence of the resistivity of the film F1 in the presence of a high magnetic field (8.5 T) is exponential, consistent with hopping conduction. In the case of film F2 it may be seen from the logarithmic plots in Fig. 9 that the data in the high- temperature range follow the same T-dependence for the two highest fields, given approximately as $R \sim \log(1/T)$. While even higher fields are needed to decide unambiguously that the normal state has been reached it is likely that the observed logarithmic dependence approximates the true normal state. Very recently Ando *et al.* observed a similar logarithmic dependence of the resistivity in LSCO with $x = 0.08$ and $x = 0.13$ in a perpendicular magnetic field of 61 T [38]. They conclude that the observed behavior is inconsistent with weak localization, which would predict a logarithmic T-dependence of conductivity rather then resistivity. Thus the ground state seems to be genuinely different from that of a conventional disordered 2D metal.

As was discussed before, the amplitude-fluctuation tuned SI transition was observed in conventional PbBi/Ge films [9]. Local fluctuations of the density of Cooper pairs, enhanced by the magnetic field, were suggested as the origin of the transition. A similar effect may occur in LSCO. However, in this case the unusual ground state with its strong Coulomb repulsion may have a strong impact on the SI transition. This is because the low concentration of carriers, short coherence length, and strong Coulomb repulsion are likely to lead to even greater local variations of the density of Cooper pairs from coherence volume to coherence volume. The wider distribution of local energy gaps may be responsible for the fact that the scaling relation (3) is not fulfilled in LSCO whereas it is observed to be followed in PbBi/Ge.

The local fluctuations of the amplitude of the superconducting order parameter might be enhanced if there are inhomogenities in the sample, either of structural origin or of electronic origin. While our samples seem to be of exceptional structural quality, with very good lattice matching to the substrate [36], one has to remember that LSCO is a material with natural chemical disorder resulting from the random distribution of Sr ions in the lattice. Moreover, there are theoretical predictions that even in the absence of structural disorder, the charge carriers in the sample may form some type of droplets or stripes [39]. Simultaneous ordering of charge and spin has been found to occur in the $La_{1-x}Sr_xNiO_4$ system [40], and it is expected that similar, but dynamical ordering may be present in LSCO. A related effect might be the presence of considerable spin scattering in the underdoped samples. It is known from various resonance experiments that with increasing doping of holes into an antiferromagnetically (AF) ordered insulating matrix of Cu-spins the long-range AF order breaks into smaller mesoscopic domains. The samples just on the insulating side of the SI transition show "cluster-spin-glass" ordering, in which the freezing occurs in the vicinity of $T_f = 4$ K as a result of mutual interaction of domains [41]. It is not clear if these domains survive on the superconducting side of the transition, and how their mutual interaction changes. It is quite plausible that they survive, although with smaller dimensions, and that their mutual interaction might be reduced, thus leaving uncompensated magnetic moments of the domains, which may act as scattering centers for carriers.

This scenario could be confirmed by a direct study of the superconducting density of states by, for example, tunneling experiments performed in the presence of a magnetic field. Unfortunately, tunneling experiments on high-T_c materials still suffer from irreproducibilities and smearing of the spectra, related to the instabilities of the cleaved surfaces or other difficulties associated with surface preparations. Besides, there are indications that the superconducting order parameter in high-T_c materials might be of the d-wave type, thus further contributing to the uncertainties in the interpretation of the tunneling data. To our knowledge no study of the influence of magnetic field on the tunneling spectra of LSCO has been done so far.

3.2 Upper critical field anomaly

It follows from the above discussion that the maximum in $R(T)$ may be used as an approximate measure of the upper critical field, H_{c2}. The dependence of H_{c2} on temperature has been the subject of renewed interest recently because of the discovery of anomalies in the behavior of H_{c2} in high-T_c systems. The studies of two overdoped systems, Tl-based [13], and Bi-based [14] revealed that H_{c2} seems to diverge as the temperature is lowered to the mK region. This is in sharp disagreement with the expectations of the classical Werthamer-Helfand-Hohenberg (WHH) theory [42] which predicts a linear increase of H_{c2} with a decrease of T in the vicinity of the zero-field transition temperature T_c, followed by saturation of H_{c2} as T approches zero.

The low-T divergence of H_{c2} has so far been unambiguously identified only in these overdoped samples. This was possible because the magnetic field causes almost parallel shifts of the superconducting transitions in the overdoped samples. It is then irrelevant with which point on the $R(T)$ curve one associates the critical field. In contrast, in the underdoped or optimally doped materials the transitions broaden in the magnetic field, with the top of the transition affected by the field much less then the bottom [43,44]. This broadening, which was studied by Suzuki and Hikita [43] in LSCO, is usually attributed to flux flow effects. It is found that the shape of the curve of H_{c2} versus T is in this case strongly dependent on the point with which H_{c2} is identified.

The samples which we investigate in the present study are in the strongly underdoped regime. However, it is seen from the logarithmic plots of the superconducting transitions in Fig. 9 that the behavior which we observe is quite different from the results described in Ref.[43]. On increasing the magnetic field the resistive transition first broadens appreciably. Then, for higher fields and at temperatures less then 1 K, the transition sharpens again. Because the sharpening effect was not observed in the previous studies, which were done at much higher temperatures, we tentatively associate the sharpening with the presence of very low temperatures, rather then the high magnetic fields. We note that if the broadening of the transition is indeed related to flux flow effects the lowering of the temperature should inhibit activated motion of fluxoids, and consequently lead to a sharpening of the transition at very low temperatures.

The sharpening of the transition encourages the attempt to extract the dependence of H_{c2} on T from the resistance data. We find, however, that we have to restrict the definition of the "critical field" to the points in the vicinity of the top of the resistive transition, otherwise the scatter of the data becomes too large for any quantitative analysis. We note that there exist experiments on the dc magnetization of YBCO [45], and on tunneling in $Ba_{1-x}K_xBiO_3$ [46] which support the identification of H_{c2} with the field near the top of the resistive transition. It has also been postulated in some theoretical models that the bottom of the resistive transition represents the irreversibility line, while H_{c2} should be identified with some point near the top of the resistive transition [47]. Remembering that this point is still to be investigated further, we restrict ourselves here to the definition of the critical field H^* as the field at which the resistance reaches 90% of the maximum of $R(T)$. We note also that it might be more sensible to use the normal-state resistance value instead of the resistance maximum. However, the dependence of the normal state resistance on the field and temperature is still not well known.

In Fig. 10 we show H^* as defined in this way as a function of temperature. Note that the zero-field point at 10.8 K is not T_c, but is close to the onset of superconductivity in the absence of the field. It may be seen that there is large scatter of the data for $T > 1$ K, related to the broadening of the transition at high T, which was discussed above. However, for $T < 1$ K the scatter is reduced, and it is possible to infer that H^* increases sharply without any sign of saturation as the temperature is lowered.

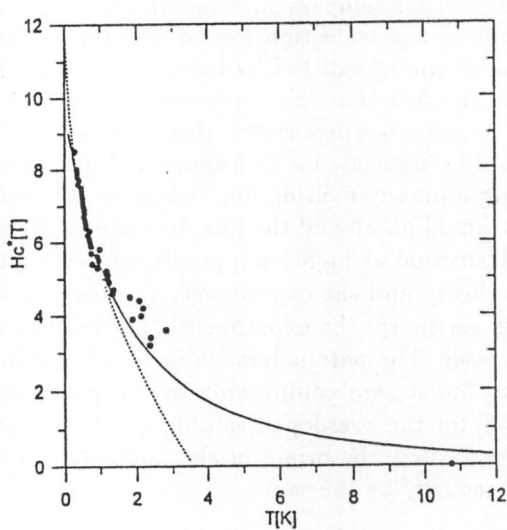

Fig. 10 Points: the critical field $H^*(T)$ as defined in the text from the resistivity at 90% of the maximum value. Solid curve: localized boson model [49]. Dotted curve: melting line resulting from the proximity to a zero temperature quantum critical point [50].

This divergent behavior is qualitatively the same as that observed for overdoped systems. Obviously, the origin of the anomalies does not have to be the same for overdoped and underdoped samples. It is possible that there are charge and spin inhomogenities in the underdoped specimens, and hence related spin scattering. The coupling of the superconducting charge carriers to magnetic moments may cause anomalies in the conventional $H_{c2}(T)$ behavior, as is known to occur in the ternary compounds RRh_4B_4 or RMo_6X_8 (R = rare earth, X = S, Se) [48] where superconducting carriers couple to the rare earth moments. On the other hand, it is also possible that the origin of the anomalies is the same for both types of samples. In this case it cannot be related to the impurity or spin scattering since the overdoped samples are in the clean limit. The divergent behavior of $H_{c2}(T)$ is predicted by Alexandrov et al. [49] in a model of bipolarons localized in a random potential. The fit of this prediction to our data is shown in Fig.10 as a continuous line. We can reproduce the shape of the curve at low temperatures but this is done assuming the quite unreasonable value of the coherence lenght $\xi = 340$ Å. Osofsky et al. [13] noted similar difficulties in analizing their data for overdoped Bi-Sr-Cu-O.

Recently it has been proposed by Kotliar and Varma [50] that the anomaly may result from a proximity to a zero temperature quantum critical point, which they identify with melting of the Abrikosov lattice. In the vicinity of the critical point, the melting field, H_m, should depend on temperature as

$$\frac{H_m(T)}{H_{c2}(0)} = 1 - \left(\frac{T}{T^*}\right)^{\frac{2}{5}}, \tag{4}$$

where T^* is a characteristic temperature dependent on the material parameters. The fit of the Eq.(4) to our data is shown in Fig. 10 by the dotted line. It is seen that the data are fitted quite well for temperatures below 1K, where the transition is sharp. The broadening of the transition for temperatures higher then 1K causes deviations from the relation predicted by theory. The origins of these deviations may be related to the fact that the experimental points were picked from 90% of the resistive maximum whereas the theory describes the melting line which most likely should be associated with a somewhat lower resistance. In the dirty superconducting films the melting line should be separated from the $H_{c2}(T)$ line by the vortex liquid phase and the flux flow effects present in this phase should broaden the transition at higher temperatures causing the observed discrepancy between the theory and the experiment. However, at low temperatures, when the transitions are sharp, the experimental data follow the predicted $H_m(T)$ dependence very well. The parameters which we obtain from the fit, $H_{c2}(0) = 13$ T and $T^* = 3.6$ K, are comparable to the parameters obtained by Kotliar and Varma [50] for the overdoped sample of $Tl_2Ba_2CuO_6$ [12]. This similarity strongly suggests that the origin of the anomalies in the overdoped and underdoped specimens may be the same.

Conclusions

The studies of the LSCO system provide two important results. First, the magnetic field tuned SI transition is qualitatively different from the SI transition tuned by the phase fluctuations which was observed in some conventional and some high-T_c systems, such as α-InO$_x$, MoGe, YBCO, or NCCO. In particular, in the temperature range which we study, we find no evidence of the critical field in agreement with the bosonic model for 2D superconducting systems. Instead, the magnetic field induces SI transition near the H_{c2}. The proximity to the H_{c2} suggests that the transition may be related to the fluctuations of the amplitude of the superconducting order parameter. It seems plausible that the combination of the low superconducting carrier density, short coherence length, strong Coulomb repulsion, and possibly spin scattering, contribute to the distribution of the local energy gaps, which are further enhanced by the magnetic field.

The second interesting observation is the low-temperature sharpening of the superconducting transitions in the underdoped LSCO specimens, broadened initially by the presence of the magnetic field. The sharpening, presumably related to the freezing of the fluxoid flow, allows for the observation of the divergence of the upper critical field as the temperature is lowered. This anomaly is qualitatively the same as in the overdoped high-T_c systems and is well described by the theory which associates anomaly with the proximity to a zero temperature quantum critical point [50].

Our observation of the unusual SI transition in LSCO, presumabely tuned by the amplitude fluctuations, brings up an important question: is this type of transition a property of LSCO, or is it a more general feature of high-T_c systems? Indeed, one may expect that the above mentioned combination of factors which seem to lead to an amplitude-tuned transition are present in all high-T_c systems. Why then did the previous studies of YBCO [11] and of NCCO [12] reach quite different conclusions? While the experiment on NCCO was done at relatively high temperatures the experiment on the oxygen deficient YBCO was performed in the milikelvin temperature range and it reveals clearly the SI transition which is qualitatively different from the transition in LSCO. One possible explanation of this difference may be related to the fact that the distribution of the local energy gaps is likely to be strongly influenced by the microscopic properties specific for each compound, such as the evolution of the electronic structure in the vicinity of the SI transition, the method of doping of the carriers, or the type of disorder. Further studies are needed to answer the question to what extent these specific properties affect the nature of the SI transition.

Acknowledgements

We are pleased to acknowledge useful and illuminating discussions with G. Kotliar, and in particular the turning our attention to the problem of the upper critical field anomaly. We also thank G. Xiao for letting us use his experimental setup in the initial stages of this experiment.

References

1. S. Uchida, T. Ido, H. Takagi, T. Arima, Y. Tokura, and S. Tajima, Phys. Rev. B **43**, 7942 (1991).
2. H. Eskes, M. B. Meinders, and G. A. Sawatzky, Phys. Rev. Lett. **67**, 1035 (1991).

3. L. Forro, in "Proceedings of the 5th Int. Conference on Hopping and Related Phenomena", Glasgow, Scotland, U.K. 1993".

4. G. Kotliar, E. Abrahams, A. E. Ruckenstein, C. M. Varma, P. B. Littlewood, and S. Schmitt-Rink, Europhys. Lett. **15**, 655 (1991).

5. P. W. Anderson and Z. Zou, Phys. Rev. Lett. **60**, 132 (1992).

6. A. F. Hebard and M. A. Paalanen, Phys. Rev. Lett. **65**, 927 (1990).

7. A. F. Hebard, in Los Angeles Symposium 1993 "Strongly correlated electronic materials", ed. K. S. Bedell, Z. Wang, D. E. Meltzer, A. V. Balatsky, and E. Abrahams, Addison-Wesley Publishing Company, 1994, p. 251.

8. A. Yazdani and A. Kapitulnik, Phys. Rev. Lett. **74**, 3037 (1995).

9. S-Y. Hsu, J. A. Chervenak, and J. M. Valles, Jr., Phys. Rev. Lett. **75**, 132 (1995).

10. M. P. A. Fisher, Phys. Rev. Lett. **65**, 923 (1990).

11. G. T. Seidler, T. F. Rosenbaum, and B. W. Veal, Phys. Rev. B **45**, 10162 (1992).

12. S. Tanda, S. Ohzeki, and T. Nakayama, Phys. Rev. Lett. **69**, 530 (1992).

13. A. P. Mackenzie, S. R. Julian, G. G. Lonzarich, A. Carrington, S. D. Hughes, R. S. Liu, and D. C. Sinclair, Phys. Rev. Lett. **71**, 1238 (1993).

14. M. Osofsky, R. J. Soulen, Jr., S. A. Wolf, J. M. Broto, H. Rakoto, J. C. Ousset, G. Coffe, S. Askenazy, P. Pari, I. Bozovic, J. N. Eckstein, and G. F. Virshup, Phys. Rev. Lett. **71**, 2315 (1993).

15. P. A. Lee and T. V. Ramakrishnan, Rev. Mod. Phys. **57**, 287 (1985).

16. D. Belitz and T. R. Kirkpatrick, Rev. Mod. Phys. **66**, 261 (1994).

17. D. B. Haviland, Y. Liu, and A. M. Goldman, Phys. Rev. Lett. **62**, 2180 (1989).

18. J. M. Graybeal and M. A. Beasley, Phys. Rev. B **29**, 4167 (1984).

19. A. F. Hebard and S. Nakahara, Appl. Phys. Lett. **41**, 1132 (1982).

20. S. J. Lee and J. B. Ketterson, Phys. Rev. Lett. **64**, 3078 (1990).

21. A. F. Hebard and M. A. Paalanen, Phys. Rev. Lett. **54**, 2155 (1985).

22. A. E. White, R. C. Dynes, and J. P. Garno, Phys. Rev. B **33**, 3549 (1986); H. M. Jaeger, D. B. Haviland, A. M. Goldman, and B. G. Orr, Phys. Rev. B **34**, 4920 (1986).

23. A. F. Hebard and A. T. Fiory, Phys. Rev. Lett. **50**, 1603 (1983); A. F. Hebard and G. Kotliar, Phys. Rev. B **39**, 4105 (1989).

24. A. T. Fiory, A. F. Hebard, P. M. Mankiewich, R. E. Howard, Phys. Rev. Lett. **61**, 1419 (1988); N. -C. Yeh and C. C. Tsuei, Phys. Rev. B **39**, 9708 (1989); D. H. Kim, A. M. Goldman, J. H. Kang, R. T. Kampwirth, Phys. Rev. B **40**, 8834 (1989).

25. D. Mandrus, L. Forro, C. Kendziora, and L. Mihaly, Phys. Rev. B **44**, 2418 (1991).

26. Y. Iye, in "Physical Properties of High Temperature Superconductors III", edited by D. M. Ginsberg, (World Scientific, 1992).

27. M. C. Cha, M. P. A. Fisher, S. M. Girvin, M. Wallin, and A. P. Young, Phys. Rev. B **44**, 6883 (1991).

28. S. Tanda, M. Honma, and T. Nakayama, Phys. Rev. B **43**, 8725 (1991).

29. S. Ullah and A. T. Dorsey, Phys. Rev. B **44**, 262 (1991); R. J. Troy and A. T. Dorsey, Phys. Rev. B **47**, 2715 (1993).

30. H. Takagi, T. Ido, S. Ishibashi, M. Uota, S. Uchida, and Y. Yokura, Phys. Rev. B **40**, 2254 (1989); M. Suzuki, Phys. Rev. B **39**, 2312 (1989).

31. N. P. Ong, Z. Z. Wang, J. Clayhold, J. M. Tarascon, L. H. Greene and W. R. McKinnon, Phys. Rev. B **35**, 8807 (1987); Y. Hidaka and M. Suzuki, Nature **388**, 635 (1989).

32. Y. Hidaka, Y. Enomoto, M. Suzuki, M. Oda, and T. Murakami, J. Crystal Growth **91**, 463 (1988); I. Tanaka and H. Kojima, Nature **337**, 21 (1989); T. Kimura, K. Kishio, T. Kobayashi, Y. Nakayama, N. Motohira, K. Kitazawa, and K. Yamajuji, Physica C **192**, 247 (1992).

33. H. L. Kao, J. Kwo, R. M. Fleming, M. Hong, and J. P. Mannaerts, Appl. Phys. Lett. **59**, 2748 (1991); T. R. Lemberger, in "Physical Properties of High Temperature Superconductors III", edited by D. M. Ginsberg, (World Scientific, 1992), p.471.

34. H. Takagi, B. Batlogg, H. L. Kao, J. Kwo, R. J. Cava, J. J. Krajewski, and W. F. Peck, Jr., Phys. Rev. Lett. **69**, 2975 (1992).

35. H. Y. Hwang, B. Batlogg, H. Takagi, H. L. Kao, R. J. Cava, J. J. Krajewski, and W. F. Peck, Jr., Phys. Rev. Lett. **72**, 2636 (1994).

36. M. Z. Cieplak, M. Berkowski, S. Guha, E. Cheng, A. S. Vagelos, D. J. Rabinowitz, B. Wu, I. E. Trofimov, and P. Lindenfeld, Appl. Phys. Lett. **65**, 3383 (1994).

37. I. E. Trofimov, L. A. Johnson, K. V. Ramanujachary, S. Guha, M. G. Harrison, M. Greenblatt, M. Z. Cieplak, and P. Lindenfeld, Appl. Phys. Lett. **65**, 2481 (1994).

38. Y. Ando, G. S. Boebinger, T. Kimura, and K. Kishio, unpublished.

39. V. J. Emery and S. A. Kivelson, Physica C **209**, 597 (1993).

40. J. M. Tranquada, D. J. Buttrey, V. Sachan, and J. E. Lorenzo, Phys. Rev. Lett. **73**, 1003 (1994).

41. F. C. Chou, F. Borsa, J. H. Cho, D. C. Johnston, A. Lascialfari, D. R. Torgeson, and J. Ziolo, Phys. Rev. Lett. **71**, 2323 (1993).

42. E. Helfand, N. R. Werthamer, and P. C. Hohenberg, Phys. Rev. **147**, 295 (1966).

43. M. Suzuki and M. Hikita, Phys. Rev. B **44**, 249 (1991).

44. Y. Iye, T. Tamegai, T. Sakakibara, T. Goto, N. Miura, H. Takeya, and H. Takei, Physica C **153-155**, 26 (1988); T. T. M. Palstra, B. Batlogg, L. F. Schneemeyer, and J. V. Waszczak, Phys. Rev. Lett. **61**, 1662 (1988); A. T. Fiory, M. A. Paalanen, R. R. Ruel, L. F. Schneemeyer, and J. V. Waszczak, Phys. Rev. B **41**, 4805 (1990).

45. U. Welp W. K. Kwok, G. W. Crabtree, H. Claus, K. G. Vandervoort, B. Dabrowski, A. W. Mitchell, D. R. Richards, D. T. Marx, and D. G. Hinks, Physica C **156**, 27 (1988); U. Welp, W. K. Kwok, G. W. Crabtree, K. G. Vandervoort, and J. Z. Liu, Phys. Rev. Lett. **62**, 1908 (1989).

46. G. Roesler, P. M. Tedrow, E. S. Hellman, and E. H. Hartford, IEEE Trans. Appl. Supercond. (USA) **3**, 1280 (1993).

47. D. S. Fisher, M. P. A. Fisher, and D. A. Huse, Phys. Rev. B **43**, 43 (1991); J. R. Clem, Phys. Rev. B **43**, 7837 (1991); M. Tinkham, Phys. Rev. Lett. **61**, 1658 (1988).

48. "Superconductivity in Ternary Compounds II", edited by M. B. Maple and O. Fisher (Springer-Verlag, New York, 1982).

49. A. S. Alexandrov, J. Ranninger, and S. Robaszkiewicz, Phys. Rev. B **33**, 4526 (1986).

50. G. Kotliar and C. M. Varma, unpublished.

On the "s" and "d" Wave Symmetry in High–T_c Cuprate Superconductors

K.Alex Müller

   University of Zürich, Physics Department, CH–8057 Zürich, Switzerland;
   and
   IBM Zürich Research Laboratory, CH–8803 Rüschlikon, Switzerland.

Abstract. It is shown that the best recent tunneling data in the superconducting cuprates can quantitatively be accounted for by assuming that two order parameters of s– and d-symmetry are present. This assumption also comprises two coherence lengths that differ by nearly an order of magnitude. Owing to this difference, specific tunneling experiments clearly differentiate between the two order parameters. Furthermore, it is also pointed out that other experiments either integrate or differentiate between the two components.

   A longer summary of the above presentation has recently appeared in *Nature* **377**, 133 (1995).

Properties of Infinite–Layer and T'-Phase Electron–Doped Copper–Oxide Superconductors

John T. Markert, Jonathan L. Cobb, Christopher L. Kuklewicz, Beom–hoan O, and Ruiqi Tian

Department of Physics, The University of Texas, Austin, TX 78712 USA

Abstract. An overview of recent research on electron–doped infinite–layer compounds (e.g., $Sr_{1-x}La_xCuO_2$) and electron–doped T'-phase compounds (e.g., $Ns_{2-x}Ce_xCuO_4$) is presented. Studies of steric effects indicate that superconductivity disappears in both systems for values of the in–plane lattice constant below a critical value, $a_{cr} \approx 3.92$ Å. Data from the $Nd_{1-y}Y_y)_{2-x}Ce_xCuO_4$ system confirm that steric effects are responsible for the loss of superconductivity (i.e., not magnetic interactions, as has been recently proposed). Previous Mössbauer work has emphasized the presence of a distribution of antiferromagnetic clusters in Nd-Ce-Cu-O, as well as the strong 2-D nature of the magnetism in that system; thus, we have modelled our anisotropic resistivity data (Montgomery method) using a spin–fluctuation scattering model. The extracted parameters indicate a gap to spin excitations ($\Delta \approx 3.4$ meV), reasonable scattering lengths ($l_\parallel \approx 200$ Å; $l_{perp} \approx 4$ Å), and a mass anisotropy $m_\perp/m_\parallel \approx 570$. A remarkably linear dependence of the extracted Fermi wavevector on the residual resistivity provides values of the impurity scattering cross section and impurity concentration. Also discussed are properties of $La_{2-x}Y_xCuO_4$, the first T'-phase compound without a magnetic rare–earth element. Various other investigations of these two known electron–doped copper–oxide structures are presented or reviewed.

1 Introduction

Of all the many superconducting copper–oxide materials, only two types are generally accepted as being "electron–doped" superconductors. These are: 1) the T'-phase compounds, i.e., the lanthanide series $L_{2-x}Ce_xCuO_{4-\delta}$, with $L = $ Pr, Nd, Sm, or Eu [1,2] and $L_{2-x}Th_xCuO_{-\delta}$, with $L = $ Pr, Nd, or Sm [2–4], and 2) the so–called "infinite–layer" compounds, $A_{1-x}M_xCuO_2$, with $A = $ Sr, $Sr_{1-y}Ca_y$, or $Sr_{1-y}Ba_y$, and where M is a rare–earth, most commonly La or Nd. In the former case, electron doping is achieved by substituting tetravalent Ce or

Th for the trivalent lanthanide host ions; in the latter case, substitution of trivalent lanthanides for the divalent host alkaline-earth ions occurs. The T'-phase compounds are structurally more complex, with fluorite-like layers separating the CuO_2 planes, and non-ideal site occupancy a norm [5]; the infinite-layer electon-doped materials, on the other hand, have only single alkaline/rare-earth ion layers between the CuO_2 planes, and can be prepared in an apparently ideal structure [6]. The latter is not to be confused with its hole-doped counterpart, which is riddled with intergrowths and inhomogeneities; indeed, it is generally believed that no simple hole-doped infinite-layer structure exists [7].

The T' phase has some well-known peculiarities: superconductivity can most often be induced only in a narrow dopant concentration regime, often near x = 0.15, and a reduction anneal is required to remove some of the as-prepared oxygen [8]. Neutron diffraction indicates that partial occupancy of the apical (interstitial) oxygen site occurs, as well as some vacancies in the two usual oxygen sites [5]. The reduction anneal predominantly affects the interstitial occupancy. The as-prepared oxygen content apparently becomes larger with decreasing size of the L ion [9]; the removal of excess oxygen appears to be hampered by cerium doping and smaller lattice sizes. Thus, although it is evident from x-ray absorption spectra that each Ce ion donates one electron to a copper atom [10], the concentration range for superconducting behavior can exhibit appreciable variability [9,11], due to such host-size-dependent oxygen content and oxygen site occupancy. However, as argued below, universal and purely steric effects, in addition to the oxygen occupancy discussed above (which itself is related to steric effects), have strong influence on superconducting behavior.

Appreciable insight into the microscopic origin of many physical properties of the T'-phase materials, as well as how they relate to the oxygen defects just described, is provided by emission Mössbauer studies [12–15]. The Mössbauer spectra are characterized by a magnetically split species and a paramagnetically relaxed one; after the deoxygenation procedure, the proportion of the latter grows at the expense of the former. The magnetically split species can be associated with extraneous apical oxygen occupancy (which localize doped electrons and nucleate regions of quasi-static magnetic order); the amount of paramagnetically relaxed species correlates with the superconducting volume fraction [14]. Most peculiar is the cerium concentration dependence: the extraneous oxygen can be easily removed only for cerium concentrations near the optimum value (near x = 0.16 for the Nd host), while both above and below that concentration, the deoxygenation procedure causes little change. The observation that even after detailed attempts at optimizing the deoxygenation procedure there exist appreciable non-superconducting regions with antiferromagnetic spin clusters that are static on the Mössbauer timescale ($\sim 10^{-8}$ s) is in agreement with typical low-field magnetization studies [16], where the superconducting volume fraction rarely exceeds 60%.

The infinite-layer superconductors have been much less studied, no doubt because of their later advent on the high-T_c scene [17], as well as their more complicated synthesis requirements: typically, they must be reacted at 25 kbar or higher pressure. Some of the physical properties of the electron-doped infinite-layer materials are reviewed below; for now, suffice it to say that no structural irregularities are readily evident [6].

With these interesting behaviors of the dopant and oxygen content of the T' phase in mind, as well as the underlying antiferromagnetism common to all of the copper oxides, we now discuss a number of specific experimental results. First, some steric effects are described which appear to be quite similar in both electron-doped systems. Then, certain properties of the infinite-layer materials are discussed. Various results of studies of the T'-phase materials are presented next, followed by a brief summary.

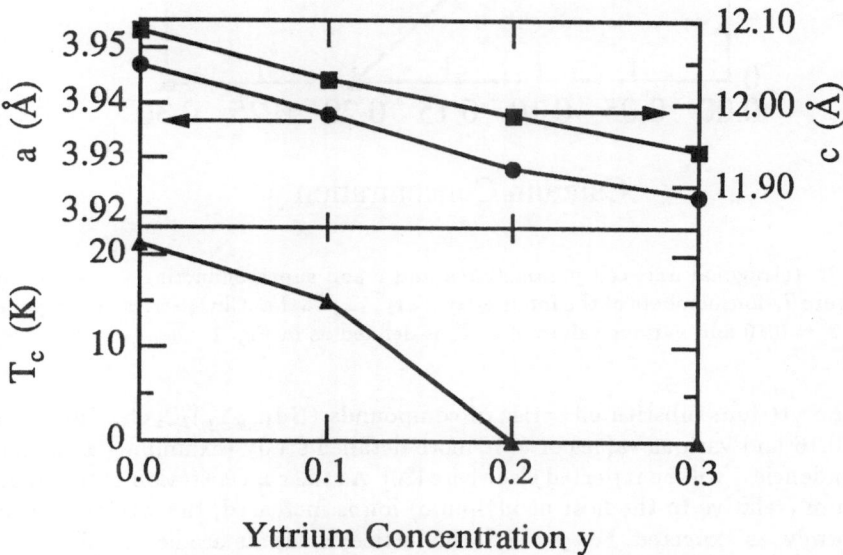

Fig. 1. Tetragonal unit cell parameters a and c and superconducting transition temperature T_c for members of the T'-phase $(Nd_{1-y}Y_y)_{2-x}Ce_xCuO_{4-\delta}$ series of compounds with $x = 0.16$ and various values of y. Here, T_c is defined as the temperature at which the zero-field-cooled magnetization data reaches 5% of ideal diamagnetism.

2 Steric Effects on T_c in Electron-Doped Superconductors

The decreasing volume fraction and eventual disappearance of superconductivity in the $L_{2-x}Ce_xCuO_{4-\delta}$ series as the rare-earth host ion size decreases (in the sequence $L = $ Pr, Nd, Sm, Eu, and Gd) has generally been assumed to be a steric effect, as has the pressure dependence of T_c [18]. Recent propositions [19] that the disappearance of superconductivity for the $L = $ Gd system was due to magnetic pairbreaking are easily laid to rest. In Fig. 1 are shown lattice constants

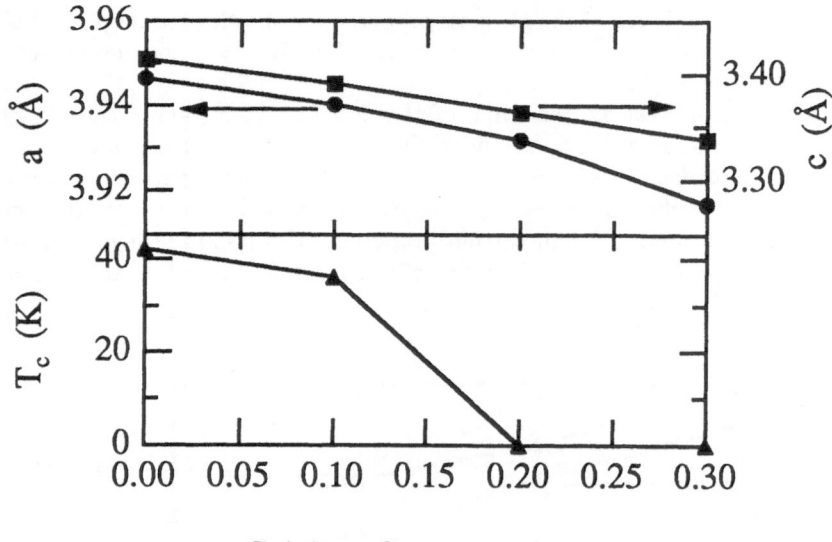

Calcium Concentration y

Fig. 2. Tetragonal unit cell parameters a and c and superconducting transition temperature T_c for members of the infinite-layer $Sr_{1-x-y}Ca_yLa_xCuO_2$ series of compounds with $x = 0.10$ and various values of y. T_c is defined as in Fig. 1.

for the yttrium-substituted series of compounds $(Nd_{1-y}Y_y)_{2-x}Ce_xCuO_{4-\delta}$ for $x = 0.16$ and various values of y. A more detailed study (examining x, y, and δ dependencies) will be reported elsewhere [20]. As the concentration of the smaller yttrium (relative to the host neodymium) ion is increased, the lattice contracts uniformly, as expected. For $y = 0.3$, the lattice constants are nearly identical to those of the smallest rare earth which forms the structure, $L = Gd$.

Also shown in Fig. 1 are values of T_c determined from SQUID magnetization data, as defined in the figure caption. Other measures of T_c give essentially the same result. It is evident that superconductivity diminishes with decreasing lattice constants and disappears around $a \approx 3.92$ Å and $c \approx 11.94$ Å, corresponding to an interplanar distance of $c/2 \approx 5.97$ Å. Obviously, substitution of non-magnetic yttrium has the same effect as decreasing the rare-earth host ion size; thus, steric effects are responsible for the loss of superconductivity, not magnetic ones. Of course, the physical mechanism behind the steric effect remains an open question; it may be related to the apical-oxygen-induced electron localization effects discussed above, or may simply be a reflection of a rapid variation of exchange interaction strength with bond length. A comparison with the infinite-layer materials may shed some light on this issue.

Figure 2 shows data similar to Fig. 1, but for a series of infinite-layer compounds $Sr_{1-x-y}Ca_yLa_xCuO_2$, with $x = 0.10$ and various values of y. Because of the difficulties involved with high-pressure synthesis, these data reflect an optimization of properties, attained from over thirty sample syntheses in this series. After pre-reaction of the starting materials under ambient conditions, the samples were sealed in platinum capsules and reacted at 25 kbar pressure at $\sim 1030°C$. Data from the best single-phased materials are shown.

It is evident from Fig. 2 that steric effects similar to those at work in the T'-phase materials are also occuring in the infinite-layer materials. It is also evident that superconductivity disappears near a universal value of the intraplanar lattice constant, $a \approx 3.92$ Å. The interplanar distance (for a constant value of a), on the other hand, seems to have the opposite effect on T_c, with a decrease in c from ~ 6 Å to ~ 3.4 Å corresponding to an increase in T_c from 24 to 42 K. Such behavior is not unlike that of artificially layered superconductors, where T_c decreases from a maximum when the separation of the two-dimensional layers is increased [21]. It should be pointed out that in the electron-doped infinite-layer case the steric effects on T_c probably result from a simple physical mechanism, like the enhanced exchange interactions discussed above, since no structural oddities (e.g., interstitial oxygen, as in the T' phase) have been reported in this system.

3 Some Physical Properties of Infinite-Layer Materials

It was mentioned above and should be emphasized here that the electron-doped infinite-layer materials are strucurally quite simple [6], whereas attempts at hole doping result in highly defected structures. Steric effects described above for Ca-substituted $Sr_{1-x}La_xCuO_2$ are characterized by lower T_c's; we have also investigated Ba substitutions, but find that at 25 kbar, such syntheses result in multiphased materials; this is consistent with other work at 50–60 kbar [22]. Thus for larger average alkaline-earth ion size, an apparently higher pressure is required for synthesis; at the other extreme are the first-synthesized infinite-layer parent compound with the smallest average alkaline-earth ion size, $Sr_{0.15}Ca_{0.85}CuO_2$, which can be synthesized at ambient pressure [23].

Like the T'-phase, the T_c of the infinite-layer materials have been reported to be relatively insensitive to pressure; unlike the small negative pressure derivatives of the T' phase, however, a small positive value, $dT_c/dP = +0.06$ K/kbar, has been reported for $Sr_{1-x}Nd_xCuO_2$ [24]. A similarly small change with pressure has been reported for $Sr_{0.9}La_{0.1}CuO_2$ with $dT_c/dP = +0.02$ K/kbar [25]. Such differences with the generally negative pressure derivatives observed for the T' phase [18,26-28] may indicate that the balance between interplanar and intraplanar effects has been altered for the infinite-layer materials, as discussed in Ref. 24. Since the changes in lattice constants for ~ 15 kbar pressure are generally appreciable in the cuprates, it is evident that external pressure has much different effects than those of "chemical pressure" caused by the Ca substitution discussed above.

One very appealing property of the infinite-layer materials is that they are the only electron-doped copper-oxide superconductors which contain no magnetic rare-earth elements; thus, various techniques (NMR, μSR, etc.) can study the copper magnetism without hindrance from competing rare-earth moments, and more reliable data can be obtained from other techniques (e.g., dc magnetization, because it is not necessary to subtract a large paramagnetic background). All superconducting T'-phase materials to date contain magnetic rare-earth elements (our efforts to overcome this limitation, successful as yet only in the parent compound, are described in Sec. 4).

Uemura *et al.* have highlighted the role of the interplanar distance c_{int} in determining T_c from their muon spin relaxation studies [29]; in a study of infinite-layer $Sr_{1-x}Nd_xCuO_2$ (with $c_{int} = 3.4$ Å), they found substantial differences in the ratio of spin relaxation rate to T_c as compared to other superconducting copper-oxide systems (with $c_{int} \sim 6$ or 12 Å).

In studies of superconducting $Sr_{0.9}La_{0.1}CuO_2$, Imai *et al.* [30,31] found that at high temperatures, the ^{63}Cu NMR spin-lattice relaxation time T_1 behaved in manner quantitatively quite similar to the hole-doped compounds. Specifically, the temperature dependence of T_1, which reflects the q-averaged spin susceptibility, is spin-fluctuation-enhanced and is Curie-Weiss-like. At very high temperatures, $1/T_{1a} \approx 2.5 \times 10^3$ s, identical to the hole-doped compounds. Also, as is the case for the hole-doped materials [32], no Hebel-Slichter coherence peak is observed just below T_c.

The superconducting parameters of the electron-doped infinite-layer compound can be determined from reversible magnetization measurements [33]. In Fig. 3(a) is shown the temperature dependence of the magnetization in the reversible regime for various magnetic fields applied perpendicular to the CuO_2 planes for an aligned powder sample of $Sr_{0.9}La_{0.1}CuO_2$. From these data, the thermodynamic critical field H_c and the Ginzburg-Landau parameter $\kappa_\perp = \lambda_\parallel/\xi_\parallel$ are extracted at each temperature, according to the theory of Hao, Clem, *et al.* [34]. The deduced values of H_c as a function of T^2 are shown in Fig. 3(b); a least squares fit to the relation $H_c(T) = H_c(0)(1-(T/T_c)^2)$ yields the value $H_c(0) = 126$ mT. The magnetization data as a function of applied field (both in units of $H_c(T)$) are shown in Fig. 3(c), where the fit for $\kappa_\perp = 63$ is shown. The high quality of the fit is apparent; also, the deviation of the data and fit from the (high-field) linear Abrikosov result is evident. Standard relations [33] may then be used to obtain the various superconducting parameters: $\lambda_\parallel = 290$ nm; $\xi_\parallel = 46$ Å; $H_{c1\parallel}(0) = 4.2$ mT; and $H_{c2\parallel}(0) = 23.8$ T. Parameters for fields parallel to the CuO_2 planes have also been reported [33]; these show appreciably more variability than those just mentioned, perhaps due to inhomogeneity, penetration depth, and misorientation effects [35].

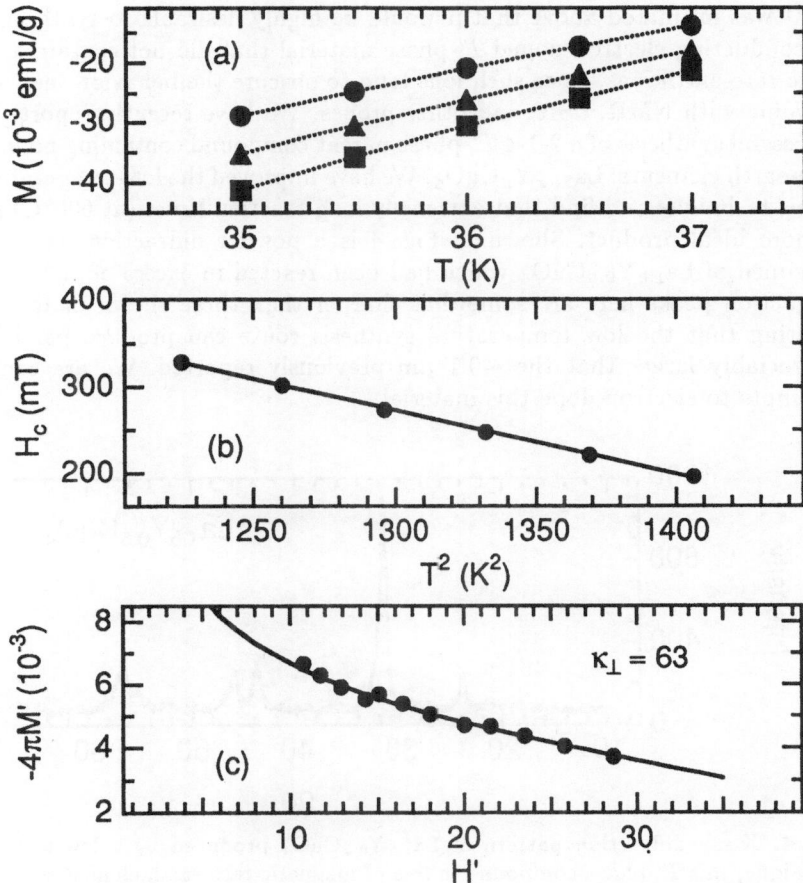

Fig. 3. (a) Reversible dc magnetization as a function of temperature for $Sr_{0.9}La_{0.1}CuO_2$ for various fields appled perpendicular to the CuO_2 planes. (b) Values of the thermodynamic critical field $H_c(T)$ deduced from the data of (a) plotted as a function of the square of the absolute temperature. (c) Normalized reversible dc magnetization as a function of normalized magnetic field (both in units of $\sqrt{2}H_c(T)$). Solid curve is a fit to the theory of Ref. 34 using the deduced value of $\kappa_\perp = 63$.

4 Some Physical Properties of T'-Phase Materials

Several interesting features of the T'-phase 2-1-4 materials were discussed above, particularly with regards to structure and composition. Here we will discuss some of the superconducting and transport behavior of these compounds. Before doing so we briefly mention some recent substitutional studies.

It was intimated above that it would be highly desirable to synthesize a superconducting electron-doped T'-phase material that did not contain any magnetic rare-earth ions, since such ions tend to obscure the behavior one wishes to examine with NMR, μSR, and other probes. We have recently reported [36] a successful synthesis of a 2-1-4 T'-phase parent compound containing no magnetic rare-earth elements: $La_{2-x}Y_xCuO_4$. We have improved the low-temperature synthesis technique and find that extremely long reaction times (at 600°C) provide a more ideal product. Shown in Fig. 4 is a powder diffraction pattern for a specimen of $La_{1.5}Y_{0.5}CuO_4$ which had been reacted in excess of 150 days. The diffraction peaks here are somewhat sharper than those shown in Ref. 36, indicating that the low temperature synthesis route can produce particle sizes appreciably larger that the $\sim$0.1 μm previously reported. We are continuing attempts to electron-dope this material.

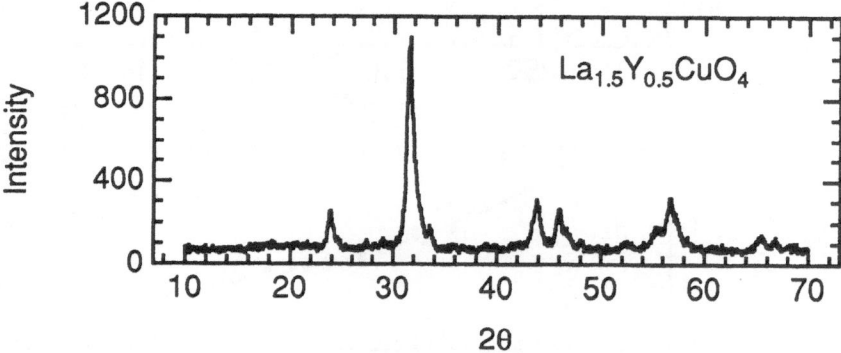

Fig. 4. X-ray diffraction pattern of $La_{1.5}Y_{0.5}CuO_4$ produced by a low-temperature technique; this T'-phase compound is free of magnetic rare-earth elements.

The superconducting properties of the T'-phase materials have attracted considerable interest. Although much more anisotropic than $La_{2-x}Sr_xCuO_4$ or $YBa_2Cu_3O_{7-\delta}$, the electron-doped compound $Nd_{2-x}Ce_xCuO_4$ appears to behave much more like a conventional BCS superconductor. For example, surface resistance measurements [37], which provide the temperature dependence of the penetration depth, indicate that the order parameter seems to be isotropic, unlike the more exotic behavior of some hole-doped compounds [38]. More recently, polarized Raman experiments [39] on high quality single crystals of $Nd_{2-x}Ce_xCuO_4$ have unambiguously identified an electronic response below T_c; these experiments clearly show the formation of an almost isotropic gap below T_c with a value $2\Delta \approx 4.1$–$4.9kT_c$.

At first glance, transport behavior may also seem conventional. Although many sometimes exotic theories have been applied to explain the "anomalous" linear in-plane resistivity of the hole-doped copper oxides, the nearly quadratic behavior observed for the T' phase has often been termed "conventional" and attributed to well-established electron-electron scattering in a Fermi liquid. However, few workers point out that the coefficient of the T^2 term is anomalous, about 10^6 times larger than in conventional metals. Such large enhancements occur in theories incorporating itinerant magnetism, but such theories permit many free parameters. Here we adopt a more localized picture [40]. Because the primary effect of electron doping is merely magnetic dilution (unlike the magnetic frustration of the hole-doped materials), long range spin correlations are maintained at high concentrations; the large quasistatic antiferromagnetic spin clusters observed by Mössbauer spectroscopy even in superconducting materials were discussed above. Since local moments are correlated on length scales much longer than the electron mean free path, and are quasistatic on time scales much longer than electron scattering times, it is natural to consider a simple model with a nearly cylindrical Fermi surface and scattering by anisotropic spin fluctuations. We previously reported anisotropic resistivity data (Montgomery method) for $Nd_{2-x}Ce_xCuO_{4-y}$ [40]. Standard anisotropic Boltzmann transport equations were solved, yielding for the temperature-dependent part of the resistivity tensor components:

$$\rho_\alpha(T) = \frac{a^2 c^3 C_\alpha}{4\pi^2 k_B T e^2 \hbar^2} \int_0^{2k_F} dq_\parallel \int_0^{2\pi/c} dq_\perp \frac{[(2k_F)^2 - q_\parallel^2]^{-1/2}[(2\pi/c) - q_\perp]q_\alpha^2 |g_0|^2}{[\exp(\hbar\omega_q/k_B T) - 1][1 - \exp(-\hbar\omega_q/k_B T)]}, \tag{1}$$

where $\alpha = \parallel$ or $\perp$, corresponding to the a and c lattice constants, $|g_0|$ is an electron-magnon scattering matrix element, and the coefficients C_α are $C_\parallel = m_a^2/k_F^4$ and $C_\perp = 9m_c^2/(2(\pi/c)^4)$. The full resistivities are given by $\rho_\alpha = \rho_{0\alpha} + \rho_\alpha(T)$, where the $\rho_{0\alpha}$'s are the residual resistivity components. We assume isotropic impurity scattering, where $\rho_{0\perp} = (m_c/m_a)(C_\perp/2C_\parallel)^{1/2}\rho_{0\parallel}$. The magnon dispersion is well known to have a gap to out-of-plane excitations Δ which may be due interplanar coupling, in-plane anisotropy, or other effects. We adopt the magnon dispersion

$$\omega_{q\pm} = \omega_\parallel[(1 + (\Delta/2J_\parallel))^2 - ((\cos q_x a + \cos q_y a)/2 \pm (\Delta/2J_\parallel)\cos q_z c)^2]^{1/2}, \tag{2}$$

where $J_\parallel \approx 980$ K is the in-plane exchange coupling. Some anisotropic resistvitiy data are shown in Fig. 5. Our previous attempts to fit similar data resulted in some scatter in the extracted parameters [40]. There are five parameters; two are determined by the residual resistvity, so only three parameters are used to fit the full temperature dependence of both components of the resistivity. There is appreciable scatter if all parameters float; however, if we fix the spin excitation gap and the electron-magnon coupling at their average fit values, $\Delta = 39$ K $= 3.4$ meV and $|g_0'| \equiv \sqrt{\hbar/2J_\parallel}(m_a/m)|g_0| = 180$ meV, and refit the data to Eq. (1), we obtain the fitted curves in Fig. 5 and the parameters shown in Fig. 6.

Figure 6 shows the effective mass ratio, scattered about a value of $m_c/m_a = 570$, and the Fermi wavevector as a function of the in-plane residual resistivity. Such values of k_F and m_c/m_a provide estimates of the electron mean free paths $l_\parallel$ and $l_\perp$; for layer separation $d = c/2$, one has $l_\parallel = \hbar d/(e^2 k_F \rho_\parallel) \approx 200$ Å and $l_\perp = l_\parallel(\rho_\parallel/\rho_\perp)(m_c/m_a)(k_F/(\pi/c)) \approx 4$ Å. Thus the in-plane scattering length is comparable to the size of the antiferromagnetic clusters and the out-of-plane

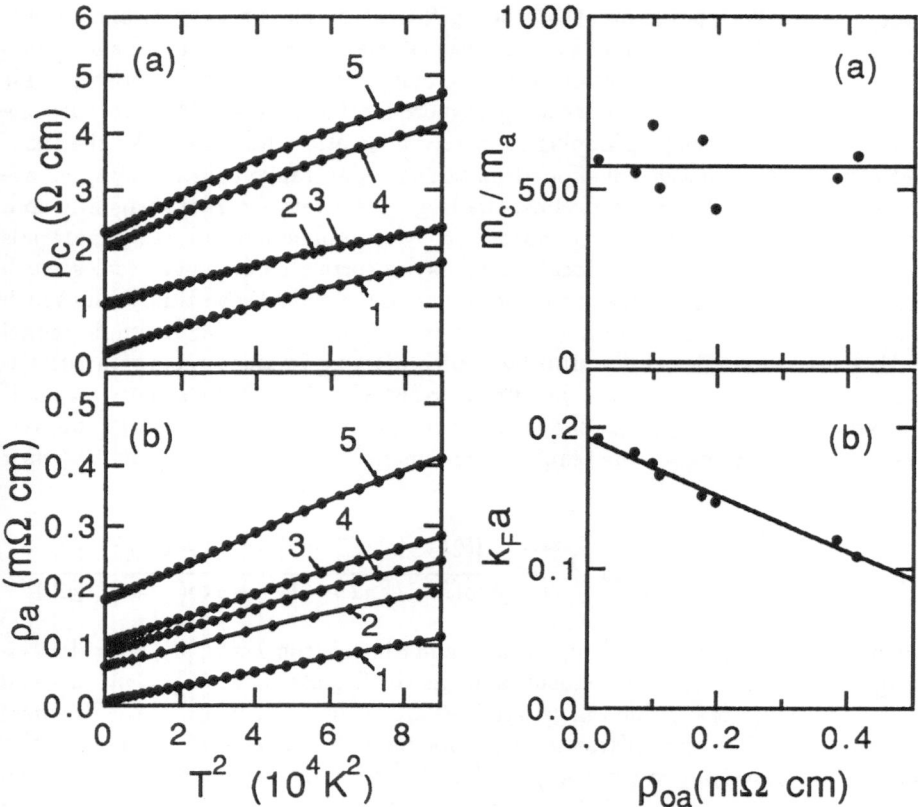

Fig. 5. Normal-state anisotropic resisitivity of $Nd_{2-x}Ce_xcuO_{4-y}$ as a function of the square of the absolute temperature for various crystals. Symbols are data; curves are fits to a spin fluctuation scattering model.

Fig. 6. Extracted values of the effective mass ratio, m_c/m_a, and the Fermi wavevector, k_F, as functions of the inplane residual resistivity.

scattering length is comparable to the layer separation. The Fermi wavevector shows a remarkable linear decrease with increasing impurity scattering, indicating that defects, possibly the oxygen interstitials discussed above, are trapping electrons and decreasing the conduction electron density. Quantitatively, for a density n_i of impurities of scattering cross section σ, one has (from $\sigma n_i = 1/l$) that $n_i = k_F e^2 \rho_{0a}/(hd\sigma)$. If one assumes that each O^{2-} defect traps two electrons, i.e., $n_e = n_{0e} - 2n_i$, then one can linearize the usual 2D expression for the Fermi wavevector to obtain

$$k_F a = \sqrt{2\pi da^2 n_{0e}} - (e^2 a/(\hbar\sigma))\rho_{0a}. \tag{3}$$

From the slope of the line in Fig. 6 one obtains $\sigma \approx 470$ Å^2, implying an impact parameter of 12 Å, indicating a charged impurity, as assumed. One also obtains $n_i \approx 0.05$ per formula unit, very close to the typical oxygen defect concentration.

5 Summary

We have reported or reviewed various results for infinite-layer and T'-phase electron-doped copper-oxide compounds. Steric studies indicated that superconductivity disappears in both systems for values of the in-plane lattice constant below a critical value, $a_{cr} \approx 3.92$ Å. Data from the $(Nd_{1-y}Y_y)_{2-x}Ce_xCuO_4$ system confirmed that steric effects are responsible for the loss of superconductivity (as opposed to magnetic effects). We modelled our anisotropic resistivity data (Montgomery method) using a spin-fluctuation scattering model. The extracted parameters indicate a gap to spin excitations ($\Delta \approx 3.4$ meV), reasonable scattering lengths ($l_{\parallel} \approx 200$ Å; $l_{\perp} \approx 4$ Å), and a mass anisotropy $m_{\perp}/m_{\parallel} \approx 570$. A remarkably linear dependence of the extracted Fermi wavevector on the residual resistivity provides values of the impurity scattering impact parameter (12 Å) and impurity concentration (~ 0.05 per formula unit). Also discussed was $La_{2-x}Y_xCuO_4$, the first T'-phase compound without a magnetic rare-earth element. Reversible magnetization measuremnts of the infinite-layer compound were also reported, providing values of the in-plane coherence length ($\xi_{\parallel} = 46$ Å), the in-plane penetration depth ($\lambda_{\parallel} = 290$ nm), and other parameters. Various other results (Mössbauer, NMR, Raman, μSR, pressure, etc.) were reviewed.

Acknowledgements

Assistance from A. L. Barr, B. C. Dunn, M. I. Larkin, J. M. McGuirk, and A. Morosoff is appreciated. This work was supported by the U. S. National Science Foundation under Grant No. DMR-9158089 and the Robert A. Welch Foundation under Grant No. F-1191.

References

1. Y. Tokura, H. Tagaki, and S. Uchida, Nature **337**, 345 (1989); H. Tagaki, S. Uchida, and Y. Tokura, Phys. Rev. Lett. **62**, 1197 (1989).
2. J. T. Markert, E. A. Early, T. Bjørnholm, S. Ghamaty, B. W. Lee, J. J. Neumeier, R. D. Price, and M. B. Maple, Physica C **158**, 178 (1989).
3. J. T. Markert and M. B. Maple, Solid State Commun. **70**, 145 (1989).
4. E. A. Early, N. Y. Ayoub, J. Beille, J. T. Markert, and M. B. Maple, Physica C **160**, 320 (1989).
5. P. G. Radaelli, J. D. Jorgensen, A. J. Scholtz, J. L. Peng, and R. L. Greene, Phys. Rev. B **49**, 15,322 (1994).

6. J. D. Jorgensen, P. G. Radaelli, D. G. Hinks, and J. L. Wagner, Phys. Rev. B **47**, 14,654 (1993).

7. Z. Hiroi, M. Azuma, M. Takano, and Y. Takeda, Physica C **208**, 286 (1993); S. Tao and H.-U. Nissen, Phys. Rev. B **51**, 8638 (1995).

8. J. T. Markert, N. Y. Ayoub, T. Bjørnholm, E. A. Early, C. L. Seaman, P. K. Tsai, and M. B. Maple, Physica C **162–164**, 957 (1989).

9. Y. T. Zhu and A. Manthiram, Phys. Rev. B **49**, 6293 (1994).

10. G. Liang, Y. Guo, D. Badresingh, W. Xu, Y. Tang, M. Croft, J. Chen, A. Sahiner, B.-H. O, and J. T. Markert, Phys. Rev. B **51**, 1258 (1995).

11. M. Brinkmann, T. Rex, H. Bach, and K. Westerholt, Phys. Rev. Lett. **74**, 4927 (1995).

12. V. Chechersky, N. S. Kopelev, B.-H. O, M. I. Larkin, J. L. Peng, J. T. Markert, R. L. Greene, and A. Nath, Phys. Rev. Lett. **70**, 3355 (1993).

13. V. Chechersky, N. S. Kopelev, B.-H. O, M. I. Larkin, J. L. Peng, J. T. Markert, R. L. Greene, and A. Nath, Hyperfine Interactions **93**, 1721 (1994).

14. A. Nath, N. S. Kopelev, V. Chechersky, B.-H. O, M. I. Larkin, J. T. Markert, J. L. Peng, and R. L. Greene, Science **265**, 73 (1994).

15. V. Chechersky, N. S. Kopelev, A. Nath, B.-H. O, M. I. Larkin, J. T. Markert, J. L. Peng, and R. L. Greene, J. Radioanalytical and Nuclear Chemistry **190**, 391 (1995).

16. N. Y. Ayoub, J. T. Markert, E. A. Early, C. L. Seaman, L. M. Paulius, and M. B. Maple, Physica C **165**, 469 (1990).

17. M. G. Smith, A. Manthiram, J. Zhou, J. B. Goodenough, and J. T. Markert, Nature **351**, 549 (1991).

18. J. T. Markert, J. Beille, J. J. Neumeier, E. A. Early, C. L. Seaman, T. Moran, and M. B. Maple, Phys. Rev. Lett. **64**, 80 (1990).

19. H. A. Blackstead and J. D. Dow, Superlattices and Microstructures **14**, 231 (1993).

20. C. E. Kuklewicz and J. T. Markert, Physica C, in press.

21. N. R. Werthamer, Phys. Rev. **132**, 2440 (1963).

22. M. Azuma, Z. Hiroi, M. Takano, Y. Bando, and Y. Takeda, Nature **356**, 775 (1992).

23. T. Siegrist, S. M. Zahurak, D. W. Murphy, and R. S. Roth, Nature **334**, 231 (1988).

24. C. L. Wooten, B.-H. O, J. T. Markert, M. G. Smith, A. Manthiram, J. Zhou, and J. B. Goodenough, Physica C **192**, 13 (1992).

25. J. L. Cobb, A. Morosoff, L. Stuk, and J. T. Markert, Physica B **194-196**, 2247 (1994).

26. J. Beille, A. Gerber, Th. Grenet, M. Cyrot, J. T. Markert, E. A. Early, and M. B. Maple, J. Less-Comm. Met. **164-165**, 800 (1990).

27. S. L. Budko, A. G. Gapotchenko, A. E. Luppov, E. A. Early, M. B. Maple, and J. T. Markert, Physica C **168**, 530 (1990).

28. J. Beille, A. Gerber, Th. Grenet, M. Cyrot, J. T. Markert, E. A. Early, and M. B. Maple, Solid State Commun. **77**, 141 (1991).

29. Y. J. Uemura, A. Keren, G. M. Luke, W. D. Wu, Y. Kubo, T. Manako, Y. Shimakawa, M. Subramanian, J. L. Cobb, and J. T. Markert, Nature **364**, 605 (1993).

30. T. Imai, C. P. Slichter, K. Yoshimura, M. Katoh, K. Kosuge, J. L. Cobb, and J. T. Markert, Physica C **235-240**, 1627 (1994).

31. T. Imai, C. P. Slichter, J. L. Cobb, and J. T. Markert, J. Phys. Chem. Solids, in press.

32. J. T. Markert, T. W. Noh, S. E. Russek, and R. M. Cotts, Solid State Commun. **63**, 847 (1987).

33. J. L. Cobb and J. T. Markert, Physica C **226**, 235 (1994).

34. Z. Hao, J. R. Clem, M. W. McElfresh, L. Civale, A. P Malozemoff, and F. Holtzberg, Phys. Rev. B **43**, 2844 (1991); Z. Hao and J. R. Clem, Phys. Rev. Lett. **67**, 2371 (1991).

35. J. L. Cobb, A. L. Barr, and J. T. Markert, Physica C **235–240**, 1547 (1994).

36. B.-H. O, R. Tian, and J. T. Markert, Physica C **235-240**, 551 (1994).

37. D. H. Wu, J. Mao, S. N. Mao, J. L. Peng, X. X. Xi, T. Venkatesan, R. L. Greene, and S. M. Anlage, Phys. Rev. Lett. **70**, 85 (1993).

38. D. A. Wollman, D. J. Van Harlingen, W. C. Lee, D. M. Ginsberg, and A. J. Leggett, Phys. Rev. Lett. **71**, 2134 (1993).

39. B. Stadlober, G. Krug, R. Nemetschek, R. Hackl, J. L. Cobb, and J. T. Markert, Phys. Rev. Lett. **74**, 4911 (1995).

40. B.-H. O and J. T. Markert, Phys. Rev. B **47**, 8373 (1993); *ibid.*, Synthetic Metals **71**, 1579 (1995).

Magnetically Modulated Microwave Absorption MMMA in Investigation of Superconductors

J.Stankowski, B.Czyżak, J.Martinek, and B.Andrzejewski

Institute of Molecular Physics, Polish Academy of Sciences, Smoluchowskiego 17, PL–60-179 Poznań, Poland

1 Introduction

Discovery of high–temperature superconductivity near the end of the 1980's was followed by the appearance of a new method for investigation of high temperature superconductors. It was based on a microwave technique commonly used in the a studies [1–3]. The absorption of microwave power in the vicinity of zero magnetic field discovered in polycrystalline samples of high–temperature superconductors was related to Josephson dissipation [2] ans was described in terms of the rf–SQUID theory [4]. This interpretation was supported by results reported by Czyżak and Martinek who studied microwave absorption in poly– and single crystals of high temperature superconductors [5-7]. We shall begin this work with a brief description of the phenomenon and its interpretation.

Dissipation of energy takes place between two superconductors separated by a layer of insulator, forming an S–I–S junction, or between two superconductors separated by a layer of a normal metal, forming an S–N–S junction. This arrangement induces a weak contact which has all the characteristics of a weak Josephson junction.

The term Josephson effect is used to describe oscillations of the Cooper pair current induced by the voltage U and occurring at a frequency defined by the relation:

$$f_{\text{Jos}} = \frac{e}{\hbar}U .\tag{1}$$

This frequency is independent of the nature of the junction as it is related to the voltage through universal constants: the charge of a Cooper pair $e = 2e_0$ and Planck constant $\hbar$. This frequency can be detected as a signal emitted from the polarized Josephson junction. If a junction emits photons of the energy $E_{\text{J}} = \hbar f_{\text{Jos}}$ we infer that it has taken the energy from a source of dc current.

Due to a complex mechanism of Cooper pair scattering of the type: Cooper pair – two quasiparticles – Cooper pair, the dissipation of energy occurs in the phase of "two quasiparticles". These quasiparticles are normal electrons losing energy according to Ohm's law; the difference between them and the normal state conduction electrons is that their energy is, by the apparent Δ, higher than that of the Fermi level. Apart from this phenomenon there occurs the so called reverse Josephson effect in which a junction exposed to microwaves gets polarized to the voltage U. In this effect (1) is also satisfied.

The frequency from the X band (9.4 GHz) corresponds to the voltage of 20 μV and exactly that value was reported by Huo Yuhua *et al.* [8] who observed it at the junction: Nb–I–HTS (Fig.1). In the range of low microwave power 1–10 mW, the voltage was observed to fluctuate between two discrete levels: $U_1 = 0$ V and $U_2 = 20$ μV. With higher powers the effects of overheating were observed, which suppressed the fluctuations related to the quantum bistable behaviour of the Josephson junction. The reverse Josephson effect was studied by Chen *et al.* [9] for standard Pb–PbO–Pb and Sn–SnO–Sn junctions. The presence of quasiparticles in a weak junction leads to junction behaviour described by Ohm's law and to the microwave loss in MMMA.

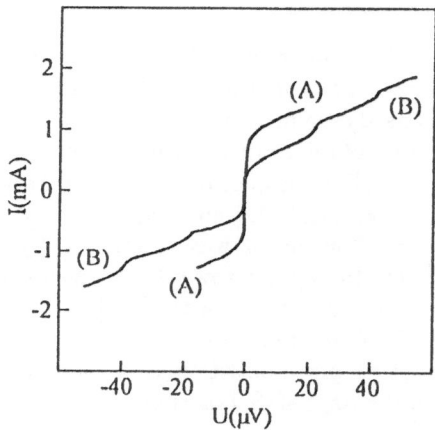

Fig. 1. *ac* and *dc* $I - V$ relations (from Ref.[6]): A) microwave power $P = 0$; B) microwave power $P \neq 0$, $f = 10$ GHz

Discovery of the unexpectedly strong microwave absorption in polycrystalline HTS samples [1–3] aroused so much interest that in the few following years it was the subject of many analyses and reports. However, the wave of enthusiasm weakened when it turned out that interpretation of the data provided by the microwave absorption technique is obscure and ambiguous. Because of the interpretation difficulties, there was much confusion about using it for determination of the fundamental properties of newly found superconductors. Nonetheless, the method soon became so popular that it was even used as a standard test for superconductivity [10,11], even though Oseroff *et al.* [12] proved that microwave absorption in low magnetic fields is not necessarily the evidence of superconductivity in the rare–earth copper oxides and fullerides.

The widespread use of the technique was undoubtedly favoured by its high sensitivity and ease of use. Moreover, the availability of EPR spectrometers which were most frequently used for microwave absorption measurements, stimulated a large number of studies in the field.

One of the most successful applications of microwave absorption measurements, arousing almost no interpretational reservations, is for determination of the critical temperature of superconductors [13,14]. Determination of the critical fields and currents in HTS with the use of microwave absorption measurements has not been so successful. To the best of our knowledge, apart from the early attempts at determination of the critical field of Josephson junctions $H_{c1\,J}$ [2,3] there have been only a few papers reporting measurements of the critical fields of ceramic grains $H_{c1\,G}$ and $H_{c2\,G}$ [15–17] and critical currents J_c [18,19]. Similarly, surprisingly few authors have been interested in applying the method based on microwave absorption measurements to investigation of phenomena related to the dynamics of vortices in HTS, $e.\,g.$ flux creep [20–22], vortex network melting [23] or determination of the irreversibility line [22,24,25]. This is even more surprising since microwave absorption in low magnetic fields is associated with the Josephson phenomenon [4], while in high fields microwave absorption is associated with viscous flow of fluxons driven by the microwave field [26–28], should be the most suitable for investigation of these phenomena.

On the other hand, microwave absorption in combination with the EPR technique has been already approved as a method enabling investigation of the chemical processes accompanying formation of the superconducting phase in the newly discovered fullerides $A_x C_{60}$ [29–32]. Moreover, increasing interest has been aroused by the application of microwave absorption to the investigation of minor superconducting phases forming during synthesis of superconducting materials [33,34] as well as reactions taking place in superconductors under the effects of various chemical agents [35–37].

In this paper we would like to describe a few specific applications of magnetically modulated microwave absorption, MMMA, developed at our laboratory. We shall subsequently describe specific methods for determination of the critical temperature T_c , and for studying dynamics of the synthesis of superconducting fullerides Rb_3C_{60} and the quality of thin films and single crystals. We shall also discuss the possibility of applying the MMMA technique for measurements of the lower Josephson critical field $H_{c1\,J}$, the lower critical field of the grains $H_{c1\,G}$ and intergrain critical current density J_c .

2   Mechanisms of Microwave Absorption

The fundamental transport and magnetic properties of high–temperature superconductors as well as their interaction with high frequency fields are closely related to the occurrence of weak Josephson junctions forming among grains of ceramic materials and thin films, and to structural defects of crystals. The Josephson junctions between copper oxide layers in structural layers of cuprates, considered on the molecular scale, also play an important role.

Measurements of magnetically modulated microwave absorption (MMMA), in addition to measurements of resistivity and permittivity, have become a very sensitive method for determination of the critical temperature of the conductor–superconductor phase transition, and are commonly used for detection of the superconducting phase. For measurements of microwave absorption a sample of the studied material is placed in a microwave resonator (a component of a standard EPR spectrometer) at the maximum of the magnetic component of the microwave field. The microwaves reflected from the resonator with the sample in place are recorded by a microwave semiconductor diode. The sample mounted in the resonator is subjected to a *dc* external magnetic field and an *ac* field of 100 kHz frequency parallel to the *dc* one, which is the so called second modulation. The signal recorded by the microwave diode is subjected to phase detection at the frequency of the field hence the signal at the output is proportional to the derivative of the absorption signal with respect to the external *dc* magnetic field. In such an experimental setup, the high frequency field losses are proportional to the magnetic susceptibility which is related to the magnetic hysteresis loop of the studied system [38]. For granular superconductors and superconducting samples with randomly distributed defects, the shape of the MMMA signal is close to that of the derivative of Lorentz line, which is a typical EPR line obtained using the modulation technique, but of the opposite phase [1–3,39]. Such a line is observed at zero applied magnetic field [40,41] where microwave absorption has a sharp minimum.

In increasing external fields, the signal slowly and linearly increases with the value of the field. On the background of a broad signal, there appear oscillations resembling noise signals which, however, are periodic and reproducible for a given sample. Such trains of signals, periodic in the magnetic field scale, have been observed for samples of twinned single crystals of copper oxide superconductors of regular shape [42–46] as well as for two small irregular pieces of low temperature superconductor pressed against each other [3,47,48]. The observed trains of peaks formed well–separated narrow lines. Distances between the lines in the spectrum are determined by macroscopic quantization of the magnetic flux which indicates formation of loops of superconducting current in the materials.

To explain the character of the spectrum Xia and Stroud [49] proposed a simple model whose main idea was to consider a system of junctions connected in loops, for which they determined the eigenenergy. Later Vichery *et al.* [44] developed this model on the basis of the works by Silver and Zimmerman [50] and explained why an MMMA signal appeared at microwave power levels greater than a certain threshold value.

A general model of microwave absorption in superconductors was proposed in the paper of Martinek and Stankowski [5]. Their model proved successful in explaining not only the basic features like periodicity of the spectrum and the threshold of microwave power but also the shape of particular lines in an MMMA signal, and changes of the shape of this spectrum *vs* temperature and external magnetic field. Basic assumption of this model was that formation of regular loops of superconducting current, including Josephson junction arrays, occurs in the superconductors. Under this assumption, the possible reasons for microwave absorption were analysed using the fundamental *rf*-SQUID model [4,50,51], taking into regard thermal fluctuations [52–56].

The magnetic field dependence of the *rf* SQUID is described by the equation:

$$\Phi + LI_c \sin\left(2\pi\Phi/\Phi_0\right) = \Phi_x ,\qquad(2)$$

where Φ_x and Φ are the external and internal magnetic flux, respectively, Φ_0 is the quantum of magnetic flux, I_c the critical current of the junction, L — inductance of the superconducting loop. For $LI_c > \Phi_0/2\pi$ the behavior becomes hysteretic, so for a given external magnetic field the system can be in different states (Fig.2). The losses in a studied system are related to the irreversible transitions between these states [38]. For this model, the shape of the MMMA spectrum and its characteristic dependences on magnetic field, microwave power and temperature were calculated. The obtained results were in good agreement with the experimental data — see Fig.3 [3,42–48].

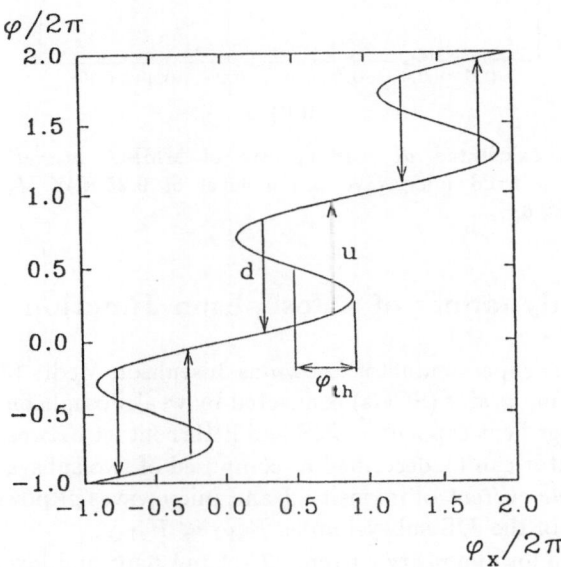

Fig. 2. Internal flux $\phi = 2\pi\Phi/\Phi_0$ *versus* applied flux $\phi_x = 2\pi\Phi_x/\Phi_0$ for a superconducting loop with a Josephson junction. Arrows indicate thermally activated transition to a state of the higher (u) and lower (d) value of the internal flux.

The shape of the microwave absorption signal in ceramic samples is determined by the distribution of superconducting current loops that are randomly orientated with respect to an external magnetic field. Assuming that the absorption signal from a single loop is a typical periodic one and disregarding interaction among the loops, a spectrum was obtained as a superposition of signals from the individual loops. The shape of this spectrum was in good agreement with experimental evidence [5–7,57].

MMMA [arb. units]

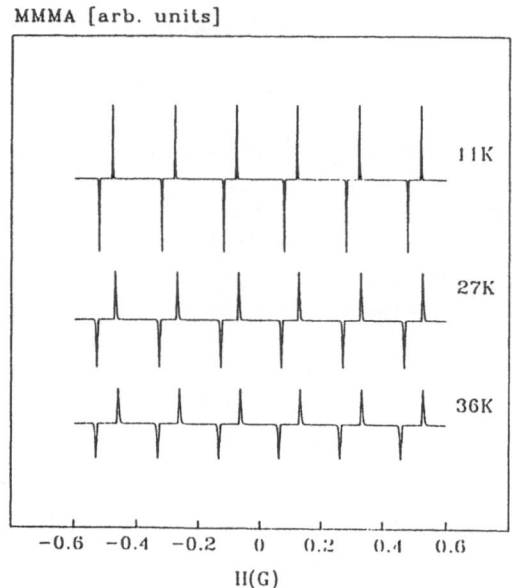

H(G)

Fig. 3. Plots of calculated *dc* scan spectra of MMMA at various temperatures as indicated at a fixed microwave power level of 0.25 mW. $I_c = 3 \times 10^{-5}$ A, $\beta = 2\pi L I_c / \Phi_0 = 2.6$

3 Thermodynamics of a Josephson Junction System

A polycrystalline superconductor known as Josephson Media (JM) is composed of superconducting grains (SCGs) connected by weak Josephson junctions (JJS). Because of a large heat capacity of JJS and little contact between them, a granular superconductor can be described as composed of two subsystems, grains and JJS (Fig.4). A *dc* current of intensity J and microwaves of power P can at first affect the state of the JJS subsystem as $H_{c1J} \ll H_{c1G}$.

In the case of low–intensity currents $J < 1$ mA/cm^2 and low microwave powers $P < 10$ mW, the sample temperature T_s is equal to the local temperature of a Josephson junction T^* : $T_s = T^*$. The value of $T_c^{(o)}$ determined by MMMA for the sample Y–123 is the same on heating and on cooling. This means that the disturbance to JJS is small enough for JM to be characterised by a single temperature T_s. With increasing current flowing through the JM, JJS heats up faster than SCGs, which results in the appearance of MMMA at $T_{eff} < T_c^{(o)}$. An explanation can be given in terms of the above proposed two–subsystem model JM (Fig.4). The critical value of the current intensity is first reached in weak junctions which rapidly heat up. The temperature T^* becomes higher than the sample temperature which is proportional to the square of intensity of the current flowing through the sample $T \sim J^2$. This relation, reported for the first time in [58], has led to formulation of a more general linear relationship between the

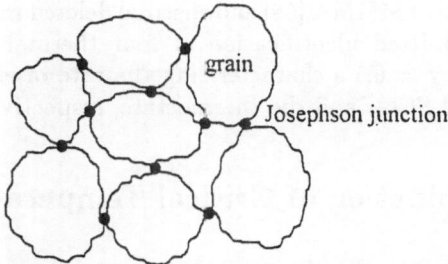

Fig. 4. Two subsystem model of a granular superconductor. The first subsystem is composed of superconducting grains, whereas the other is composed of intergrain Josephson junctions (shaded areas).

temperature T_{eff} at which MMMA appears and the microwave power P supplied to the JJS system:

$$T_{eff} = T_c^{(o)} - aP . \tag{3}$$

Determination of T_{eff} in the case of very strong currents is difficult because the transition from the normal to superconducting state becomes broadened. Experimental evidence supporting relation (3) has been provided by measurements of T_{eff} *versus* the microwave power applied [59]. Analysis of this dependence shows that the transition from the normal to superconducting state in the sample Y-123 becomes more clearly defined with increasing power P. Thus, a more exact determination of T_{eff} can be made at higher levels of applied microwave power. Figure 5 presents the experimental data obtained for $P = 1$, 15.8, and 30 mW; as seen, T_{eff} of the whole sample is evidently shifted.

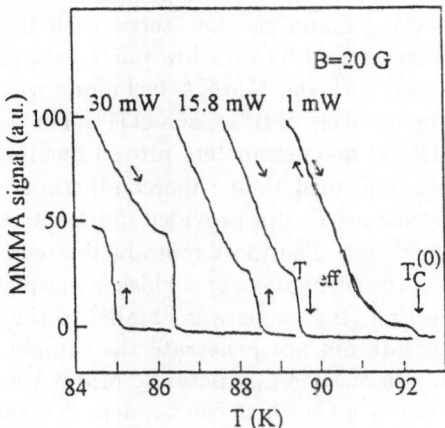

Fig. 5. MMMA signal of $YBa_2Cu_3O_{6.9}$ ceramic *versus* temperature for the microwave powers $P = 1$, 15.8 and 30 mW for A) cooling and B) heating the sample

The studies of MMMA [59] and thermal deleted microwave absorption TDMA [40] have permitted identification of four thermal reservoires and two times $\tau = 0.8$ s and $\tau = 0.1$ s characterizing the rate of energy transfer between JJS and SCGs and SCGs and the thermostate, respectively.

4 Determination of Critical Temperature

As one of its first applications, MMMA was used to measure the critical temperature T_c of a superconductor. This was achieved thanks to the fact that a transition into the superconducting phase leads to a strong microwave dissipation in HTS. The MMMA method permits determination of the transition temperature T_c with an accuracy to 0.1 K. Moreover, it is contactless which is of considerable importance in high pressure studies [60]. These advantages of the MMMA technique inspired Stankowski *et al.* to work out a new method for determination of the temperature of transition to the superconducting state [61]. The novelty of the solution was to replace the *dc* magnetic field in which the sample is placed with a magnetic field of a constant intensity but periodically reverse4d direction. These periodic changes of the magnetic field sense imply a certain characteristic shape of the MMMA signal which permits accurate determination of the critical temperature T_c (Fig.6). Measurements should be made using the minimum microwave power to prevent overheating of the Josephson junction system [59] and with carefully chosen values of the magnetic field intensity to ensure the maximum MMMA signal.

We have also designed and built an HTS tester [61], which permits measurements of microwave absorption versus magnetic field intensity, sweep rate and amplitude of the 100 kHz modulation.

MMMA was also used to study the quality of superconducting materials — especially thin films and monocrystals. Generally speaking, it was found out that the best superconducting materials, *i.e.*, those with the fewest defects in the crystal lattice, are characterised by very low microwave absorption [62]. An outstanding success achieved with the MMMA technique was its application to very small superconducting whiskers of BiSrCaCuO [11]. For these small samples, even the sensitivity of SQUID magnetometers proved insufficient [63]. MMMA studies of these materials confirmed their superconductivity, permitted determination of critical temperatures T_c and provided much interesting information about defects in their structures. The most reproducible and periodic MMMA signals were observed when the (a, b) plane of a whisker was placed perpendicular to the magnetic field direction (the c axis was parallel to the field) (Fig.7). For small fields, the magnetic flux did not penetrate the sample and consequently no microwave absorption was observed. A periodic MMMA signal appeared for magnetic field intensities over 1.5 Oe which can be identified with a lower Josephson critical field $H_{c1\,J}$. The periodic character of the MMMA line has been found to be related to nucleation of vortices and quantization of magnetic flux on the single crystal twin boundaries. As a result, the MMMA measurements provide

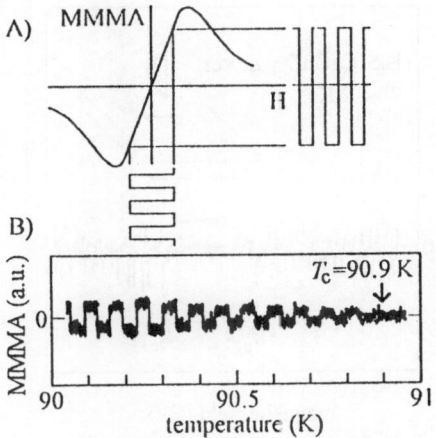

Fig. 6. Measuring the critical temperature T_c by means of MMMA.
A) The superconductor is placed in a magnetic field H which changes its direction periodically.
B) In the superconducting state, a characteristic "step–like" (square wave) MMMA signal can be observed. Its disappearance upon warming indicates a transition into the normal state.

information about defects in the crystal structure, including their size and orientation with respect to the crystal walls [42,44,64]. The area of defects can be estimated from the condition of magnetic flux quantization $\phi_o = SB$, where ϕ_o is a quantum of magnetic flux, S is the area of a superconducting loop restricted by the crystal lattice defects and B is the period of the MMMA signal. The periodicity measured in the whiskers to be 0.19 and 3.75 Oe corresponds to the defects areas 100 μ^2 and 5 μ^2, respectively. The SEM photographs of BiSrCaCuO whiskers confirmed the occurrence of low–angle boundaries in the planes (a, b) and (a, c) at which the flux quantization may take place. However, the visible area of the interfaces formed on the boundaries is much greater than the calculated value. This result is consistent with the presence of boundaries forming interfaces in the (b, c) plane, whose existence have been indicated by many authors [65,66]. Hence, the regular pattern of the MMMA line could be ascribed to the surfaces formed in the (a, b) plane by intersection of (a, b), (a, c) boundaries and limited by the (b, c) defects.

5   Kinetics of the Reaction Leading to Formation of Superconducting State

Thanks to its high sensitivity, MMMA was used to study dynamics of chemical processes in superconductor synthesis [29–33]. We studied the processes of diffusion responsible for formation of the superconducting phase in $Rb_x C_{60}$ fullerides [32] using alkali metal azides RbN_3 as a source of the alkali metal. The

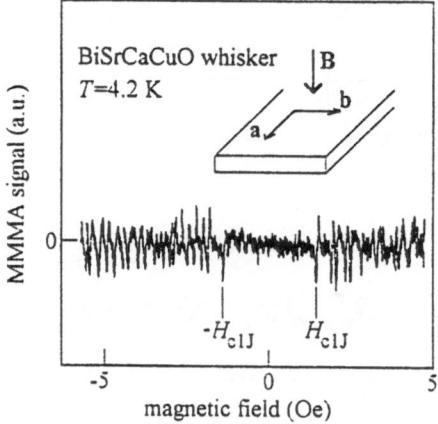

Fig. 7. The regular MMMA signal observed for superconducting whiskers. In the low field range microwave absorption does not occur, which proves the fact that for $H < H_{c1\,J}$ magnetic flux does not penetrate into the Josephson junctions.

sample preparation procedure and experimental details have been described in [29,32]. Kinetics of the reaction was already determined by the chosen synthesis procedure as the fullerene grains were covered homogeneously with the alkali rubidium immediately after thermal decomposition of the azide. The fullerene grains coated with the alkali metal are the nuclei of the product phase. As a result of rubidium diffusion, the thickness of the product phase on the grain surface increases as: $u = -dr/dt$ (r is the radius of the unreacted grain core). Since the number of fullerene grains N remains the same, the reaction is described by the Avrami formula [67]:

$$\alpha = \exp\left[-\frac{\pi}{3}NU^3t^3\right], \quad \text{where} \quad \alpha = \frac{V_s(t)}{V_s(0)}, \tag{4}$$

where a is a coefficient of the reaction progress, $V_s(0)$ and $V_s(t)$ specify the amount of substrate at the beginning of the reaction and after time t. As can be seen in Fig.8, (4) gives a correct description of the time dependence of the MMMA signal intensity as the latter is directly proportional to the number of Josephson junctions formed, and thus to the amount of superconducting phase [34]. Microwave absorption can only occur to the depth of magnetic field penetration λ into a superconductor. The intensity of the MMMA signal will stop growing when the thickness of the product phase U becomes greater than λ. Saturation of the signal was observed 13 hours after the beginning of the synthesis. In order to estimate λ we can resort to the definition of the coefficient of reaction progress:

$$\alpha_{(t=13\,h)} = 1 - \left(1 - \frac{\lambda}{r_G}\right)^3. \tag{5}$$

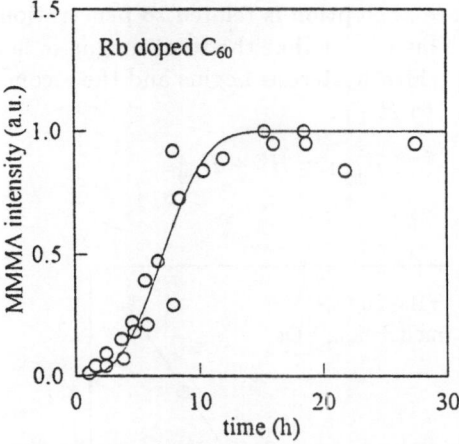

Fig. 8. An increase in MMMA intensity as a function of time elapsed from the onset of the synthesis of $Rb_x C_{60}$. The solid line represent the best fit obtained on the basis of (4).

Substituting the value of α determined from additional EPR measurements and assuming that the mean radius r_G of fullerene particle of the order of 5000 Å, λ was estimated to be 1500 Å. A recent application of MMMA worth mentioning was observation of two superconducting phases in $K_x C_{60}$ for $x \approx 1$ [68].

6 Determination of Critical Fields $H_{c1\,J}$, $H_{c1\,G}$

The application of MMMA for determination of critical fields $H_{c1\,J}$ and $H_{c1\,G}$ has been our recent interest [69]. In order to do that we first cooled a ceramic superconductor YBaCuO down to $T = 4.2$ K in zero magnetic field, the so–called zero field cooling. Then the width of the hysteresis loop ΔH for MMMA was measured as a function of the amplitude of the magnetic field sweep H_{max}. The results, shown in Fig.9, indicate that the penetration of magnetic flux occurs in two stages. Above the lower Josephson critical field $H_{c1\,J}$, the flux penetrates the intergrain areas which is accompanied by an increase in the MMMA hysteresis width. The plateau in Fig.9 corresponds to the established intergrain critical state. At the second stage, with fields higher than the lower critical field of the ceramic grains $H_{c1\,G}$, the width of the hysteresis increases again. Now the hysteresis loop width which is related to penetration of the magnetic flux into individual grains and establishment of the critical state in each of them. The lower Josephson critical field $H_{c1\,J}$ can be found from the data given in Fig.10. As can be seen in this figure, for $H_{max} < 7$ Oe, the width of the MMMA hysteresis increases linearly. We found, by extrapolation, that hysteresis appears when the magnetic field sweep amplitude H_{max} is higher than a certain threshold value H^*, which depends on the amplitude of the second modulation. The appearance of hysteresis in microwave absorption is related to penetration of magnetic flux into a superconductor. This means that the total magnetic field which is a sum of the sweep field H^* at which hysteresis begins and the second modulation field $H mod$, is actually equal to H_{c1J} :

$$H_{c1J} = H^* + H_{\mathrm{mod}} .\tag{6}$$

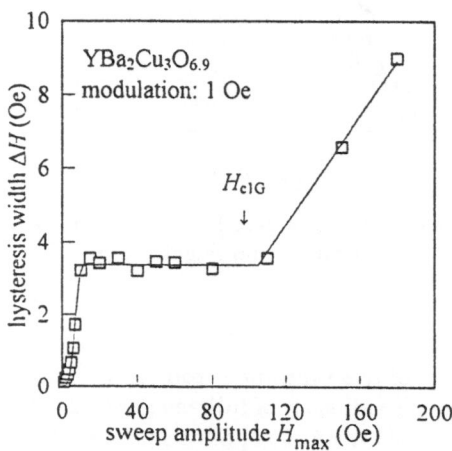

Fig. 9. The relation between the MMMA hysteresis width ΔH and the amplitude of the magnetic field sweep H_{max} . The solid line shows the best fit between the empirical data and (9).

The measurements of the lower Josephson critical field H_{c1J} performed in the above described way, gave $H_{c1J} = 1.7 \pm 0.1$ Oe [70].

Parameters of the critical state can be obtained interpreting the experimental data in terms of the generalized model of a critical state (GCSM) [71]. The intergrain critical state equation can be expressed as:

$$\frac{\mathrm{d}H_{\mathrm{int}}}{\mathrm{d}r} = J_{c\,\mathrm{int}}(H) , \quad \text{where} \quad J_{c\,\mathrm{int}}(H) = \frac{J_{c\,\mathrm{int}}(0)}{\left(1 + \dfrac{H}{H_0}\right)^n} .\tag{7}$$

$J_{c\,\mathrm{int}}(0)$ is the intergrain critical current density in the absence of the field, and n and H_0 are critical state parameters which depend on the mechanism of vortex pinning. The solution of (7) is:

$$H_{\mathrm{int}}(r) = \left[(H_{\mathrm{eff}} + H_0)^{n+1} \pm (n+1)J_{c\,\mathrm{int}}(0)H_0^n(R-r)\right]^{\frac{1}{n+1}} - H_0 ,\tag{8}$$

where R is the radius of a cylindrical sample, H_{eff} is the effective magnetic field, and the signs $+$ and $-$ correspond to the solutions in which $H(r)$ decreases and increases, respectively.

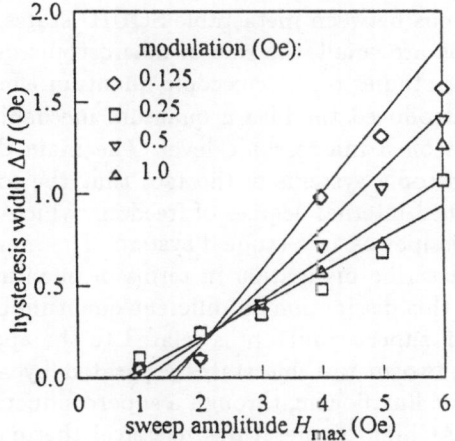

Fig. 10. The changes in the width of the MMMA hysteresis ΔH in the low range of the magnetic field sweep H_{max} and for the various modulation values H_{mod}.
The straight lines have been determined by the least squares method to fit the points on the linear part of the graph, which extends up to about 6–7 Oe. The lines cross over the H_{max}–axis at points H^*.

Assuming that the width of a hysteresis loop ΔH is proportional to the average intensity of the field trapped in a superconductor, in each magnetic field cycle we have:

$$\Delta H = p\langle H\rangle_{trap} \, , \tag{9}$$

where p is a coefficient dependent on the structure and geometry of the superconductor. The mean value of intensity of the magnetic field trapped in a superconductor can be found from (8):

$$\langle H\rangle_{trap} = \frac{2\pi}{S} \int_S H(r)\, ds \, , \tag{10}$$

where S is the area of the base of the cylindrical sample. The best fit of the model (9) to the experimental points, shown in Fig.9, has been obtained for the intergrain critical current density of $J_c = 400$ A/cm^2, the critical exponent $n = 1$ and a characteristic field $H_0 = 3$ Oe. These values are typical of granular YBaCuO superconductors and are comparable to the results of magnetic measurements [72,73].

7 Investigation of Macroscopic Quantum Effects

A recent proposition that has enjoyed much interest is the application of a periodic MMMA signal to the investigation of macroscopic quantum effects.

In ref.[5] the *rf*-SQUID model was applied to interpret microwave absorption at temperatures much above 1 K taking into account the influence of thermal fluctuations on transitions between metastable SQUID states. At low temperatures where fluctuations are small, the role of macroscopic quantum tunneling is more important. The studies on macroscopic quantum effects were initiated by Leggett [74] who introduced the idea a quantum mechanism could be used to describe phenomena on a macroscopic level. The main difference between macroscopic and microscopic systems is the fact that the former are coupled with external or undefined internal degrees of freedom, which unavoidably leads to the appearance of dissipation in a studied system. The main theoretical task was to find a way to describe dissipation in terms of quantum mechanics and determine the effect of this dissipation on different quantum phenomena.

Low field MMMA in superconductors is related to the appearance of losses at transitions between two metastable states separated by an energy barrier, associated with magnetic flux flowing through a superconducting loop including a Josephson junction. At high temperatures, classical thermal activation is responsible for transitions between these states but at low temperatures, quantum tunnelling becomes important. The shape of the MMMA signal is closely related to the shape of the probability distribution for transition between the states [5] and the latter depends on which of the two kinds of processes is involved. Bamas et al. [75] reported experimental results which indicated that in the system they studied macroscopic tunnelling could have the most important influence on the MMMA signal. The calculated shape of the MMMA signal and the temperature dependence of this shape, obtained assuming model [76], proved that the observed effect is related to macroscopic quantum tunnelling.

Acknowledgments

Work partly supported by National Committee for Scientific Research under grant No.2P03B 009 08.

Mr. Bartłmiej Andrzejewski is a holder of a scholarship awarded by the Foundation for Polish Science.

References

1. R.Durny, J.Hautala, S.Ducharme, B.Lee, O.G.Symko, P.C.Taylor, D.J.Zheng, and J.A.Xu, *Phys.Rev.*B **36**, 2361–2363 (1987).

2. J.Stankowski, P.K.Kahol, N.S.Dalal, and J.S.Moodera, *Phys.Rev.*B **36**, 7126–7128 (1987).

3. K.W.Blazey, K.A.Müller, J.G.Bednorz, W.Berlinger, G.Amoretti, E.Buluggiu, A.Vera, and F.C.Matacotta, *Phys.Rev.*B **36**, 7241 (1987).

4. A.Barone and G.Paterno, *Physics and Applications of the Josephson Effect* (Wiley, New York 1982).

5. J.Martinek and J.Stankowski, *Phys.Rev.*B **50**, 3995 (1994).

6. B.Czyżak, *Physica* C **243**, 327 (1995).

7. B.Czyżak, *Supercond.Sci.Technol.* **9**, 25 (1996).

8. Huo Yuhua, Shao Fangwu, Hu Xujie, Fen Honghui, Sun Fengtong, Yang Xiaoming, and Yan Jie, *Solid State Commun.* **73**, 131 (1990).

9. J.T.Chen, R.J.Todd, and Y.W.Kim, *Phys.Rev.*B **5**, 1843 (1972).

10. A.F.Hebard, M.J.Rosseinsky, R.C.Haddon, D.W.Murphy, S.H.Glarum, T.T.M.Palstra, A.P.Ramirez, and A.R.Kortan, *Nature* **350**, 600 (1991).

11. B.Czyżak, B.Andrzejewski, L.Szcześniak, N.Danilova, and J.Stankowski, *Appl.Magn.Res.* **8**, 25 (1995).

12. S.B.Oseroff, D.Rao, F.Wright, D.C.Vier, S.Schultz, J.D.Thompson, Z.Fisk, S-W.Cheong, M.F.Hundley, and M.Tovar, *Phys.Rev.*B **41**, 1934 (1990).

13. B.Czyżak, T.Żuk, and J.Stankowski, *Acta Phys.Pol.*A **78**, 769 (1990).

14. R.Pragasam, V.R.K.Murthy, B.Viswanathan, and J.Sobhanadri, *phys.stat.sol.*(a) **142**, 465 (1994).

15. J.S.Ramachandran, M.X.Huang, and S.M.Bhagat, *Physica* C **234**, 173 (1994).

16. T.Yuan, H.Jiang, A.Widom, C.Vittoria, A.Drehman, D.Chrisey, and J.Horwitz, *Physica* C **242**, 197 (1995).

17. M.Požek, I.Ukrainczyk, and A.Dulčić, *Appl.Magn.Res.* **8**, 99 (1995).

18. A.Morimoto, T.Matsuki, T.Minamikawa, and T.Shimizu, *Physica* C **220**, 332 (1994).

19. V.V.Srinivasu, S.V.Bhat, G.K.Muralidhar, G.Mohan Rao, and S.Mohan, *Pramana-J.Phys.* **40**, 119 (1993).

20. A.Nishida, K.Shiiyama, T.Fujita, H.Shibayama, K.Iwahashi, and K.Horai, *Physica* C, **185–189**, 2153 (1991).

21. A.Nishida and K.Horai, *Solid State Commun.* **86**, 447 (1993).

22. E.J.Pakulis, *Phys.Rev.*B **39**, 9618 (1989).

23. F.J.Owens, *Physica* C **178**, 456 (1991).

24. A.Nishida, K.Shiiyama, T.Fujita, H.Shibayama, K.Iwahashi, and K.Horai, *Solid State Commun.* **79**, 259 (1991).

25. J.Dumas, A.Neminsky, B.P.Thrane, and C.Schlenker, *Mol.Phys.Rep.* **8**, 51 (1995).

26. J.I.Gittleman and B.Rosenblum, *J.Appl.Phys.* **39**, 2617 (1968).

27. Y.W.Kim, A.M.de Graaf, J.T.Chen, E.J.Friedman, and S.H.Kim, *Phys.Rev.*B **6**, 887 (1972).

28. M.Mahel and Š.Beňačka, *Solid State Commun.* **83**, 615, (1992).

29. F.Bensebaa, B.Xiang, and L.Kevan, *J.Phys.Chem.* **96**, 6118 (1992).

30. N.Kinoshita, Y.Tanaka, M.Tokumoto, and S.Matsumiya, *Solid State Commun.* **83**, 883–886 (1992).

31. J.A.Schlueter, U.Welp, H.H.Wang, U.Geisler, J.M.Williams, M.J.Bauer, J.M.Cho, J.L.Smart, and S.A.Taha, *Physica* C **216**, 305 (1993).

32. J.Stankowski, L.Kevan, B.Czyżak, and B.Andrzejewski, *J.Phys.Chem.* **97**, 10 430 (1993).

33. J.Stankowski, W.Kempinski, and L.Piekara–Sady, inL: *Proc.of 178th Meeting of The Electrochemical Society, Inc.*, Reno '95, in press.

34. R.S.de Biasi and S.M.V.Araujo, *Phys.Rev.*B **51**, 8645 (1995).

35. A.A.Zakhidov, I.I.Khairullin, P.K.Khabibullaev, V.Yu.Sokolov, K.Imaeda, K.Yakushi, H.Inokuchi, and Y.Achiba, *Synth.Met.* **55–57**, 2967 (1993).

36. A.D.Shengelaya, H.Drulis, J.Klamut, A.Zygmunt, and N.M.Suleĭmanov, *Solid State Commun.* **89**, 875 (1994).

37. A.Rastogi, V.Sivasubramanian, V.R.K.Murthy, M.S.Hegde, and S.V.Bhat, *Appl.Phys.Lett.* **66**, 1995 (1995).

38. J.Martinek and J.Stankowski, *Mol.Phys.Rep.*, in press (1996).

39. J.Stankowski and B.Czyżak, *Appl.Magn.Res.* **2**, 465 (1991).

40. J.Stankowski, J.Martinek, and B.Czyżak, *Phys.Rev.*B **48**, 3383 (1993).

41. B.Czyżak, J.Stankowski, and J.Martinek, *Physica* C **201**, 379 (1992).

42. K.W.Blazey, A.M.Portis, K.A.Müller, and F.H.Holtzberg, *Europhys.Lett.* **6**, 457 (1988).

43. A.Dulčić, R.H.Crepeau, and J.H.Freed, *Physica* C **160**, 223 (1989).

44. H.Vichery, F.Beuneu, and P.Lejay, *Physica* C **159**, 823 (1989).

45. K.Kish, S.Tyagi, and C.Kraft, *Phys.Rev.*B **44**, 225 (1991).

46. A.Poppl, L.Kevan, H.Kimura, and R.N.Schwartz, *Phys.Rev.*B **46**, 8559 (1992).

47. J.E.Drumheller, Z.Trybula, and J.Stankowski, *Phys.Rev.*B **41**, 4743 (1990).

48. A.S.Kheifets and A.I.Veinger, *Physica* C **165**, 491 (1990).

49. T.K.Xia and D.Stroud, *Phys.Rev.*B **39**, 4792 (1989).

50. A.H.Silver and J.E.Zimmerman, *Phys.Rev.* **157**, 317 (1967).

51. J.E.Zimmerman, P.Thiene and J.T.Harding, *J.Appl.Phys.* **41**, 1572 (1970).

52. K.K.Likhariev and B.T.Urlich, *Josephson Junction Circuits and Applications* (Izd.Mosk.Universiteta, Moscow 1978).

53. J.Kurkijarvi, *Phys.Rev.*B **6**, 832 (1972).

54. J.Kurkijarvi and W.W.Webb, in: *Proc.1972 Appl.Supercond.Conf.*, 581 (1971).

55. L.D.Jackel and R.A.Buhrman, *J.Low.Temp.Phys.* **19**, 201 (1975).

56. K.K.Likhariev, *Dynamics of Josephson Junctions and Circuits* (Gordon and Breach, 1984).

57. J.Martinek and J.Stankowski, *Appl.Magn.Res.* **8**, 83 (1995).

58. J.Stankowski, S.Waplak, S.Hutton, and J.Martinek, in: *Proc.IWCCCL Zaborów'91* (World Sci., Singapore 1992) p.372.

59. J.Stankowski, B.Czyżak, and J.Martinek, *Phys.Rev.*B **42** 10 255 (1990).

60. J.Stankowski, B.Czyżak, M.Krupski, J.Baszyński, T.Datta, C.Almasan, Z.Z.Sheng, and A.M.Hermann, *Physica* C, **160**, 170–176 (1989).

61. J.Stankowski and B.Czyżak, *Rev.Sci.Instrum.* **64**, 2930 (1993).

62. G.Grüner, in: *Phenomenology and Applications of High–Temperature Superconductors*, ed.by K.S.Bedell (Addison–Wesley, NY 1992) p.82.

63. H.Szymczak, private information.

64. P.Bele, H.Brunner, D.Schweitzer, and H.J.Keller, *Solid State Commun.* **92**, 189 (1994).

65. J.Jung, J.P.Franck, S.C.Cheng, and S.S.Sheinin, *Jpn.J.Appl.Phys.* **28**, 1182L (1989).

66. S.Shi Lei, L.Fanqing, J.Yunbo, Z.Guien, G.Yunlong, and Z.Yuheng, *Physica* C **203**, 398 (1992).

67. M.Avrami, *J.Chem.Phys.* **7**, 1103 (1939).

68. W.Kempiński and J.Stankowski, *Acta Phys.Pol.*A **88**, 549 (1995).

69. B.Andrzejewski, B.Czyżak, and J.Stankowski, *Physica* C, **235–240**, 2044 (1994).

70. B.Andrzejewski, B.Czyżak, J.Stankowski, and L.Szcześniak, *Appl.Magn.Res.* **8**, 35 (1995).

71. Ming Xu, Donglu Shi, and R.F.Fox, *Phys.Rev.*B **42**, 10 773 (1990).

72. P.Fournier and M.Aubin, *Phys.Rev.*B **49**, 15 976 (1994).

73. P.Fournier and M.Aubin, *Physica* B **194–196**, 1833 (1994).

74. A.J.Leggett, *Progr.Theor.Phys.* **69**, 801 (1980).

75. P.Bamas, R.Leband, B.Dessertenne, K.Bouzehonane, H.Bouchiot, B.Reulet, and P.Monod, *Physica* C **235–240**, 2024 (1994).

76. J.Martinek and J.Stankowski, to be published.

Emerging Applications
of High Temperature Superconductors
in Electric Power

Paul M. Grant

Electric Power Research Institute, Palo Alto, CA 94304, USA

Abstract. As the 20th century draws near its end, electricity is now found to pervade virtually all human societies and cultures. That modern civilization could not exist without electricity has become self–evident. In 1911, at the very dawn of the commercialization of electricity, a discovery occurred which has since held great promise for the advancement of this technology. This discovery was superconductivity — the ability of certain materials, under appropriate conditions, to transmit electric power losslessly. In the spring of 1986 a new class of materials, copper oxide perovskites, were shown to exhibit superconductivity at substantially higher temperatures than any previously known. In this paper we discuss the renewed promise held out for electric power technology by these recent revolutionary discoveries. Central to the realization of this promise is the development of a practical wire technology, especially to address the need for cable and magnet wire operable at and above liquid nitrogen in fields greater than 1 T and critical currents above 10^6 A/cm^2. We will focus on a new approach, oriented thick films of $YBa_2Cu_3O_{7-y}$ deposited on flexible nickel alloy tape substrates, with potential to satisfy this need.

1 Introduction

From the very earliest days following the experiments revealing zero resistance in mercury metall cooled to the boiling point of liquid helium by Kamerlingh Onnes in 1911, a result he immediately and aptly named "superconductivity", the potential for this astounding phenomena to revolutionize our electric technology has been recognized. However, as the century of its discovery now comes to a close — many worthy and well–planned efforts notwithstanding — this potential remains essentially unfulfilled. By and large, the principal impediment to wide–spread application, especially those requiring operation over large distances or volumes, has been the necessity to employ complex and expensive refrigeration systems to produce and maintain the surrounding ultra–low temperature environment. For example, it has remained simply impractical to construct and operate economically superconducting electric power transmission cables over kilometer–scale lengths. Likewise, the application of superconducting magnet technology to electric energy storage at levels comparable to pumped hydroelectric has eluded practical realization for similar reasons of refrigeration difficulties. It is true, on the other hand, that superconductivity has achieved modest commercial success in certain specialized markets such as medical magnetic resonance imaging. Nonetheless, such markets, as well as new opportunities for electric power generation, transmission, distribution and storage, would greatly expand were superconductors to be found that operate at substantially higher temperatures then those currently employed, and be amenable to development as practical and commercializable electric wire. We emphasize this last remark as vital to any future deployment of superconducting technology in the electric power industry.

A significant step in this direction took place with a series of discoveries beginning in 1986 [2] which continue to the present time. Previous to these events, the highest temperature below which superconductivity existed was about 21 K (-252^0C), requiring a maximum operating temperature of 12 – 15 K. The newest [3] of these new materials, the so–called "high temperature superconductors (HTSC)", becomes superconducting at 133 K (-140^0C), with an operating temperature near 90 K. Thus, we are now at the point where liquid nitrogen — cheap, plentiful, and environmentally friendly — or cryocoolers whose technology is in principle no more complicated than that employed in the ubiquitous and highly reliable residential refrigerator, can be employed for the application of superconductivity. What remains to be seen is whether these discoveries are capable of leading to cheap and practical wire as just emphasized.

2   Overview of HTSC Materials

Figure 1 charts progress in superconducting materials in terms of transition temperature since their initial discovery in 1911 up to the present HTSC era. There are now well over 100 separate HTSC compounds differentiated by atomic structure and elemental content. The unifying feature of all HTSC compounds is the presence of a clearly identifiable square–planar network of copper and oxygen ions, generically termed "layered copper oxide perovskites" [4]. This network has turned out to be a necessary, and almost sufficient, condition for the occurrence of high temperature superconductivity in copper oxides. In the last several years, three non–copper oxide compounds have been discovered with transition temperatures between 20 – 30 K, but so far have not shown promise of being extendible to higher temperatures [5]. Regarding future discoveries, the opening statement of the Bednorz–Müller paper [2], "At the extreme forefront of research in superconductivity is the empirical search for new materials...", still remains the most cogent advice to those seeking the next materials breakthrough.

The principal copper oxide material systems which carry potential for power application are shown in Figs.2 and 3. Figure 2 represents the crystal structure of $YBa_2Cu_3O_{7-y}$, the now-famous "123," or "YBCO." YBCO was the first HTSC to display a transition temperature above the boiling point of liquid nitrogen [6]. In Fig.3 we show the atomic arrangement for $Bi_2Sr_2Ca_2Cu_3O_{10}$, Bi-2223 or BSCCO. In both cases the square–planar aspect of the copper oxide coordination can be seen. The latter structure is host to several other HTSC compounds where bismuth is replaced by thallium, mercury and partially by lead. In some cases, the double layer of these metal oxides can be reduced to a single one ("1223").

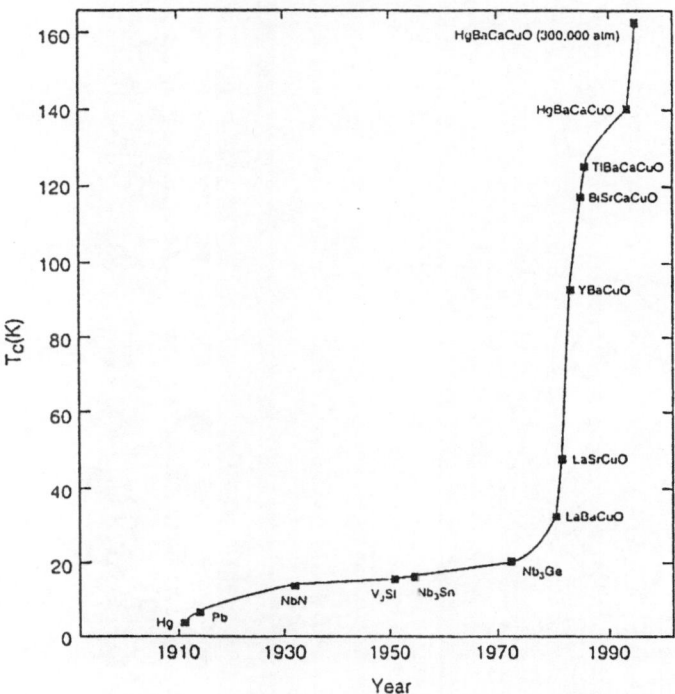

Fig. 1. Progress in raising the superconducting transition temperature, T_c from the year of its discovery to the present

The presence of this intermetallic oxide zone separating the copper oxide complex results in the BSCCO family having quite micaseous, or clay–like, mechanical properties, important for wire processing as will be discussed subsequently. On the other hand, in YBCO, the CuO planar complex is separated by a secondary, and, with regard to superconductivity, passive, linear copper oxide structure shown in Fig.2 between the barium ions. These copper oxide "chains" provide much stronger bonding (*i.e.*, increased dimensionality) between the active CuO planes in YBCO, than the intermetallic oxides in BSCCO. This structural aspect is prosaically, yet strikingly, manifested when one grinds BSCCO and YBCO in a mortar and pestle — the former yields a "greasy" graphitic feel and the latter is more akin to attempting to crush sand.

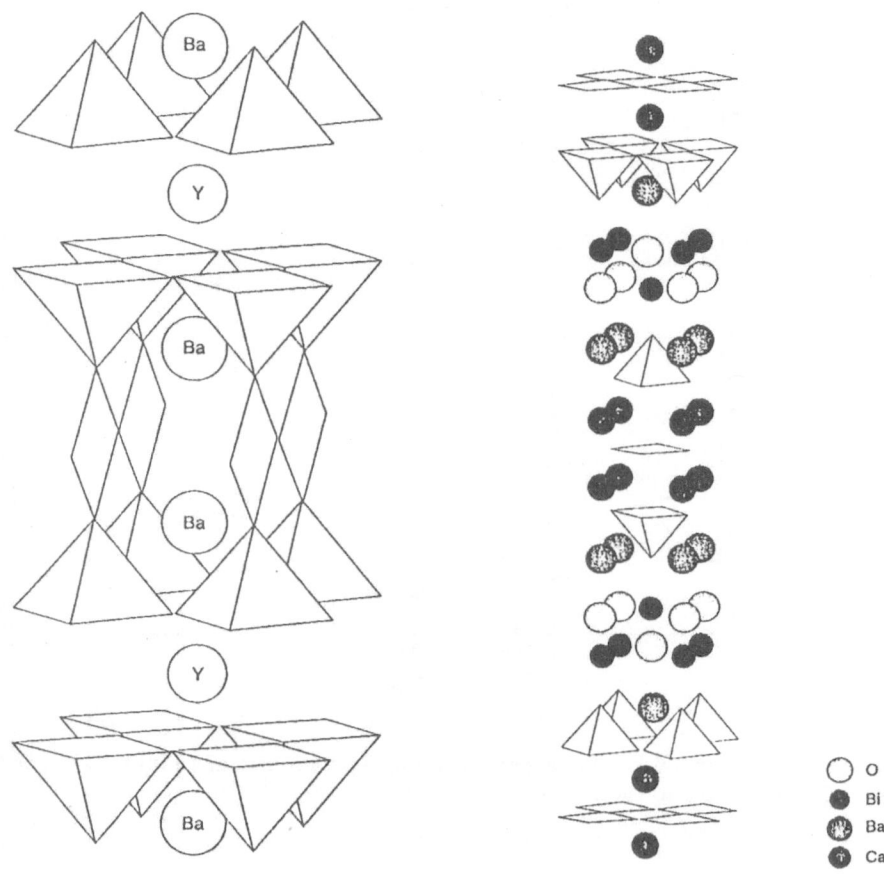

Fig.2. Crystal structure of
$YBa_2Cu_3O_{7-y}$ (YBCO) (See Ref.4)

Fig.3. Crystal structure of
$Bi_2Sr_2Ca_2Cu_3O_{10}$ (BSCCO) (See Ref.4)

These dimensional aspects profoundly influence transport properties in the presence of a magnetic field. This is most clearly seen in Fig.4, where we plot the irreversibility lines for several HTSC compound [7]. We see a wide variation in behavior, and especially note that those materials with the highest transition temperatures do not necessarily yield the highest irreversibility field at liquid nitrogen temperature, the operating point of choice for many applications. This variation is a consequence of unit cell dimensionality. Unlike low temperature type II superconductors, almost all of which have cubic and isotropic structures, HTSC materials are highly anisotropic due to the two–dimensional character of the CuO planes. This 2D character greatly affects the overall vortex pinning properties. Imagine the cylindrical vortex tube as a long length of dry spaghetti, but with "wet" segments between the copper oxide planar complexes. The entire tube could be pinned simply by an impurity or defect located at one of the "dry" regions. The wet segments would represent the regions of the BiO planes in BSSCO in Fig.3 and the YBCO copper oxide chains in Fig.2. However, for the structural bonding reasons discussed earlier, these segments would be "wetter" or softer in BSCCO than in YBCO. The irreversibility characteristics of HTSC materials is determined by the temperatures and fields required to break these weak segments in the vortex tube, resulting in little dry pieces of vortex spaghetti which themselves are unpinned and whose motion creates resistance. Although overly simplistic, the "wet spaghetti" picture nonetheless fundamentally illustrates why the irreversibility line of YBCO is so robust — because its crystal structure is the most 3D of all cuprate superconductors.

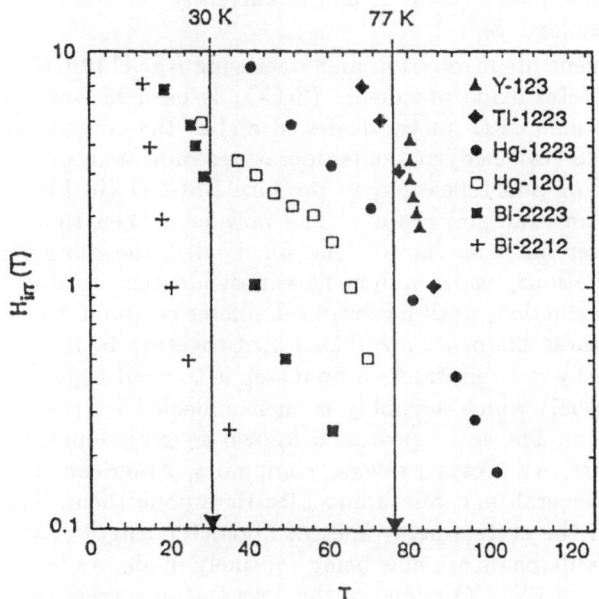

Fig. 4. Irreversibility field *vs.* temperature for a number of HTSC materials (See Ref.6).

3 Current Status of HTSC Wire Technology

We re–emphasize that essentially every application of superconductivity to electric power technology depends on the successful development of suitable wire. It required many years to bring the current low temperature wire, based on NbTi and, more recently, Nb_3Sn, to commercialization. Progress toward this end using the new HTSC materials has occured quite rapidly despite what at first appeared to be an improbability given the universal non–ductility of ceramic materials. Such progress has occurred for a number of reasons, not the least of which has been the extraordinarily large number of materials scientists drawn to HTSC research, as well as some unanticipated "gifts of nature." In the latter category falls the ease of processing BSCCO in the presence of silver. In common with low temperature superconductors, the granularity and poor ductility of the HTSC ceramics require the use of either a binder or sheath as an element of wire embodiment. Unfortunately, otherwise attractive common metals such as copper and aluminum have unfavorable phase equilibrium properties in contact with copper oxide materials. Only silver has been found to have suitable equilibrium properties at temperatures which the cuprate perovskites must be processed as well as possessing high oxygen diffusivity, necessary for wire processing as will be seen shortly. Indeed, Ag also appears to aid in the stabilization of crystallographically single phase BSCCO, and is currently the material of choice for HTSC wire technology [8].

The central elements involved in manufacturing HTSC BSCCO wire are outlined in Fig.5. Metal oxide precursors ($SrCO_3$ is used because of the general instability of alkaline earth metal oxides in air) of the constituent cations are mixed and reacted (calcined) at the temperatures indicated, undergoing several cycles of regrinding and reheating, to produce BSCCO (in Fig.5, the Bi-2212 form is used as an example) powder. The powder is then tightly packed into a cylindrical silver billet as shown. The filled billet then undergoes repeated draw/swage operations, using equipment almost identical to that employed for common wire production, until an overall diameter of about 1 mm is obtained. Several tens of these filaments are then rolled together to form a tape of the order 6 mm wide by < 1 mm thick. A final step is to wind a given length of tape on a mandril (spool) which assembly is then annealed for several days under oxygen atmosphere. The entire process is known as oxygen–powder–in–tube, or "OPIT," for short. At present, several companies, American Superconductor, Intermagnetics General, and Sumitomo Electric among them, have refined the OPIT process to the state where tapes of kilometer length containing several hundred BSCCO filaments are now being routinely made. As hinted earlier, the micaseous nature of BSCCO is one of the keys to the success of the OPIT approach, as the draw/swage/roll process shears the material along its bismuth oxide intergrowth like spreading out a deck of playing cards, producing an unexpectedly high degree of crystallographic alignment of the cuprate planes from an originally random powder. This reduction in randomness results in significantly improved critical current.

Present levels of HTSC wire performance at 77 K in zero magnetic field are summarized in Fig.6 [9]. Engineering current density, J_E (critical current I_c divided by the total areal cross–section of the wire which includes both silver and the BSCCO filaments — J_c would involve only the area of active superconductor), is plotted against the product of critical current, I_c, times wire length L in meters. This unusual choice of abscissa units reflects the dependence of J_E on total wire length due to the accumulation of defects which interrupt the supercurrent path, and to place in perspective the needs of various wire applications which demand differing values of critical current and wire length. For example,

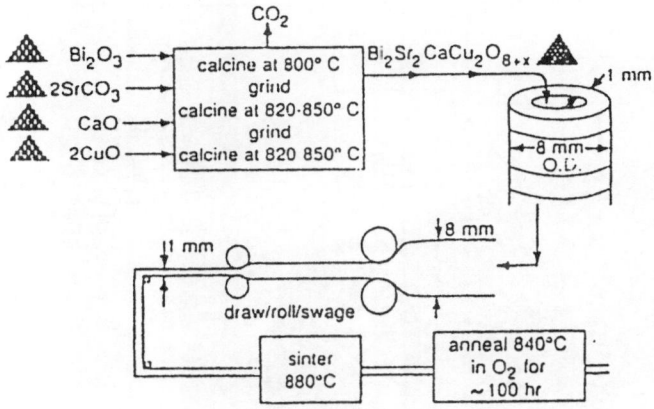

Fig. 5. Sketch of a typical BSCCO OPIT wire manufacturing process.

it takes considerably longer continuous wire to wind a MRI coil than needed by a 500 m underground cable. Ideally, J_E vs I_cL would circumscribe a rectangle whose area would encompass all the application boxes shown in Fig.6. However, as can be seen, progress has been rapid and some applications are being closed in upon.

One that already falls within the "magic rectangle" is a prototype 30 m underground superconducting ac transmission/distribution cable whose development was announced in the fall of 1994 by a consortium comprising the Electric Power Research Institute (EPRI), Pirelli Cable Corporation, American Superconductor Corporation (ASC) and the US Department of Energy (DOE). A schematic of the cable is given in Fig.7. This figure shows the prototype cable design as a three–phase circuit enclosed in an 8 in (203 mm) steel tube, the most common deployment of urban underground transmission/distribution systems in the United States. The basic cable design follows closely the current conventional size and construction to allow as much use of present manufacturing plant as possible without extensive retooling. The same outside cable diameter as conventional ($\approx$ 85 mm) was chosen to permit retrofit of present room temperature cable with a superconducting model as an initial market entry strategy. The cable design shown in Fig.7 represents essentially a direct replacement of the present copper conductor and its nearest layer of electrical insulation with superconducting and cryogenic components. The performance specifications of the 30 m prototype have been set at a 115 kV voltage level to carry 2000 A of current, or a power delivery capacity of 400 MVA within a three-phase circuit. The central conductor comprises a hollow, flexible stainless steel former, approximately 1 in (25.4 mm) diameter, spirally wound with 150 tapes (10 layers, 15 tapes per layer, tape dimensions 4.5 mm wide by 0.25 mm thick, each layer counterwound at 45 degrees) of silver–sheathed BSCCO wire

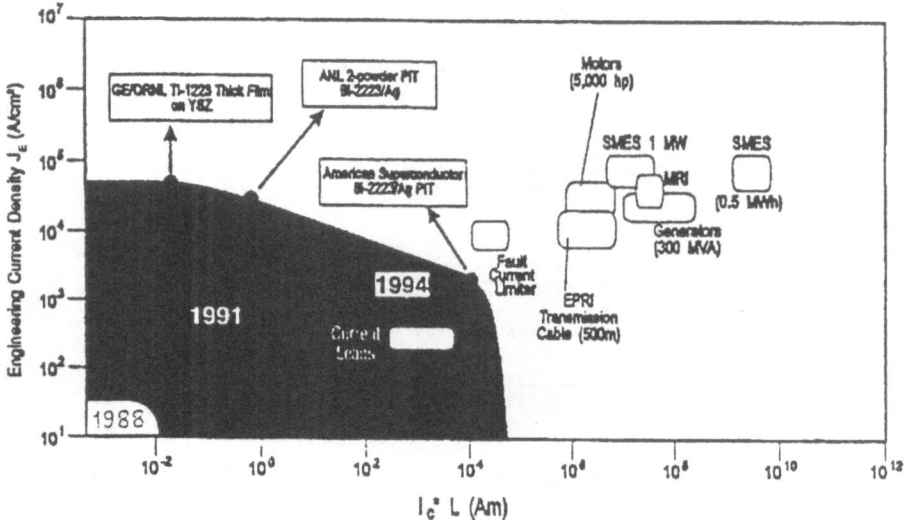

Fig. 6. HTSC wire requirements for several electric power applications in terms of engineering critical current density, J_E (A/cm^2), *vs.* the product of critical current times wire length, $I_c L$ (Am). BSCCO OPIT wire performance characteristics shown are at 77 K in 0 T magnetic field; however, magnet applications would require temperature operation in the 30 K range (See Ref.9).

manufactured as described above. To meet the stated cable specification, each individual tape must sustain a critical current of around 15 A in lengths of approximately 100 m (as of the writing of this paper, critical currents in excess of 30 A have been obtained). The multi–tape conductor is enclosed in two concentric corrugated stainless steel tubes, containing between the inner and outer a "super–insulation" layer, about 12 mm thick, of crinkled aluminized mylar tapes which is vacuum–pumped to around 10^4Torr. Liquid nitrogen is pumped at a rate of several l/s through the central stainless steel former of one phase and returned through those of the two co–phases. Total losses calculated for the design are of the order of 700 W/km/MW (which includes 14 W/W of cooling power). This figure is about 30than a convential 300 MVA copper conductor cable, but the power delivered is equivalently more as well. These losses are approximately equally proportioned between eddy current losses induced in the steel pipe conduit and hysteretic losses arising from self–field in the BSCCO conductor. Not included in the loss estimate are the hysteretic losses induced by the magnetic fields of the respective co–phases on each other.

This cable design is not ideal. Its configuration was dictated by a desire to utilize current conventional cable production tools and to fit the space constraints of an 8 in (203 mm) steel conduit with retrofit applications in mind. The lack of a superconducting shield results both in eddy current losses in the conduit and, as just mentioned, exposure to co–phase induced hysteretic losses. Nonetheless,

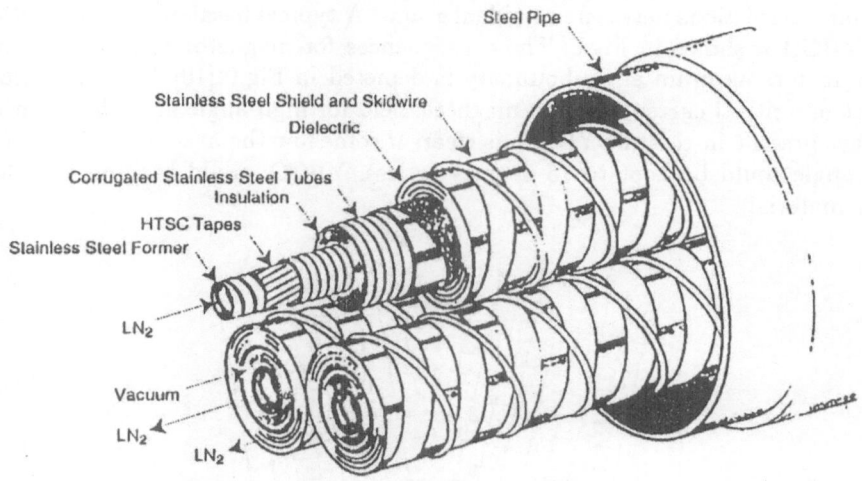

Fig. 7. Schematic drawing of the EPRI/Pirelli Wire/American Superconductor/US-DOE superconducting *ac* underground cable project. The HTSC tapes employed are manufactured with the BSCCO OPIT process.

it is a good choice for the initial market insertion of HTSC into the present high power underground transmission/distribution system.

4   Possibilities for Advanced Performance HTSC Wire

As just discussed, a major drawback of the current BSCCO wire technology is its poor performance at 77 K in even moderate magnetic field rendering unfeasible the construction of practical electromagnets for operation at this temperature. As also pointed out previously, this limitation is ultimately due to the high two–dimensionality of its crystal structure resulting in the almost non–existent irreversibility field response at 77 K as shown in Fig.4. Certain kinds of extrinsic defects can be introduced through high-energy particle bombardment which vastly improve pinning and raise the vortex lattice "melting point" substantially above intrinsic levels. This is more or less the case for all HTSC materials, so it nonetheless remains important to start from as high a value of the intrinsic irreversibility field as possible. A glance at Fig.4 reveals YBCO to satisfy this criteria better than any other HTSC at liquid nitrogen temperatures.

This feature of YBCO has not gone unnoticed. Many attempts have been made to develop a YBCO wire technology. Essentially, the difficulty has been that the same aspect of its crystal structure that results in good magnetic field performance, that is, its lower degree of two–dimensionality, creates problems when attempting to process it by the OPIT method. The lack of a micaseous layer, as present in BSCCO, results in a random arrangement of copper–oxygen planar intersections between individual grains. A typical basal–plane intersection for YBCO is shown in Fig.8. The consequences for magnetotransport response with increasing grain angle boundary is depicted in Fig.9 [10]. The deleterious effect on critical current at high magnetic field for high angle grain boundaries, always present in ceramic YBCO, is clear. If somehow the average grain boundary angle could be kept to 15 degrees or less, YBCO would become the ideal wire material.

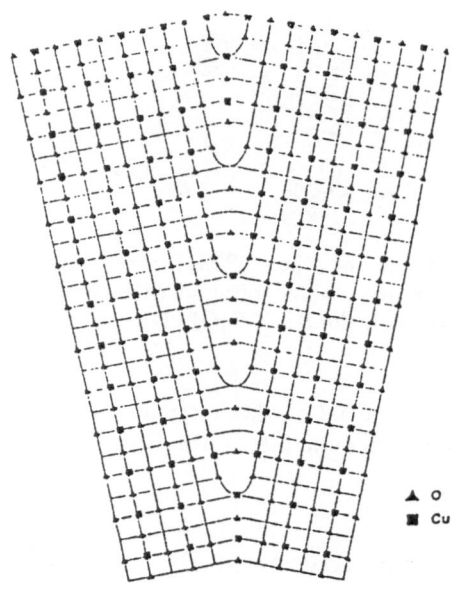

Fig. 8. Atomic lattice positions in a tilted grain boundary in the $a - b$ basal plane of YBCO of about 23 degrees.

We should mention that it is not yet entirely clear why large angle grain boundaries result in the observed "weak link" behavior in superconducting properties. There is evidence that some oxygen depletion occurs in the CuO "chain" components of YBCO (see Fig.2). There are also indications that significant strains exist at grain boundaries which may impede the passage of a supercurrent. In addition, one might speculate that if the symmetry of the superconducting pair state of an HTSC is indeed d–wave, decreasing overlap of the superconducting wave functions between adjacent grains with increasing grain boundary angle, and hence forming an ever weaker link, may be an intrinsic property of these materials. It is important to understand the source of the wide angle grain boundary weak link, as clues may be found to ameliorate its negative effect on superconducting magnetotransport properties. At present, the only path for improvement is to explore methods to keep the angle between adjacent YBCO grains small.

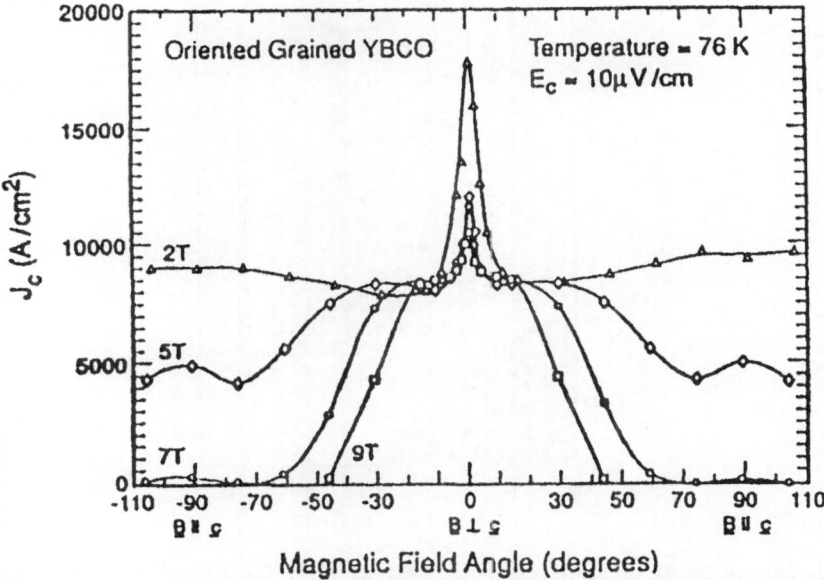

Fig. 9. Transport critical current density J_c as a function of the angle between magnetic field and the $a - b$ plane of bulk grain–oriented YBCO at 77 K. (See Ref.10).

One procedure for grain alignment in YBCO that works well is to "melt process" the ceramic form using methods analogous to zone refining of semiconductors. An initial region of the material is brought to its melting point and this region is drawn slowly along its remaining length. This method produces very good J_c characteristics in high magnetic field, but is painfully slow — of the order of a few mm/hr — hardly practical for manufacture on a kilometer-length scale. The best magnetotransport characteristics are achieved in epitaxial films of YBCO which are essentially single grain structures. Figure 10 summarizes the improvement of YBCO transport properties as a function of processing. Once again, however, epitaxial films require rigid (and expensive) single crystal substrates, not a practical embodiment for a wire technology. Nonetheless, the successes of both melt–processing and epitaxial film growth demonstrate extraordinary results can be obtained for YBCO once high angle grain boundaries can be eliminated and suggest possible paths to bring this about, especially those that might be quasi–epitaxial in nature.

In the mid–1970s, a film deposition technique was developed by workers in the IBM Research Division whereby the material being deposited could be preferentially oriented in a given crystallographic direction [11]. An auxiliary heavy ion source was used to bombard a buffer layer along a particular crystallographic axis as it was being laid down. This had the effect of microscopically "scribing" or "scratching" the forming film much in the manner of a diffraction grating. A subsequently post–deposited film on such a prepared buffer layer would pos-

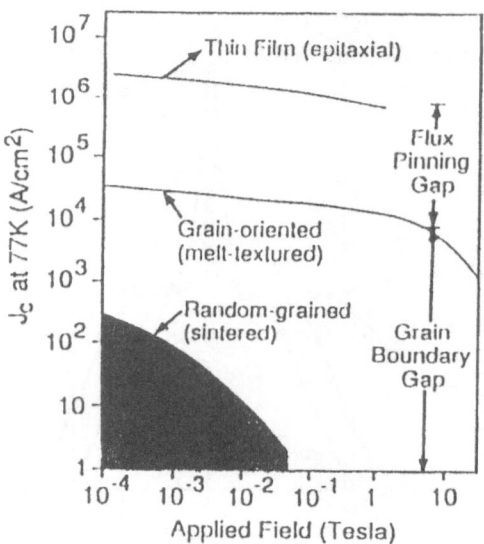

Fig. 10. Improvement of J_c in magnetic field at 77 K for YBCO as a function of synthesis and processing.

sess a high degree of orientation with subsequent small angle boundaries between adjacent crystallites (formally termed "texturization" or "quasi–epitaxy"), providing other conditions, such as closely matching lattice constants and thermal expansion factors, were adequately satisfied. This technique was applied by Iijima *et al.* [12] and by Reade, Berdahl and Russo, [13] to the deposition of yttrious-stabilized zirconia (Y_2O_3–stabilized ZrO_2, or YSZ) on thin flexible metal tapes, such as stainless steel or Ni alloys (Hastalloy), in the hope of providing a platform for post–deposition of oriented YBCO. The basic layer configuration with typical thickness values for the components is shown in Fig.11. Note that, unlike OPIT–produced wire, these tapes can be wound with the superconductor in compression, a very attractive feature for electromagnet applications.

These initial efforts showed great promise, and further work is now ongoing at Stanford [14], Los Alamos National Laboratory (LANL) [15] and Lawrence Berkeley Laboratory (LBL) [16] in the United States, as well as at Siemens [17] in Germany and Sumitomo [18] in Japan. Figure 12 shows recently published results of Wu and co–workers at LANL. The performance in magnetic field is truly spectacular and compares favorably with melt–textured YBCO (at date of writing, short length samples have been made which exhibit J_cs of 1.2×10^6 A/cm^2 in zero field and 3×10^5 A/cm^2 at 1 tesla at a temperature of 75 K. These figures are approaching levels usually only found for epitaxial films). Moreover, short tape specimens have shown excellent electromechanical properties, with some withstanding as much as 1degradation in critical current performance.

A major question, of course, is whether the IBAD–buffered, YBCO thick

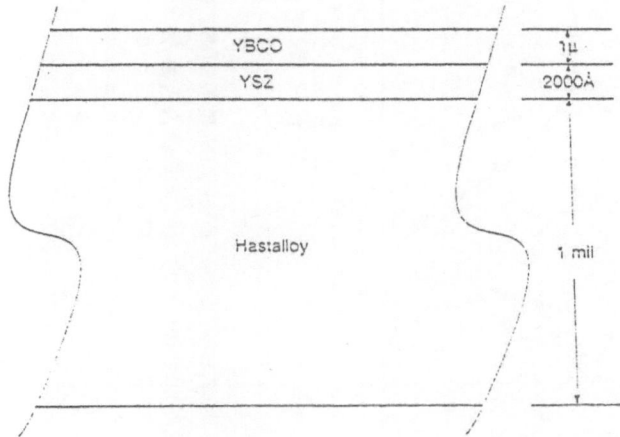

Fig. 11. Lengthwise cross–section of an envisioned YBCO/YSZ/Hastalloy tape wire.

film approach is scaleable to manufacturing levels of thousands of kilometers per year at a competitive cost. A preliminary analysis using as a basis a 1970s study proposing the manufacture of Nb_3Sn tape by thin film deposition methods, properly adjusted for current cost–of–funds, labor costs, materials procurement and capital plant, indicate that production costs comparable to presently marketed volumes of low temperature superconducting wire (NbTi and Nb_3Sn) are achievable. This study assumed critical current levels now within reach ($\approx 3 \times 10^6$ A/cm^2) thin film thicknesses of the order given in Fig.11 produced at deposition rates obtainable in volume production–size vacuum coating chambers such are used today in the magnetic disk storage industry. One possible manufacturing visualization is given in Fig.13. This figure shows how either slower deposition rates, or the need for thicker films should the target J_c prove impractical, can be managed by employing a multiple pass arrangement. The successful commercialization of textured YBCO tape would indeed revolutionize the superconductor wire industry.

5   Conclusion

We have emphasized from the beginning of this paper that the core embodiment underlying deployment of superconductivity in electric power technology — from transmission/distribution cables to rotating machinery to energy storage and power conditioning — is wire and the temperature, current and magnetic fields at which it can be operated. The advent of high temperature superconductivity has pushed this operation window to new levels of performance. The OPIT process of manufacturing Ag–sheathed BSCCO tape provides at present wire in

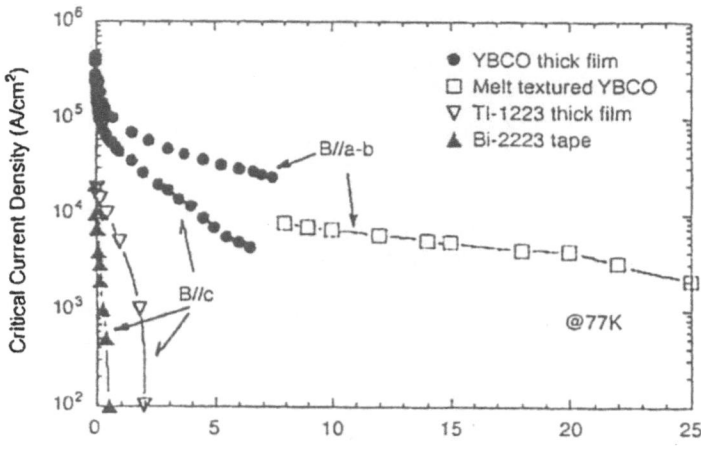

Fig. 12. Comparison of potential HTSC wire technologies and their transport properties in magnetic field at 77 K (See Ref.15).

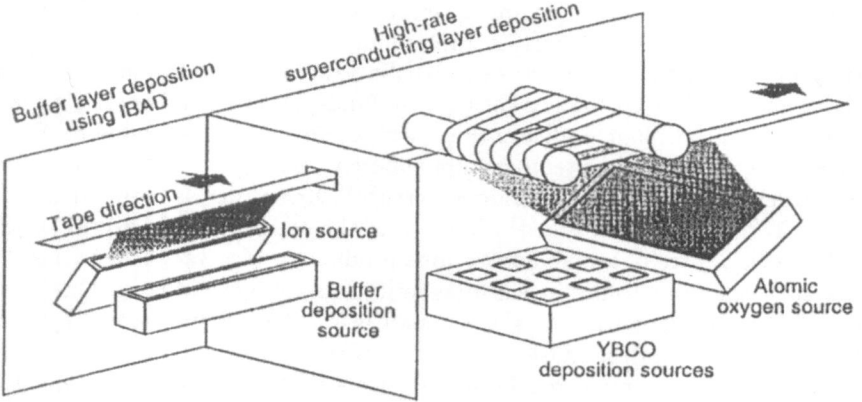

Fig. 13. Possible high–rate manufacturing process for YBCO wire tape using ion–beam–assisted–deposition (IBAD) (See Ref.14).

kilometer lengths capable of electromagnet application at 25 K and in linear conductors for cables cooled by liquid nitrogen. This development has spawned new commercial activity in both small and large enterprises that would not otherwise have occurred. Our new age of superconductivity and electric power is presaged by the construction and initial testing of a prototype 30 m *a.c.* underground transmission cable, a project already underway as this paper is being written.

Yet the full potential of the discovery of the cuprate superconductors for wire application remains to be realized. We have pointed out that YBCO, the first material to be found superconducting above liquid nitrogen temperature, still retains title to the most optimum properties in that temperature range, even though materials with higher T_c have been found. The development of the IBAD–textured buffer layer technique holds out the promise that the next round of HTSC wire technology will greet the new century as the watershed event enabling the truly widespread electric power application of superconductivity that has for so long remained illusive.

Acknowledgements

The author would like to acknowledge the support and encouragement of his colleagues at EPRI in the writing of this paper, particularly Tom Schneider, Don Von Dollen and Fritz Kalhammer. He would also like to affirm the contributions of the many contractors in support of the EPRI program in superconductivity over the years. Those that have had a direct impact on the contents of this paper were David Larbalestier, Tom Sheahan,19 Robert Hammond, Ted Geballe and Xin Di Wu.

Rererences

1. H.K.Onnes, *Commun.Leiden Lab.* Nos. 120b, 122b, 124c (1911).

2. J.G.Bednorz and K.A.Müller, *Z.Phys.*B **64**, 189 (1986).

3. A.Schilling, *et al.*, *Nature* **363**, 56 (1993).

4. H.Shaked, *et al.*, "Crystal Structures of the High–T_c Superconducting Copper Oxides," (Elsevier, Amsterdam, 1994); obtainable from Argonne National Laboratories.

5. R.J.Cava, *et al.*, *Nature* **367**, 252 (1994),   A.F.Hebard, *et al.*, *Nature* **352**, 600 (1991).

[6. M.K.Wu, *et al.*, *Phys.Rev.Lett.* **58**, 908 (1987).

7. K.Isawa, *et al.*, in: *Proc.of the 6th Int.Symp.on Superconductivity* (26-29 October 1993, Hiroshima).

8. Y.Feng, *et al.*, *Physica* C **192**, 293 (1992).

9. "Superconductivity Program for Electric Power Systems: FY 1994-1998 Multi-Year Plan," (U.S. Department of Energy, 1995).

10. J.W.Ekin, K.Salama, and V.Selvamanickam, *Appl.Phys.Lett.* **59**, 360 (1991).

11. Lock See Yu, J.M.E.Harper, J.J.Cuomo, and D.A.Smith, *Appl.Phys.Lett.* **47**, 932 (1985).

12. Y.Iijima, N.Tanabe, O.Kohno, and Y.Ikeno, *Appl.Phys.Lett.* **60**, 769 (1992).

13. R.P.Reade, P.Berdahl, and R.E.Russo, *Appl.Phys.Lett.* **61**, 2231 (1992).

14. K.B.Do, *et al.*, *Bull.Am.Phys.Soc.II* 40 (1), 13 (1995).

15. S.R.Foltyn, P.Tiwari, R.C.Dye, M.Q.Le, and X.D.Wu, *Appl.Phys.Lett.* **63**, 1838 (1993);

X.D.Wu, *et al.*, in: *Proc.of the Applied Superconductivity Conf.*, Boston, October, 1995 (to be published).

16. P.Berdahl, *et al.*, in: *Proc.of the Applied Superconductivity Conf.*, Boston, October, 1995 (to be published).

17. W.Schmidt, G.Endres, U.Haerlen, H.-W.Neumüeller, K.J.Schmatjko, S.Klöck, G.Reichel, and G.Saemann-Ischenko, in: *Proc.of the European Conference on Applied Superconductivity* (4-9 October 1993, Göttigen), ed.by H.C.Freyhardt (DGM Informationgesellshaft mbH), p.637.

18. S.Takano, US Patent 4 921 833 (May 1, 1990).

19. T.P.Sheahen, "Introduction to High-Temperature Superconductivity," (Plenum, New York, 1994).

Spin Fluctuations and $d_{x^2-y^2}$ Pairing in the Cuprate Superconductors: A Progress Report

David Pines

Department of Physics, University of Illinois,
1110 West Green Street, Urbana, IL 61801-3080 USA

Abstract. I present an overview of recent progress using a nearly antiferromagnetic Fermi liquid description of normal state behavior on three key issues for high temperature supercondujctivity in the cuprate superconductors: the physical characteristics and physical origin of the anomalous spin and charge behavior in the normal state; the mechanism for high temperature superconductivity, and the superconducting pairing state. I select four topics for detailed discussion: characterizing pseudogap and pseudoscaling behavior in the normal state; calculating magnetotransport in YBa$_2$Cu$_3$O$_7$, demonstrating the relationship between the effective pairing potential for superconductivity and $d_{x^2-y^2}$ pairing; and recent evidence for $d_{x^2-y^2}$ pairing from NMR experiments on the Knight shift and the spin-echo decay time in YBa$_2$Cu$_4$O$_8$.

1 Introduction

In preparing this written account of my lecture at Dushniki, I decided to highlight progress which has been made since the Grenoble meeting in July, 1994 on three key issues for high temperature superconductivity in the cuprate superconductors: the physical characteristics and physical origin of the anomalous spin and charge behavior in the normal state; the mechanism for high temperature superconductivity, and the superconducting pairing state. I begin with a brief overview of the approach our group has taken to these issues: that because of the close approach of the superconducting cuprates to antiferromagnetism, it is the magnetic interaction between planar quasiparticles which is responsible for their anomalous normal state behavior and the transition, at high temperatures, to a superconducting state with $d_{x^2-y^2}$ pairing. This led to a nearly antiferromagnetic Fermi liquid (NAFL) description of normal state behavior (Pines, 1994a), based on the ansatz that the effective magnetic interaction between planar quasiparticles mirrors the highly anisotropic momentum dependence of the spin-spin response function measured in NMR and neutron scattering experiments. The spin-spin response function, and by inference the effective magnetic quasiparticle interaction, is strongly peaked at wave vectors in the vicinity of the commensurate AF wave vector, $\mathbf{Q} \equiv (\pi, \pi)$, and explicit calculations for YBa$_2$Cu$_3$O$_7$ (Monthoux and Pines, 1993; 1994a) show how this peaking leads to both the anomalous normal state spin and charge response of the planar excitations and the transition at high temperatures to a superconducting state with $d_{x^2-y^2}$ pairing. I then consider four topics on which further progress has recently been made:

characterizing pseudogap and pseudoscaling behavior in the normal state; calculating magnetotransport in $YBa_2Cu_3O_7$, demonstrating the relationship between the effective pairing potential for superconductivity and $d_{x^2-y^2}$ pairing; and recent evidence for $d_{x^2-y^2}$ pairing from NMR experiments on the Knight shift and the spin-echo decay time in $YBa_2Cu_4O_8$.

2 A Brief Overview of Earlier Work

At Grenoble (Pines, 1994b) I reviewed the considerable progress which has been made in understanding the anomalous spin and charge response of the planar excitations in the normal state. Their anomalous low frequency magnetic behavior is seen clearly in NMR experiments on the Knight shift, spin-lattice relaxation rates, and the ^{63}Cu spin-echo decay time (Slichter, 1994). Within a one-component description of magnetic behavior based on the hyperfine Hamiltonian of Shastry (1989), and Mila and Rice (1989), which explains the temperature-dependent Knight shifts found in all underdoped cuprates,the strikingly different temperature dependence of the spin-lattice relaxation rates for adjoining planar ^{17}O and ^{63}Cu nuclei *requires* the presence of strong antiferromagnetic correlations between the barely itinerant Cu^{2+} spins. A quantitative account of the NMR experiments is obtained by using a phenomenological expression for a spin-spin correlation function peaked at Q,

$$\chi(q,\omega) = \chi_{MMP}(q,\omega) + \chi_{FL}(q,\omega) \tag{1}$$

where the anomalous contribution, $\chi_{MMP}(q,\omega)$, introduced by Millis, Monien and Pines (1990, hereafter MMP) may be written as,

$$\chi_{MMP}(q,\omega) = \frac{\chi_Q}{1+(Q-q)^2\xi^2 - i(\omega/\omega_{SF})} \equiv \frac{\alpha\xi^2}{1+(Q-q)^2\xi^2 - i(\omega/\omega_{SF})} \tag{2}$$

and χ_{FL} is a parametrized form of the normal Fermi liquid contribution, which is wave vector independent over most of the Brillouin zone,

$$\chi_{FL}(q,\omega) \cong \frac{\chi_0(T)}{1-i\pi\omega/\Gamma} \tag{3}$$

modified to take into account the temperature dependence of the bulk susceptibility, $\chi_0(T)$. As discussed by Barzykin and Pines (1995), for systems for which measurements have been made of the anisotropic spin-lattice relaxation time, $^{63}T_1$, and the spin-echo decay time, $^{63}T_{2G}$, the three parameters which characterize $\chi_{MMP}(q,\omega)$ can all be determined from experiment [χ_Q, the static staggered magnetic susceptibility, ξ, the antiferromagnetic correlation length which measures the fall-off of the static susceptibility as one moves away from Q, and ω_{SF}, the frequency of the relaxational mode at or near Q, which is quite generally very small compared to the quasiparticle Fermi energy or bandwidth ($\sim$ eV)].

Quite remarkably, as I discuss in more detail in the following section, the parameters ω_{sf} and ξ are not independent for a wide range of temperatures and hole concentrations, but rather display scaling behavior (Sokol and Pines, 1993),

$$\omega_{SF} = \text{const}_/' \xi^z \tag{4}$$

where one has a crossover from $z = 2$ to $z = 1$ behavior at the same temperature, T_{cr}, at which $\chi_0(T)$ begins to decrease as T is decreased

We shall also see that experiments on the temperature and hole concentration dependence of the resistivity and the Hall effect in the underdoped cuprates, when combined with measurements of the specific heat and uniform magnetic susceptibility, demonstrate unambiguously the inseparability of spin and charge behavior in the normal state; the onset of pseudoscaling and pseudogap behavior, detected magnetically, in NMR measurements of the ^{63}Cu and ^{17}O relaxation rates and Knight shifts,is accompanied by measurable changes in the charge response.

Also reviewed at Grenoble were the strong coupling (Eliashberg) calculations, carried out in collaboration with Philippe Monthoux (Monthoux and Pines, 1993;1994a), of the normal state properties and T_c for YBa$_2$Cu$_3$O$_7$ using the model experiment-based planar quasiparticle interaction introduced by Monthoux, Balatsky, and Pines (1992),

$$V_{\text{mag}}^{\text{eff}}(\mathbf{q}, \omega) = g^2 \chi(\mathbf{q}, \omega) \sigma_1 \cdot \sigma_2. \tag{5}$$

where $\chi(\mathbf{q}, \omega)$ takes the form, Eq. (2). We found that when the full structure of the quasiparticle interaction specified by Eqs. (2) and (5) is taken into account, although lifetime effects do lead to a dramatic reduction in T_c, for parameters appropriate to YBa$_2$Cu$_3$O$_7$ a superconducting transition into a $d_{x^2-y^2}$ pairing state occurs at $T_c \sim 90$K for comparatively modest values of the coupling constant g. For this same range of coupling constants our calculated resistivity and optical properties were in quantitative agreement with experiment, so that a bridge was built between the measured anomalous normal state charge response, the anomalous normal state spin response, and $d_{x^2-y^2}$ superconductivity at high temperatures.

Our calculations thus provided for YBa$_2$Cu$_3$O$_7$ a proof of concept of the proposal that the measured anomalous spin and charge properties of the normal state are those of *nearly antiferromagnetic Fermi liquids*, which find their physical origin in the novel wave vector dependent interaction and the novel magnetic low energy scale, ω_{SF}, brought about by the strong antiferromagnetic correlations between the quasiparticles.

As Monthoux and I frequently emphasized, despite this theoretical support for the magnetic mechanism, since our nearly antiferromagnetic Fermi liquid approach *predicted unambiguously that the pairing state of YBa$_2$Cu$_3$O$_7$ must possess $d_{x^2-y^2}$ symmetry*, experimental detection of that pairing state was a *necessary* condition for the magnetic mechanism to explain high T_c. Initially only NMR experiments on YBa$_2$Cu$_3$O$_7$, reviewed in Slichter (1994), showed the

characteristic $d_{x^2-y^2}$ signature of a linear dependence on T of the Knight shift
and a T^3 dependence of the ^{63}Cu spin-lattice relaxation rate we had predicted.
However, subsequent measurements, discussed at Grenoble, of the temperature-
dependence of the penetration depth of YBa$_2$Cu$_3$O$_7$, ARPES and Raman scattering experiments on Bi$_2$Sr$_2$CaCu$_2$O$_8$, and especially the direct Josephson measurements, suggested by Leggett, and carried out by the Van Harlingen group
(Wollman et al., 1993), of the change in sign of the order parameter under $\pi/2$
rotation on SNS tunnel junctions of YBCO and Pb, all supported this pairing
state assignment (Van Harlingen, 1995).

As the $d_{x^2-y^2}$ nature of the order parameter became better established, advocates of other than the magnetic mechanism have begun to find ways of obtaining the pairing state from their mechanism of choice. Thus it is natural to
inquire whether there is a "smoking gun" for the magnetic mechanism whose
principal advocates were, early on, the Schrieffer and Scalapino groups in Santa
Barbara, the Moriya group in Tokyo, and our group in Urbana. Monthoux and
I anticipated this question, and developed (Monthoux and Pines, 1994a) a candidate smoking gun: the remarkable influence of impurities which change the
local planar magnetic order on both T_c and the low temperature superconducting properties of the cuprate superconductors. We argued that quite generally,
because $\xi(T_c)$ is typically a few lattice spacings, any imperfection which destroys
local magnetic order may be expected to act as a unitary (strong) scatterer in
its influence on the low temperature properties, while for a pairing potential of
magnetic origin, because it both scatters at the unitary limit and changes the
pairing potential, such an impurity will exceed the unitary limit in its influence
on T_c. We carried out explicit strong coupling calculations for a simple model
which demonstrated these two important points, and noted that the substantially different influence of planar Zn impurities which destroy local magnetic
order and Ni impurities (which do not) (Ishida et al., 1993), on both T_c and
the low temperature superconducting properties of the 1-2-3 system provided
experimental support for our proposal. Further confirmation of these arguments
came from the Knight shift results of Ishida et al. (1994) and Takigawa and
Mitzi (1994) on Bi$_2$Sr$_2$CuCu$_2$O$_8$ who found that even the "best" samples of this
material behave like dirty d-wave superconductors. The reason: Bi$_2$Sr$_2$CuCu$_2$O$_8$
contains an intrinsic level of magnetic imperfections associated with variations
in the superstructure associated with the BiO planes.

3   Pseudogap and Pseudoscaling Behavior in the Normal State

Perhaps the strongest single constraint on candidate descriptions of normal state
behavior in the cuprate superconductors is the appearance of pseudogap and
pseudoscaling behavior. Early experiments on the Knight shift in the normal
state of the 1-2-3 system by Alloul et al. (1989) showed that for the underdoped
systems, as the temperature decreased, the uniform (bulk) susceptibility, χ_0,

rather than being temperature independent as is the case for a normal Fermi liquid, also decreased, a phenomenon which has since been called variously a spin gap or pseudogap. Subsequent experiments on both this system and the 2-1-4 system (for a recent review see Barzykin and Pines (1995)) showed that this change in character of $\chi_0(T)$ was a general property; while at high temperatures the bulk susceptibility <u>increased</u> (or was nearly T independent), below a characteristic temperature, T_χ, χ_0 begins to decrease linearly with decreasing T; moreover at a second temperature, which we call T_*, this decrease in T becomes more rapid, as depicted in Fig. 1. Loram *et al.* (1993) carried out specific heat experiments which show that this decrease in χ_0 is accompanied by a corresponding decrease in the planar quasiparticle density of states, $N_T(0)$, so that the decrease can be regarded as a quasiparticle phenomenon. Sokol and

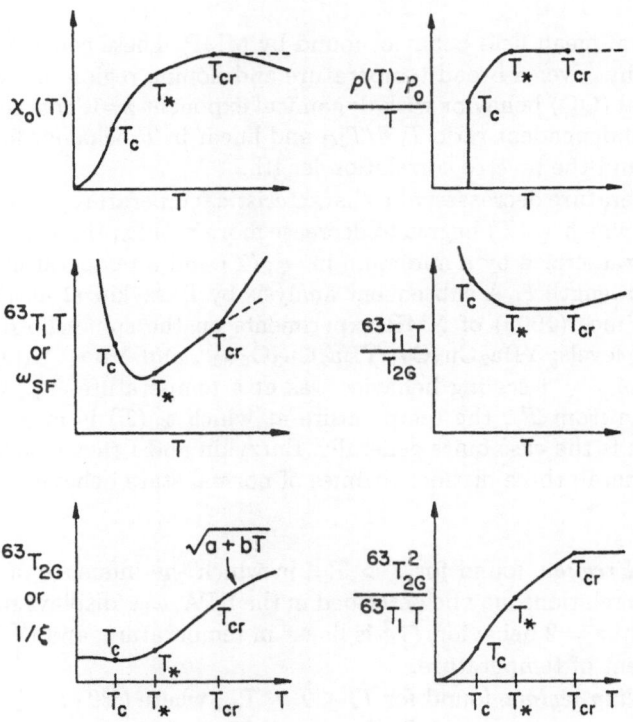

Fig. 1. A schematic depiction of measured changes in the low frequency magnetic and transport properties in the normal state as one passes from the mean field regime above T_{cr} through the pseudoscaling regime for $T_* \leq T \leq T_{cr}$, to the pseudogap regime; also shown are the changes in these quantities which occur below T_c.

Pines (1994) then applied scaling arguments to to the interpretation of measurements of the ^{63}Cu spin-echo decay rate $1/T_{2G}$ and spin-lattice relaxation rate $1/T_1$ and found a remarkable tendency toward scaling behavior. Thus, on using the MMP expression, Eq. (2), and the appropriate microscopic expressions for $^{63}T_1T$ and $^{63}T_{2G}$ (Pennington and Slichter, 1991), one has (Thelen and Pines, 1994),

$$^{63}T_{2G} \sim \alpha/\xi \tag{6}$$

$$^{63}T_1T \sim \omega_{SF}/\alpha \tag{7}$$

so that if $(^{63}T_1T/^{63}T_{2G})$ is independent of temperature, one has $z = 1$ scaling behavior with

$$\omega_{SF} = c'/\xi \tag{8}$$

while if $(^{63}T_1T/^{63}T_{2G}^2)$ is independent of temperature, one has

$$\omega_{SF} = \text{const}/\xi^2, \tag{9}$$

the non-universal mean field behavior found by MMP. These considerations led us to propose that over a broad temperature and doping region one would find quantum-critical (QC) behavior with dynamical exponent z=1, characterized by a temperature independent ratio T_1T/T_{2G} and linear in T behavior for T_{2G}, the product, T_1T, and the inverse correlation length.

As the temperature decreases, at a characteristic temperature, T_*, close to the temperature at which $\chi_0(T)$ begins to decrease more rapidly, there is a crossover to a regime characterized by a minimum in $\omega_{SF}(T)$ and a temperature independent correlation length ξ. A subsequent analysis by Barzykin et al. (1994) and Barzykin and Pines (1995) of NMR experiments on the spin-echo decay time at three doping levels: YBa$_2$Cu$_3$O$_7$, YBa$_2$Cu$_3$O$_{6.63}$, and YBa$_2$Cu$_4$O$_8$, showed that the onset of $z = 1$ scaling behavior was at a temperature, T_{cr}, which was indistinguishable from T_χ, the temperature at which $\chi_0(T)$ is maximum, and we shall see this is the case quite generally. Barzykin and I thus concluded that there are, in general, three distinct regimes of normal state behavior:

- *a mean field regime*, found for $T > T_{cr}$, in which the influence of antiferromagnetic correlations may be described in the RPA, ω_{SF} displays mean-field, non-universal $z = 2$ behavior, ξ^{-2} is linear in temperature, and $(^{63}T_1T/T_{2G}^2)$ is independent of temperature.
- *a pseudoscaling regime*, found for $T_* \leq T \leq T_{cr}$, where $(^{63}T_1T/T_{2G})$ is independent of temperature, ω_{SF} displays non-universal $z = 1$ scaling behavior (since the measured dependence of c' on doping level does not reflect the universal behavior calculated in a non-linear sigma model description of spin behavior and there has been no experimental finding of the spin waves which should accompany true scaling behavior), and both ω_{SF} and ξ^{-1} vary linearly with temperature.

— *a pseudogap regime*, found for $T_c \leq T \leq T_*$, in which $\chi_0(T)$, ξ, and ω_{SF} display behavior which is qualitatively similar to what happens when one gets a quasiparticle gap through the long range order which accompanies superconductivity or itinerant antiferromagnetism. Thus $\chi_0(T)$ changes character at T_*, and as T decreases, ω_{SF}, after passing through a minimum, increases as T decreases, while ξ and T_{2G} become independent of temperature. Since there is no long range order, "pseudogap" rather than "gap" is the proper description of quasiparticle behavior in this regime.

Barzykin and I further proposed that at T_{cr}, $\xi \sim 2$. With this ansatz, one then is able to set the scale for the temperature variation of ξ measured in T_{2G}. For $YBa_2Cu_4O_8$, confirmation of our predicted crossover from $z = 1$ to $z = 2$ behavior at $T_* \sim 470$ has come in very recent work by Curro *et al.* (1996), who find $z = 1$ behavior below $\sim 485K$, and $z = 2$ behavior from 485K to $\sim 700K$.

Figure 1 contains, in capsule form, the generic behavor of $\chi_0(T)$, $^{63}T_1T$, $^{63}T_{2G}$, $(^{63}T_1T/^{63}T_{2G})$, and $(^{63}T_{2G}^2/^{63}T_1T)$ for the three regimes of interest while the doping dependence of T_{cr} and T_* for the 1-2-3 system is shown in the magnetic phase diagram of Fig. 2 and the corresponding phase diagram which

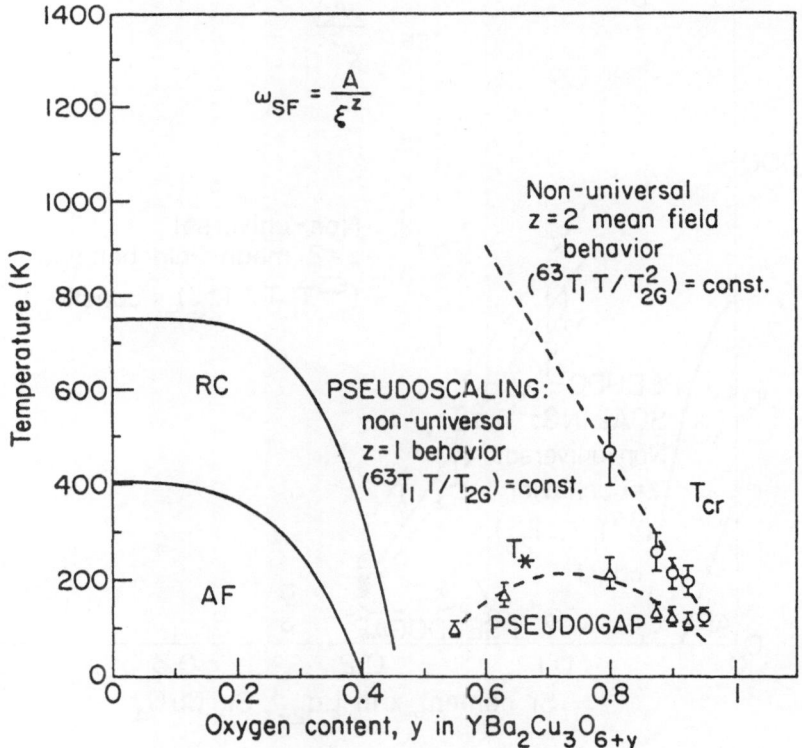

Fig. 2. Normal state phase diagram for $YBa_2Cu_3O_{6+y}$ (Barzykin and Pines, 1995)

Barzykin and I found for the 2-1-4 system is shown in Fig. 3. Note that for both systems, T_{cr} increases as the doping decreases, while T_* displays a comparatively broad maximum at intermediate doping levels.

From a Fermi liquid perspective, the only way pseudoscaling and pseudo-gap behavior can come about is through a non-linear feedback of the magnetic interaction on the quasiparticles which are in turn the source of that interaction. Monthoux and I have developed a Fermi-liquid description of pseudoscaling (Monthoux and Pines, 1994b), and have reviewed (Pines and Monthoux, 1995) the non-linear feedback effects at work above T_{cr} and when the system goes superconducting, and then considered how similar effects in the normal state could lead to pseudoscaling and pseudogap behavior below T_{cr}. However, finding a first-principles, microscopic derivation of pseudoscaling and pseudogap behavior remains a major unsolved problem.

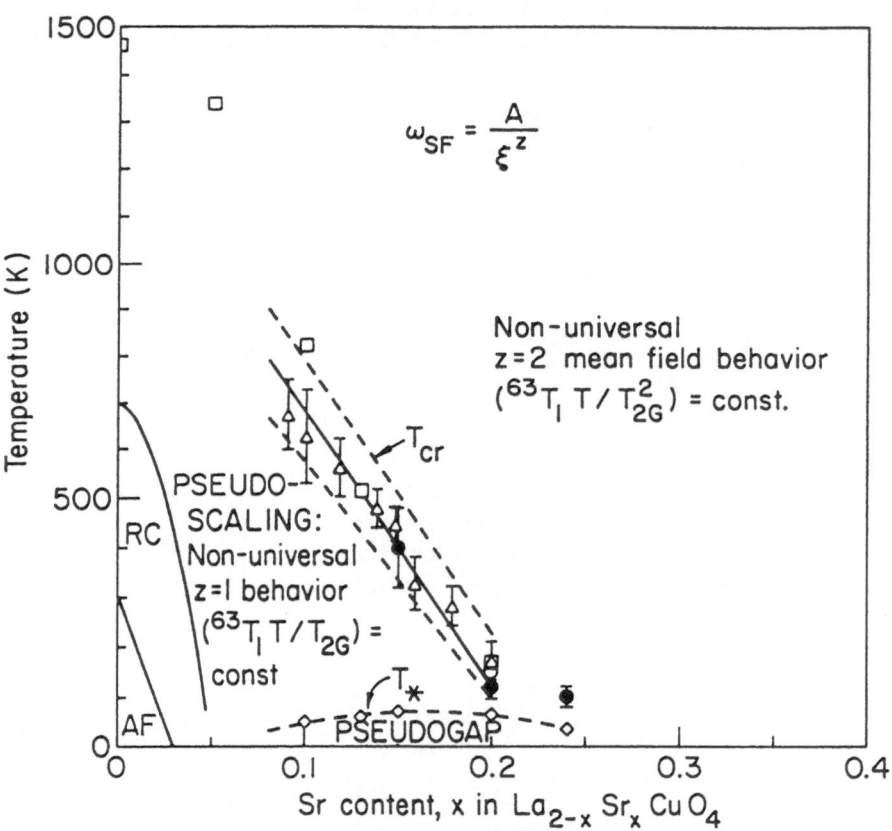

Fig. 3. Normal state phase diagram for $La_{2-x}Sr_xCuO_4$ (Barzykin and Pines, 1995)

It should be emphasized that pseudoscaling and pseudogap bahavior is not confined to the magnetic properties of the normal state, but is found as well in the charge response. Thus, Ito *et al.* (1993) showed for $YBa_2Cu_3O_{6.63}$ that below a temperature, T_ρ, the planar resistivity ceases to exhibit a linear in T behavior, while the Hall effect likewise changes character at this temperature; a similar conclusion for $YBa_2Cu_4O_8$ was reached by Bucher *et al.* (1993). Within experimental error, $T_\rho = T_*$. In the $La_{2-x}Sr_xCuO_4$ system, Hwang *et al.* (1994) have shown that for $x \geq 0.15$, the Hall coefficient, $R_H(T)$, follows scaling behavior, with a characteristic temperature, T_H, which not only exhibits the same strong dependence on doping level that is found for T_{cr}, but which possesses very nearly the same magnitude as the T_χ determined from the maximum in the spin susceptibility. For this same system, with $x \leq 0.14$, Nakano *et al.* (1994) find that below T_*, the resistivity ceases to be linear in T. Thus the crossover temperatures, T_{cr} and T_*, which Barzykin and I identify in low frequency magnetic measurements possess direct counterparts in transport measurements. A further indication that the anomalous spin and charge behavior are linked and that both originate in quasiparticle behavior comes from the work of Zha *et al.* (1996) who show that by using a temperature dependent quasiparticle density of states taken from Knight shift measurements, one can arrive at a quantitative understanding of the way in which c-axis resistivity and optical measurements on both the 2-1-4 and 1-2-3 systems reflect the behavior of planar quasiparticles.

4 Magnetotransport

While a NAFL model of normal state behavior has been shown to be consistent, both qualitatively and quantitatively, with NMR experiments, optical conductivity, resistivity, the superconducting transition temperature, and the role played by impurity scattering, a major challenge for the NAFL approach has been explaining anomalous transport in an applied magnetic field, where experiments show that the Hall resistivity is a strong function of temperature, yet the cotangent of the Hall angle has a very simple behavior: $\cot\theta_H = A + BT^2$. This has been explained in terms of spin-charge separation and has been considered as major support for that approach.

Quite recently, Stojkovic and I (Stojkovic and Pines, 1996, hereafter SP) have investigated the temperature dependence of the normal state Hall effect and magnetoresistance in $YBa_2Cu_3O_7$ using the NAFL approach. We obtain a direct (nonvariational) numerical solution of the Boltzmann equation and find that highly anisotropic scattering at different regions of the Fermi surface gives rise to the measured anomalous temperature dependence of the resistivity and Hall coefficient while yielding the quadratic temperature dependence of the Hall angle observed for both clean and dirty samples.

The starting point of our calculation is the assumption that the magnetic interaction between quasiparticles is determined by Eqs. (2) and (5), which we rewrite as

$$V^{\text{eff}}(\mathbf{q},\omega) = g^2\chi(\mathbf{q},\omega) = \frac{g^2 A}{\omega_{\text{sf}} + \omega_{\text{sf}}\xi^2(\mathbf{q}-\mathbf{Q})^2 - i\omega} \quad (10)$$

For optimally doped YBCO, fits to NMR experiments show that above $T_{\text{cr}} \sim$ 125K the dimensionless parameter $A \approx 1.1$, $\omega_{\text{SF}} = 0.55T$, and the energy $\omega_{\text{SF}}\xi^2 \approx 880$K is very nearly independent of temperature, while $g = 0.64$ eV is taken from the work of Monthoux and Pines (1994a) who found in their strong coupling (Eliashberg) calculation of T_c that with these parameters, and $g = 0.64$ eV, one found a superconducting transition temperature, $T_c \sim$ 90K. Thus all the input parameters of the calculation are fixed.

The use of standard Boltzmann theory to calculate the resistivity of $YBa_2Cu_3O_7$ was pioneered by Hlubina and Rice (1995, hereafter HR), who used a variational method to solve the Boltzmann equation. On adopting the same quasiparticle band parameters and doping level as those used by Monthoux and Pines, they found, in agreement with MP, that the scattering is very strong for points $\mathbf{k}$ and $\mathbf{k}'$ such that $\mathbf{k} - \mathbf{k}' = \mathbf{q} \approx \mathbf{Q}$. As may be seen in Fig. 4, there are only a small number of such *regions* on the FS, called "hot spots" by HR. Nevertheless, as HR have shown, for $\epsilon_k \gg T_0$ the *average* inverse quasiparticle lifetime, τ, is linear in the excitation energy ϵ_k. In addition, the k dependence of τ is found to be large only away from these hot spots in regions where the

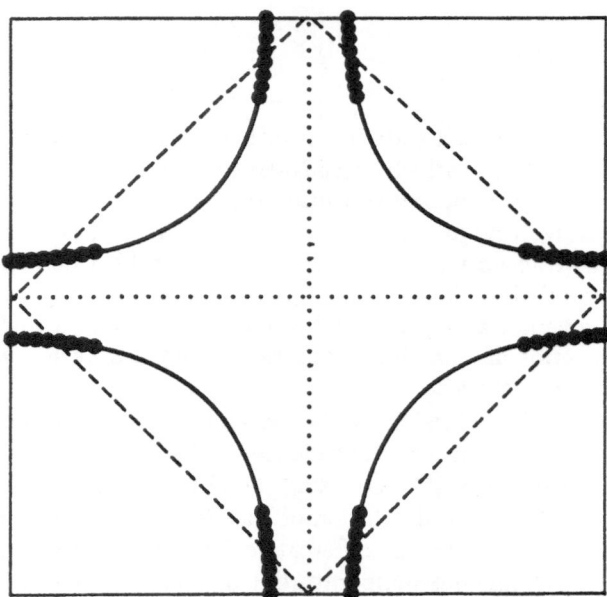

Fig. 4. The FS of YBCO (solid line) and the magnetic BZ (dashed line). The hot spots are the thick regions near the intercepts of the two lines. From Stojkovic and Pines, 1996.

scattering is no longer anomalous. SP find that is the combination of these features which is responsible for the peculiar temperature dependence of both the resistivity and the Hall angle.

As may be seen in Fig. 5, the result obtained by SP in their "non-variational" direct numerical solution of the BE for the resistivity is very similar to HR for the same doping value and quantitatively it is even close to MP at lower temperatures. Unlike MP, the calculations of both HR and SP show a non-linearity in $\rho(T)$ (see inset (a) in Fig. 5), which has not been observed in experiments, and which is due to normal FL-like scattering: for sufficiently low temperatures, in this model, the resistivity is proportional to T^2 with a crossover to linear in T resistivity. As indicated in insets of Fig. 5, if the lifetime effects are neglected, the crossover temperature is approximately equal to 200K in this model. However, SP find that a different choice of band parameters, or the hole concentration, can shift the crossover temperature to much lower values.

SP find that their results for the Hall conductivity are in qualitative agreement with the experiments for temperatures above the crossover temperature. More than *qualitative* agreement cannot be expected, since SP neglect any effects associated with CuO chains and/or planar mass anisotropy, both of which influence the experimental results. Figure 6 is their main result: it shows that $\cot\theta_H$ obeys the simple form $A + BT^2$ over the entire temperature range.

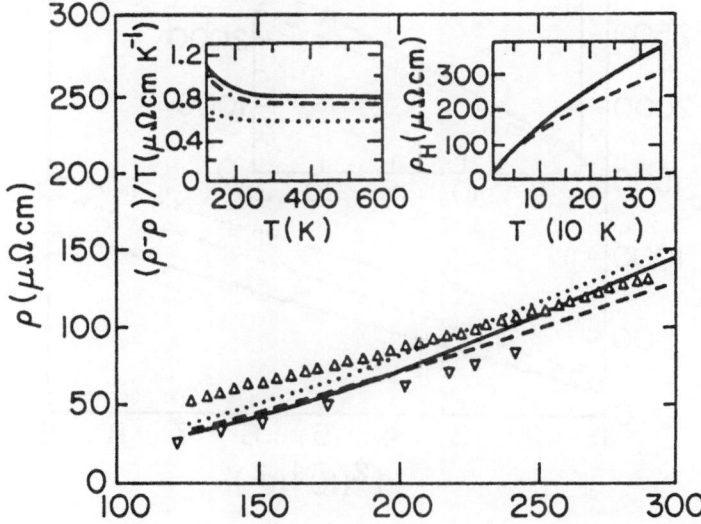

Fig. 5. A comparison of NAFL calculations of $\rho_{ab}(T)$ for YBa$_2$Cu$_3$O$_7$ with experiment. Shown in the insets are: (a) $(\rho - \rho_0)/T$, where ρ_0 is the residual resistivity, as a function of T for $\omega_{SF}\xi^2 = 880$ and 970K (dash-dotted line): (b) ρ as a function of T^2. From Stojkovic and Pines, 1996.

An important test of the soundness of the SP approach is obtained by adding weak impurity scattering to the collision term in the BE. Impurity scattering might be expected to add a temperature independent term to the collision integral, so that $\cot\Theta_H$ should be proportional to $A + BT^2$, where the slope B is unchanged from the clean case. The results of their numerical calculations show that this is indeed the case: B is the same in clean and dirty cases to within 1%, consistent with a number of experiments.

Another challenge for the NAFL model is the existence of a small orbital magnetoresistance (MR) in YBCO, which Harris et al. (1995) show does not obey the usual Köhler rule. In calculations of $\rho(B)$ at several temperatures SP have verified this behavior; the relative MR, $\Delta\rho/\rho$, is found to have very strong temperature dependence and is of order θ_H^2, as is usually the case in FLs. The values they find are somewhat higher than those obtained experimentally. Stojkovic (1996b) has quite recently extended these calculations to take into account the role played by interplanar coupling and possible incommensuration of the peaks in the spin spectrum, and finds that both effects act to improve the agreement between his theoretical results and experiment.

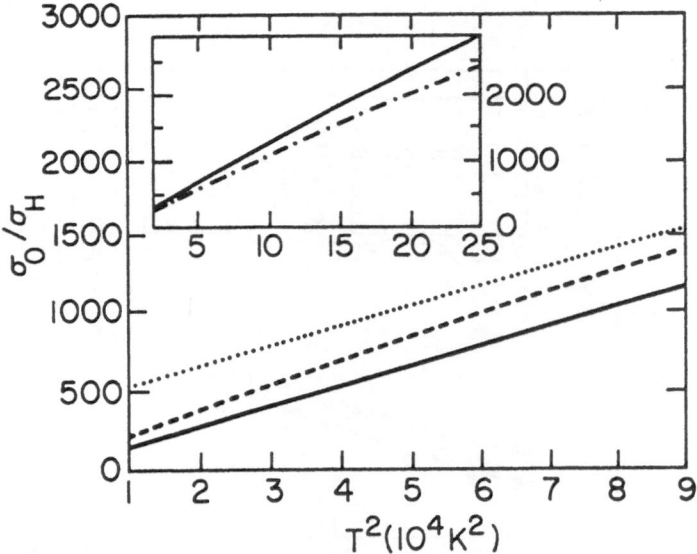

Fig. 6. Cotangent of the Hall angle as a function of temperature. The solid and dotted lines show clean and dirty samples, respectively, at doping level $n_h = 0.25$. The dashed line shows the clean result at $n_h = 0.15$. Inset: results at higher temperature for $\omega_{SF}\xi^2 = 880$ and 970K (dash-dotted line). From Stojkovic and Pines, 1996.

5 The Pairing Potential

In the NAFL description of the cuprates, the pairing potential which brings about high temperature superconductivity is the same magnetic interaction between planar quasiparticles as that which is responsible for their anomalous normal state transport behavior, with a wave vector and frequency dependence which interaction reflects directly the MMP dynamic spin susceptibility Eq. (2), which provided a fit to the NMR experiments. Zha, Barzykin and Pines (1996) have now shown how the apparently different results of NMR and neutron scattering experiments can be reconciled using a susceptibility which takes the MMP form, Eq. (2), modified to take into account the finding, in neutron scattering experiments, of four incommensurate peaks at Q_i near (π, π),

$$\chi(\mathbf{q}, \omega) = \frac{1}{4} \sum_i \frac{\chi_{Q_i}}{1 + (\mathbf{Q}_i - \mathbf{q})^2 \xi^2 - i\omega/\omega_{SF}} \tag{11}$$

The Monte Carlo Hubbard model calculations of Bulut, Scalapino, and White, (1993) carried out only on an 8×8 lattice and restricted to temperatures, $T \gtrsim J/2$, provide support for the Monthoux-Balatsky-Pines ansatz Eq. (5). These show that at temperatures $\sim 800K$, the static susceptibility, $\chi(\mathbf{Q}, 0)$ in the vicinity of (π, π) has already begun to develop peaks which are significant compared to the bulk susceptibility (being ~ 8 times larger), while the static effective quasiparticle interaction is indeed proportional to $\chi(\mathbf{Q}, 0)$. Moreover, extrapolation to 800K of the expression for the correlation length which Zha, Barzykin and Pines (1996) find provides a fit to experiment for $La_{1.86}Sr_{0.14}CuO_4$ between 80K and 325K, leads to a correlation length $\xi \sim a$, comparable to the results of Bulut, Scalapino and White at this temperature.

In considering the effective pairing potential, I follow the approach of Pines and Monthoux (1995, hereafter PM) and focus on $V_{eff}(\mathbf{q}, 0)$, since it determines the dimensionless coupling constant for superconductivity in the Eliashberg approximation. PM consider the long correlation length limit of Eq. (11), in which case $\chi(\mathbf{q}, 0)$ and $V_{eff}(\mathbf{q}, 0)$ possess either four sharp incommensurate peaks,

$$\chi(\mathbf{q}, 0) = \frac{1}{4} \sum_{i=1}^{4} \chi_{\mathbf{q}_i} \delta_{q, q_i} \tag{12}$$

where

$$\mathbf{q}_1 = \pi(1 - \delta, 1); \mathbf{q}_2 = \pi(1 + \delta, 1); \mathbf{q}_3 = \pi(1, 1 - \delta); \mathbf{q}_4 = \pi(1, 1 + \delta) \tag{13}$$

(the kind of incommensuration found in experiments on the 2-1-4 system), or, when $\delta = 0$, one sharp commensurate peak. An expression of the form, Eq. (12), with four incommensurate peaks at $\chi_{\mathbf{q}_1} = \chi_{\mathbf{q}_2} = \chi_{\mathbf{q}_3} = \chi_{\mathbf{q}_4}$, is what one obtains with the Hubbard model, in the limit of strong AF correlations. For Eq. (12), it is trivial to calculate the pairing potential in configuration space, $V_{eff}(x, 0)$. If $x \equiv (n_x, n_y)$ describes a point on the lattice, where n_x and n_y are integers, then,

$$V_{eff}(x, 0) = \frac{g^2 \chi_{q_{max}}}{2} (-1)^{n_x + n_y} [\cos(\pi\delta n_x) + \cos(\pi\delta n_y)] \qquad (14)$$

The interaction is thus maximum and repulsive at the origin, while at the nearest-neighbor sites $(\pm 1, 0)$ and $(0, \pm 1)$

$$V_{eff} = -\frac{g^2 \chi_{q_{max}}}{2} [1 + \cos(\pi\delta)]. \qquad (15)$$

Evidently, the interaction is most attractive when $\cos(\pi\delta) = 1$, *i.e.* when $\delta = 0$ and one has a commensurate spin-fluctuation spectrum. At longer distances, V_{eff} is cut off by the finite correlation length that this simple model ignores, so the main contribution to λ really comes from the nearest-neighbor term (especially for YBCO$_7$ with $\xi \simeq 2$).

Monthoux and I emphasize that it is this on-site repulsion, plus the effective attraction between the nearest neighbors, which is responsible for d-wave pairing; in this pairing state, quasiparticles, by virtue of their $l = 2$ relative angular momentum, avoid sampling the on-site repulsion while taking advantage of the attraction between nearest neighbors to achieve superconductivity. That the pairing state must be $d_{x^2-y^2}$ follows from the fact that along the diagonals in configuration space (where $n_x = \pm n_y$), the effective pairing potential, Eq. (14), is always repulsive for $n_x \lesssim \xi$, and not too large a discommensuration, δ; hence it is energetically favorable to place the nodes of the gap parameter,

$$\Delta(T) = \Delta_0(T)(\cos k_x a - \cos k_y a) \qquad (16)$$

along these same diagonals, since for that pairing state the effective repulsion, which would otherwise be deleterious for superconductivity, is rendered harmless.

The above simple model calculation explains why as long as the q_i's in Eqs. (12) and (13) span the Fermi surface, which appears to be the case for all cuprate superconductors, an effective magnetic interaction which possesses incommensurate peaks will be less effective in bringing about superconductivity than a magnetic interaction which is peaked at Q. It also explains why Hubbard calculations of T_c show a T_c which is maximum near half-filling, falling off as the doping increases, since an elementary calculation for the Hubbard model shows that $\delta = 0$ at half filling, and increases with increased doping.

6 Recent Evidence for $d_{x^2-y^2}$ Pairing

As I have noted earlier, initial evidence for $d_{x^2-y^2}$ pairing came from NMR experiments on YBa$_2$Cu$_3$O$_7$ in the superconducting state at low temperatures, which were most easily explained with the planar quasiparticle density of states, $N_T(0) \sim T$ associated with the thermal excitations in the vicinity of the nodes of the $d_{x^2-y^2}$ gap function, Eq. (16). (The spin-dependent part of the Knight shift is $\sim N_T(0)$, while spin-lattice relaxation rates are proportional to $N_T^2(0)T$; hence the measured linear in T behavior of the Knight shift and the T^3 behavior of the ^{63}Cu spin-lattice relaxation rate at low temperatures.) Subsequent ARPES experiments (Shen *et al.*, 1993) showed a highly anisotropic gap function in $Bi_2Sr_2CaCu_2O_8$ samples, with gap minima and maxima close to those expected for the $d_{x^2-y^2}$ order parameter, while the penetration depth experiments of the UBC group on $YBa_2Cu_3O_7$ (Hardy *et al.*, 1993), which showed the linear in T behavior at low temperatures expected for $N_T(0) \sim T$, made it evident that such behavior was "universal" for clean samples of this material. Still, d-wave skeptics argued that a highly anisotropic s-wave state or perhaps even a g-wave state, with a gap function whose absolute magnitude was close to Eq. (16), could explain these and related experiments.

Thus it was of interest to design experiments which were sensitive to the phase of the order parameter, either through direct measurement of that phase, or through the influence of the phase-sensitive BCS coherence factors on the property under study. The Josephson tunneling experiments suggested independently by Leggett and Sigrist and Rice (1992) are an example of the first kind of experiment, while Bulut and Scalapino (1991) demonstrated in a Hubbard model calculation that the differences between the coherence-factor dependent predictions of s-wave and d-wave pairing states for the spin-echo decay rate, $^{63}T_{2G}$, were so striking that measurements of this quantity would represent a sensitive probe of the symmetry of the order parameter. The tunnel junction SQUID experiments of the Van Harlingen group in Urbana (Wollman *et al.*, 1993) were the first to test the order parameter directly; these, and subsequent work by the Kirtley-Tsuei group at IBM and many others (reviewed in Van Harlingen, 1995) went a long way toward settling the pairing issue, since their results exhibited the change in sign of the order parameter under $\pi/2$ rotation characteristic of the $d_{x^2-y^2}$ gap function, Eq. (15). In this talk, I consider the additional evidence for $d_{x^2-y^2}$ pairing which comes from $^{63}T_{2G}$ experiments, motivated by the recent experiments of Corey *et al.* (1996), on $YBa_2Cu_4O_8$, together with the associated theoretical calculations which Piotr Wrobel and I have carried out (Pines and Wrobel, 1996, hereafter PW). This body of work not only provides strong support for $d_{x^2-y^2}$ pairing and strong antiferromagnetic correlations in $YBa_2Cu_4O_8$, but rules out anisotropic s-wave pairing.

As PW note, in the limit of the long correlation lengths found for $YBa_2Cu_4O_8$ ($\xi \gtrsim 5a$) at low temperatures, the expression for the spin-echo decay time takes the simple form, Eq. (6), which can be rewritten as,

$$\frac{1}{^{63}T_{2G}(T)} \sim \alpha \xi(T) \sim \alpha \sqrt{\chi_Q(T)} \tag{17}$$

where χ_Q is the staggered static spin susceptibility. Quite generally χ_Q may be written in the form

$$\chi_Q(T) = \frac{\tilde{\chi}_Q(T)}{1 - f_Q^a \tilde{\chi}_Q(T)} \equiv \frac{\tilde{\chi}_Q(T)}{1 - F_Q^a(T)} \tag{18}$$

where $\tilde{\chi}_Q(T)$ is the temperature dependent static irreducible particle hole susceptibility which, in the superconducting state, depends on the order parameter and the gap magnitude, and f_Q^a is the temperature-independent restoring force at Q which brings the system close to antiferromagnetic behavior. From Eqs. (17) and (18) it follows that

$$\frac{(T_{2G}(T_c))}{(T_{2G}(T))} = \sqrt{\frac{\chi_Q(T)}{\chi_Q(T)_c}} = \sqrt{\frac{[1 - F_Q^a(T_c)]\gamma_Q(T/T_c)}{1 - \gamma_Q(T/T_c)F_Q^a(T_c)}} \tag{19}$$

on introducing the normalized irreducible commensurate susceptibility in the superconducting state,

$$\gamma_Q(T/T_c) \equiv \tilde{\chi}_Q(T)/\tilde{\chi}_Q(T_c). \tag{20}$$

T_{2G} in the superconducting state is thus seen to depend only on $\gamma_Q(T/T_c)$ and the parameter, $F_{\mathbf{Q}}^a(T_c) \sim 1$, which measures at T_c, the antiferromagnetic enhancement of the commensurate irreducible particle-hole spin susceptibility.

$\gamma_Q(T/T_c)$ may be approximated by a BCS expression

$$\gamma_{\mathbf{Q}}(T/T_c) \equiv \frac{\tilde{\chi}_{\mathbf{Q}}(T)}{\tilde{\chi}_{\mathbf{Q}}(T_c)} \cong \gamma_{\mathbf{Q}}^{BCS}(T/T_c) \equiv \frac{\tilde{\chi}^{BCS}(\mathbf{Q}))_T}{(\tilde{\chi}^{BCS}(\mathbf{Q}))_{T_c}} \tag{21}$$

where $\tilde{\chi}^{BCS}(\mathbf{q}) = \tilde{\chi}^{BCS}(\mathbf{q}, \omega = 0)$ is given by

$$\tilde{\chi}^{BCS}(\mathbf{q}) = -\frac{1}{2}\sum_k (1 + \frac{\varepsilon_{\mathbf{k+q}}\varepsilon_{\mathbf{k}} + \Delta_{\mathbf{k+q}}\Delta_{\mathbf{k}}}{E_{\mathbf{k+q}}E_{\mathbf{k}}}) \frac{f(E_{\mathbf{k+q}}) - f(E_{\mathbf{k}})}{E_{\mathbf{k+q}}E_{\mathbf{k}}}$$

$$+\frac{1}{2}\sum_k (1 - \frac{\varepsilon_{\mathbf{k+q}}\varepsilon_{\mathbf{k}} + \Delta_{\mathbf{k+q}}\Delta_{\mathbf{k}}}{E_{\mathbf{k+q}}E_{\mathbf{k}}} \frac{1 - f(E_{\mathbf{k+q}}) - f(E_{\mathbf{k}})}{E_{\mathbf{k+q}} + E_{\mathbf{k}}} \tag{22}$$

where $\varepsilon_{\mathbf{q}}$ is the normal state quasiparticle energy, $E_{\mathbf{q}}$ that energy in the superconducting state, and $\Delta_{\mathbf{k}}$ is the gap function. It is through the coherence factors, $\varepsilon_{\mathbf{k+Q}}\varepsilon_{\mathbf{k}} + \Delta_{\mathbf{k+Q}}\Delta_{\mathbf{k}}$, in Eq. (22) that T_{2G} acquires its sensitivity to the phase of the order parameter. For the $d_{x^2-y^2}$ order parameter, Eq. (15), as one goes around the Fermi surface, the order parameter, $\Delta_{\mathbf{q}}$ changes sign, with $\Delta_{\mathbf{k+Q}} = -\Delta_{\mathbf{k}}$, while for an anisotropic s-wave order parameter, even though Δ may possess minima along the (π, π) and $(-\pi, -\pi)$ directions, it does not change sign. Hence for gap functions of identical magnitude, $\tilde{\chi}_{\mathbf{Q}}$ will be sensitive to the order parameter phase.

Without doing any detailed calculations, it is evident that in the limit of strong antiferromagnetic correlations, $[1 - F_{\mathbf{Q}}^a(T_c)] << 1$, unless $\gamma_{\mathbf{Q}}(T/T_c)$ is very close to unity, one will get a strong increase in T_{2G}, and a correspondingly decrease in $\xi(T)$, on going into the superconducting state. For example, Monthoux and Pines (1994a) find, on combining their strong coupling calculations for $YBa_2Cu_3O_7$ with an MMP fit to the spin-lattice relaxation rate, that $1 - F_{\mathbf{Q}}^a(T_c) \cong .025$, and a still smaller value of $1 - F_{\mathbf{Q}}^a(T_c)$ is to be anticipated for $YBa_2Cu_4O_8$ in view of the stronger AF correlations found there. For

$F_Q^a(T_c) = 0.975$, the finding of Corey *et al.* (1996), shown in Fig. 7, that T_{2G} is increased by only some 15% in the superconducting state, implies that $\gamma_Q(T/T_c)$ must change by less than 1.9% in the superconducting state. We shall see that the actual change is much smaller.

In calculating $\tilde{\chi}$ at T_c and in the superconducting state, PW used the BCS expression for $\tilde{\chi}$ and the same quasiparticle spectrum as that employed by MP as a fit to photoemission experiments for $YBa_2Cu_3O_7$,

$$\varepsilon_k = -2t(\cos k_x + \cos k_y) - 4t'\cos k_x \cos k_y, \tag{23}$$

where $t' = -0.45\,t$, and the bandwidth is $8\,t = 2$ eV. Because strong coupling effects play a significant role in determining the gap function, $\Delta(T)$, in the vicinity of T_c, they approximated these by the same functional form as that used by Monien and Pines (1990) to fit the Knight shift in the superconducting state of $YBa_2Cu_3O_7$,

$$\Delta(T) = \Delta(0)\tanh\left(\alpha\sqrt{\frac{T_c}{T} - 1}\right) \tag{24}$$

where α is an adjustable parameter. They chose $\Delta(0)$ and α to obtain the best fit, shown in Fig. 8, to the Knight shift experiments of Bankay *et al.* (1994). Their best fit parameters ($\alpha = 2.7$, $2.8k_BT_c \lesssim \Delta \lesssim 3k_BT_c$) were comparatively insensitive to the Fermi liquid corrections to $\chi_0(T)$ in the superconducting state.

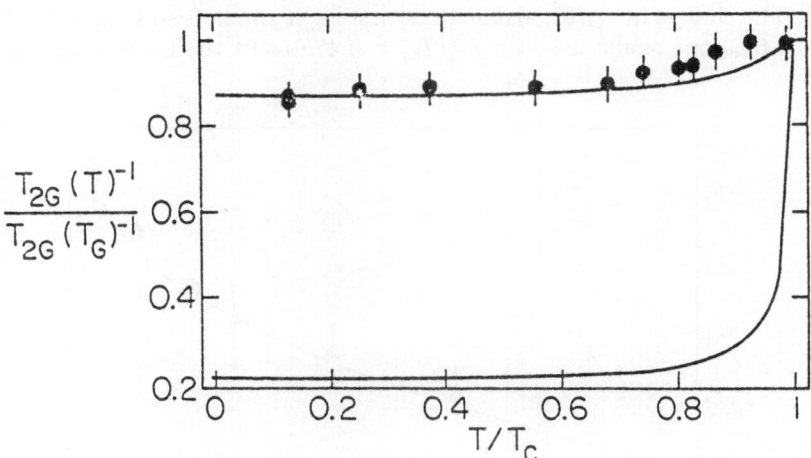

Fig. 7. A comparison of calculated (Pines and Wrobel, 1996) values of T_{2G} for $YBa_2Cu_4O_8$ with the experimental results of Corey *et al.*, 1996. The upper line is for $d_{x^2-y^2}$ pairing; the lower line for anisotropic s-wave pairing.

In order to compare the predictions of anisotropic s-wave and d-wave pairing with the experimental results of Corey *et al.* (1996), PW assumed that the gap magnitudes for the two states were the same, so that either state could provide a satisfactory fit to the Knight shift measurements. Coherence factors then provide the sole difference between the predictions of the different order parameter symmetries for $\gamma_{\mathbf{Q}}(0)$. With their Knight shift best fit values to the magnitude of the gap function, they found, for $T \lesssim 0.6T_c$,

$$\gamma_{\mathbf{Q}}(0) = 0.9979 \qquad d_{x^2-y^2} \tag{25}$$

and

$$\gamma_{\mathbf{Q}}(0) = 0.868 \qquad \text{anisotropic s-wave} \tag{26}$$

This difference may not appear substantial, but as Fig. 7 shows, because of the significant role played by the AF enhancement factor, $F_{\mathbf{Q}}^a(T_c)$ the resulting predictions differ dramatically. For $d_{x^2-y^2}$ pairing, with $F_{\mathbf{Q}}^a(T_c) = 0.993$, PW obtain excellent agreement with experiment below $0.7T_c$. Such a value, which corresponds to an AF enhancement factor in the normal state, $[1-F_{\mathbf{Q}}^a(T_c)]^{-1} \sim 143$, is not unreasonable, since Barzykin and Pines (1995) find $\chi_{\mathbf{Q}}(T_c) \sim 330$ states/eV from their analysis of the normal state NMR experiments on $^{63}T_{2G}$ and $^{63}T_1$ carried out by Corey *et al.* (1996). PW thus concluded that in the vicinity of T_c

$$\tilde{\chi}_{\mathbf{Q}}(T_c) \sim 2.3 \text{ states/eV} \tag{27}$$

a physically reasonable value.

Also shown in Fig. 7 is the fit to $^{63}T_{2G}$ obtained using the anisotropic s-wave pairing result for $\gamma_{\mathbf{Q}}(0)$, Eq. (26), and this same physically reasonable choice of $F_{\mathbf{Q}}^a(T_c)$. The change in $\gamma_{\mathbf{Q}}(0)$, although at first sight small, is so substantial that for $T \lesssim 0.6T_c$, one would need an $F_{\mathbf{Q}}^a(T_c) = 0.478$ to fit the Corey experiment.

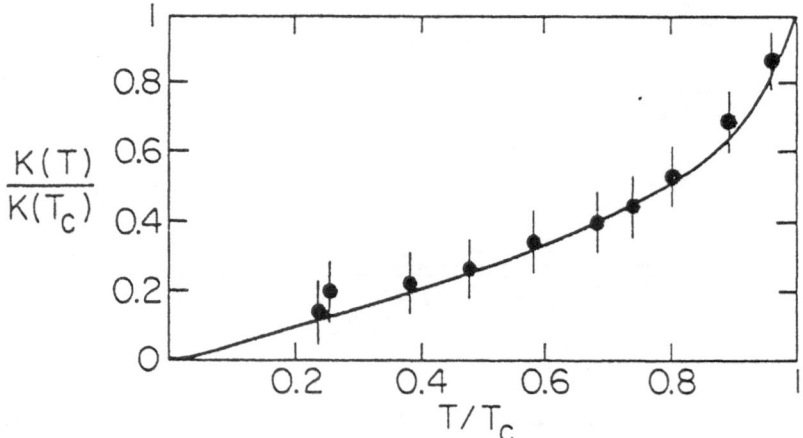

Fig. 8. Fit to $YBa_2Cu_4O_8$ Knight shift data of Bankay *et al.* (Pines and Wrobel, 1996).

This is an unrealistically low value of F_Q^a, since it would imply a normal state irreducible particle-hole susceptibility of ~ 170 states/eV. We can therefore safely conclude, with PW, that the experimental results of Corey *et al.* (1996) both rule out anisotropic s-wave pairing and strongly support $d_{x^2-y^2}$ pairing.

7 Conclusion and Acknowledgements

In this lecture, I have described progress made during the past fourteen months on the characterization of the normal and superconducting states and the extent to which a nearly antiferromagnetic Fermi liquid description of planar quasiparticle behavior enables one to explain a broad spectrum of experiments. As Monthoux and I have recently emphasized (Pines and Monthoux, 1995), the close approach of the cuprate superconductors to antiferromagnetism, the evidence from NMR (and more recently neutron) experiments that $\chi(\mathbf{q}, \omega)$ is strongly peaked near $\mathbf{Q}$, the demonstration that a magnetic interaction proportional to $\chi(\mathbf{q}, \omega)$ leads naturally to $d_{x^2-y^2}$ pairing and the anomalous normal state magnetotransport discussed here, together with our strong coupling "proof of concept" calculation for $YBa_2Cu_3O_7$, provide both back-of-the-envelope and detailed theoretical support for the view that the magnetic mechanism provides not only the simplest, but the most natural explanation for the $d_{x^2-y^2}$ pairing state which has now been observed under so many different experimental conditions. Still many important problems remain, of which perhaps the most challenging are the development of a microscopic derivation of this magnetic interaction, and the demonstration that because that interaction is of electronic origin, non-linear feedback effects can give rise to the three different normal state regimes which a cuprate superconductor traverses on its way to superconductivity.

It gives me pleasure to acknowledge here the significant contribution made to the work of our group in Urbana by my graduate students and postdocs, A. Balatsky, V. Barzykin, J.-P. Lu, H. Monien, P. Monthoux, A. Sokol, B. Stojkovic, D. Thelen, P. Wrobel, and Y. Zha. Much of the material presented in this review is drawn from articles we have written together. I should like to thank them, my STCS colleagues in Urbana, and my colleagues in the strongly correlated electron systems program at Los Alamos, for many stimulating conversations on these and related topics. I should especially like to thank Bob Corey, Nick Curro, Takashi Imai, Joe Martindale, and Charlie Slichter for communicating their NMR results to me in advance of publication, and for what has proved to be a most fruitful and enduring collaboration, and the Science and Technology Center for Superconductivity, the Frederich Seitz Materials Research Laboratory, and the National Science Foundation for their support of this work through NSF Grants DMR91-20000 and DMR89-20538.

References

Alloul, H. *et al.* 1989. Phys. Rev. Lett. **63**,1700.

Barzykin, V. *et al.* 1994. Phys. Rev. B **49**, 1544.

Barzykin, V. and D. Pines 1995. Phys. Rev. B **52**, 13585.

Bücher, B. *et al.* 1993. Phys. Rev. Lett. **70**, 2012.

Bulut, N. and D. J. Scalapino 1991. Phys. Rev. Lett. **67**, 2898.

Bulut, N., D. Scalapino and S. R. White 1993. J. Chem. Phys. Solids **54**, 1109.

Corey, R. *et al.* 1996. Phys. Rev. B, in the press.

Curro, N. *et al.* 1996. private communication.

Hardy, W. *et al.* 1993. Phys. Rev. Lett. **70**, 399.

Harris, R. *et al.* 1995. Phys. Rev. Lett. **75**, 1391.

Hlubina, R. and T. M. Rice 1995. Phys. Rev. B **51**, 9253; ibid **52**, 13043.

Hwang, H. Y. *et al.* 1994. Phys. Rev. Lett. **72**, 2636.

Ishida, K. *et al.* 1993. J. Phys. Soc. Jpn. **62**, 2803.

Ishida, K. *et al.* 1994. J. Phys. Soc. Jpn. **63**, 1104.

Ito, T. *et al.* 1993. Phys. Rev. Lett. **70**, 3995.

Loram, J. *et al.* 1993. Phys. Rev. Lett. **71**, 1740.

Mila, F. and T. M. Rice 1989. Physica C **175**, 561.

Millis, A., H. Monien, D. Pines 1990. Phys. Rev. B **42**, 167.

Monien, H. and D. Pines 1990. Phys. Rev. B **41**, 1797.

Monthoux, P., A. Balatsky and D. Pines 1992. Phys. Rev. B **46**, 14803.

Monthoux, P. and D. Pines 1993. Phys. Rev. B **47**, 6069.

Monthoux, P. and D. Pines 1994a. Phys. Rev. B **49**, 4261.

Monthoux, P. and D. Pines 1994b. Phys. Rev. B **50**, 16015.

Nakano, T. *et al.* 1994. Phys. Rev. B **49**, 16000.

Pennington, C. and C. P. Slichter 1991. Phys. Rev. Lett. **66**, 381.

Pines, D. 1995. In *High Temperature Superconductors and the C^{60} System*, (Gordon & Breach), ed. H.-C. Ren, 1.

Pines, D. 1994a. Physica B **199-200**, 300.

Pines, D. 1994b. Physica C **235-240**, 113.

Pines, D. and P. Monthoux 1995. J. Phys. Chem. Solids, in the press.

Pines, D. and P. Wrobel 1996. Phys. Rev. B, in the press.

Shen, Z. X. *et al.* 1993. Phys. Rev. Lett. **70**, 1553.

Slichter, C. P. 1995. In *Strongly Correlated Electron Systems*, ed. K. S. Bedell *et al.* (Addison-Wesley, NY), 427.

Shastry, B. 1989. Phys. Rev. Lett. **63**, 1288.

Sigrist, M. and T. M. Rice 1992. J. Phys. Soc. Jpn. **61**, 4283.

Sokol, A. and D. Pines 1993. Phys. Rev. Lett. **71**, 2813.

Stojkovic, B. and D. Pines 1996. Phys. Rev. Lett. **76**, 811.

Stojkovic, B. 1996. Phil. Mag., in the press.

Takigawa, M. and D. Mitzi 1994. Phya. Rev. Lett. **73**, 1287.

Thelen, D. and D. Pines 1994. Phys. Rev. B **49**, 3528.

Van Harlingen, D. 1995. Rev. Mod. Phys. **67**, 515.

Wollman, D. A. *et al.* 1993. Phys. Rev. Lett. **71**, 2134.

Zha, Y., S. L. Cooper, D. Pines 1996. Phys. Rev. B. in the press.

Zha, Y., V. Barzykin and D. Pines 1996. Preprint.

Crossover from BCS Superconductivity to Superfluidity of Local Pairs

R. Micnas and T. Kostyrko

Institute of Physics, A.Mickiewicz University, Grunwaldzka 6, 60-780 Poznań, Poland

Abstract. We review some recent results concerning crossover from cooperative Cooper pairing to independent bound states formation and their superfluidity, and a possible relevance of the crossover behavior to the anomalous properties of short–coherence length superconductors. Using the extended Hubbard model with on–site attraction (EHM), we analyze the behavior of collective modes, thermodynamic and electromagnetic properties versus the coupling strength and electron concentration. The normal state properties of the 2D attractive Hubbard model obtained with the conserving, self–consistent T–matrix approach, are presented. These studies also indicate possible deviations from conventional Fermi–liquid behavior, above T_c, in 2D short coherence length superconductors.

I. INTRODUCTION

A problem of crossover from a weak coupling BCS picture of Cooper pair formation and their condensation at critical temperature to Bose condensation of preformed local pairs has recently attracted great attention. A part of motivation to study the problem comes from some experimental observations regarding unusual properties of new high T_c superconductors. High temperature superconductors (HTS) (the cuprates, doped bismuthates and fullerenes) generally exhibit a low carrier density, small value of the Fermi energy ($\propto 0.1 - 0.3$ eV), short coherence length ξ_0 and they are *extreme type II superconductors* with the Ginzburg–Landau paramerter $\kappa \gg 1$ [1]. Detailed studies of HTS and also other nonconventional superconductors by the muon spin rotation (μSR) indicate that the cuprate high T_c, bismutathes, fullerenes and Chevrel phases belong to a unique group of superconductors characterized by high transition temperatures relative to the value of n_s/m^*. These materials have their T_c proportional to T_F (the Fermi temperature) or T_B (the Bose–Einstein condensation temperature) [2], with $T_B \approx (3 - 30)T_c$ and $T_F \approx (10 - 100)T_c$. Moreover, for many HTS, regardless of a specific microscopic mechanism leading to pairs, there are several universal trends in the T_c versus condensate density dependence, the T_c dependence on the pressure and the isotope effect coefficient [3]. These universal trends are consistent with a short–range, almost unretarded effective interaction responsible for pairing, and moreover they suggest that there could be a common condensation mechanism (or thermodynamic description) of short–coherence length superconductors [1].

Currently, there are several arguments showing that the fluctuations are very important in all extreme type II superconductors with a short coherence length and they play essential role in thermodynamic behavior. For various cuprate superconductors the measured thermodynamic properties reveal remarkable consistency with a universal 3D X-Y critical point behavior reminiscent of the normal fluid-superfluid transition in ^{4}He [1,3].

The above points show that these materials are clearly not in weak coupling BCS limit, but rather in the intermediate to extreme strong coupling regime. One of the central questions regarding the physics of HTS is understanding the evolution from a weak coupling limit of large Cooper pairs to the strong coupling regime of small local pairs with increasing coupling constant and description of intermediate (crossover) regime.

The problem of evolution from BCS to Bose condensation (BC) has been studied in the past for two classes of fermionic models [1,4,5]. i) For the fermions on a lattice, e.g. $U < 0$ Hubbard model, the crossover is from Cooper pairs to local pairs (composite Bosons) and BCS superconductivity transforms to superfluidity of hard-core Bosons on a lattice. ii) Continuum fermion models with attraction, generally in the dilute limit.

Many attempts to describe crossover have employed the BCS mean field theory, generalizing to allow the change in the chemical potential due to pairing away from weak coupling limit [6–9]. It has been shown that the BCS wave function for $T = 0K$ is a reasonable ansatz for any attraction strength. In continuum theories a controlling parameter is $\xi_0 k_F$ and the BCS and BC limits correspond to $\xi_0 k_F \gg 1$ and $\xi_0 k_F \ll 1$, respectively. An extension to finite temperatures has been proposed by Noziéres and Schmitt-Rink [9], who included single-pair fluctuations within a T-matrix approximation and found that T_c for the 3D continuum dilute fermion system evolves smoothly between the limits.

An essential question for the crossover is played by the system dimensionality. For 2D models with s-wave attraction in the dilute limit Miyake [10], and Randeria et al [11] have shown that an existence of a two-body bound state in a vacuum is necessary and sufficient condition for the s-wave many body instability. Further Schmitt–Rink et al [12], by extending an earlier nonconserving T–matrix calculations for continuum fermions, pointed out that in 2D systems there is always a finite number of bound fermion pairs at any T and n (coexisting bound and ionized pairs).

It is an important question whether in 2D the Fermi liquid behaviour of the normal state can be destroyed, or when 2-body bound states break Fermi liquid behaviour? The problem is of relevance for both the repulsive ($U > 0$) and attractive ($U < 0$) on–site interaction.

The model of general interest is the extended Hubbard model with on-site and intersite interaction [1], defined by the Hamiltonian:

$$\mathcal{H} = \sum_{jm\sigma} t_{jm} c_{j\sigma}^\dagger c_{m\sigma} + U \sum_j n_{j\uparrow} n_{j\downarrow} + \frac{1}{2} \sum_{jm\sigma\sigma'} W_{jm} n_{j\sigma} n_{m\sigma'} - \mu \sum_{j\sigma} n_{j\sigma} . \quad (1)$$

In above: t_{jm}, W_{jm} denote hopping integral and intersite density-density interaction, resp., between sites j and m; U - is on-site Hubbard interaction; μ - the chemical potential; the electron concentration $n = N_e/N$, N_e and N are the numbers of electrons and total lattice sites, respectively. In the case of intersite attraction but repulsive on–site interaction, the above model is one of the simplest to study competition between the anisotropic pairing of extended s or d-wave type, antiferromagnetism and phase separation [1,13]. In the case of attractive, $U < 0$ interaction it is a nontrivial model of fermions on the lattice which describes transition from weak coupling BCS like superconductivity to the strong coupling superconductor, where the superconductivity results from the condensation of hard-core composite charged bosons and is similar to superfluidity of ^{4}HeII [1,14]. Such a model has been considered as an effective model of superconductivity in CuO_2 planes [1,14,15], of doped $BaBiO_3$ [1,16] and fullerenes [17,18]. For $2Zt \gg |U|$, i.e. a large bandwidth (Z is the coordination number), the model exhibits a superconducting state of BCS-type with strongly overlapping electron pairs. In the opposite case of small bandwidth, $2Zt \ll |U|$, one has a system of local electron pairs which can undergo superfluid transition to the superconducting state with short coherence length $\xi_0 (\sim$ size of electron pair function). In this case the system of local pairs can be described by the following effective *pseudospin* Hamiltonian:

$$\tilde{\mathcal{H}} = -\frac{1}{2} \sum_{jm} J_{jm} \left(\rho_j^+ \rho_m^- + \text{H.c.} \right) + \sum_{jm} K_{jm} \rho_j^z \rho_m^z - \bar{\mu} \sum_j \left(2\rho_j^z + 1 \right), \qquad (2)$$

where: J_{jm} denotes the transfer integral for the pairs from sites $\mathbf{R}_j$, $\mathbf{R}_m$, K_{jm} is the intersite density–density interaction. Eq.(2) has the form of the anisotropic Heisenberg model with s=1/2 in an effective external field $\bar{\mu}$ in the z direction, such that the average pseudospin magnetization has a fixed value equal to $N^{-1} \sum_j \langle \rho_j^z \rangle = \bar{n} - 1/2$, $\bar{n} = n/2 = N_p/N$ is the concentration of pairs. The model (2) follows from the extended Hubbard model with on–site attraction in the strong coupling limit [1] by the second order perturbation theory (up to a constant term) with respect to hopping and we have $J_{ij} = 2t_{ij}^2/|U|$, $K_{ij} = J_{ij} + 2W_{ij}$, $\bar{\mu} = \mu + |U|/2 - W_0$, $W_0 = \sum_j W_{ij}$. In such a case we obtain the system of a tightly bound local pairs with the charge $\bar{e} = 2e$ and charge operators are the composite boson operators. The pseudospin operators ρ_i^α, $\alpha = +, -, z$, operating in a subspace of states of double occupied and empty sites, are related to the original fermion operators by: $\rho_j^+ = c_{j\uparrow}^\dagger c_{j\downarrow}^\dagger, \rho_j^z = \frac{1}{2}(n_{j\uparrow} + n_{j\downarrow} - 1), \rho_j^- = (\rho_j^+)^\dagger$, and satisfy the commutation rules of the $s = \frac{1}{2}$ operators:

$$[\rho_l^+, \rho_m^-] = 2\rho_l^z \delta_{l,m}, \quad [\rho_l^+, \rho_m^z] = -\rho_l^+ \delta_{l,m}, \quad [\rho_l^-, \rho_m^z] = \rho_l^- \delta_{l,m},$$

$$\rho_l^+ \rho_l^- = \rho_l^z + \frac{1}{2}, \quad (\rho_l^+)^2 = (\rho_l^-)^2 = 0, \quad \rho_l^+ \rho_l^- + \rho_l^- \rho_l^+ = 1.$$

The model (2) can also be used for description of the intersite local pairs [1] resulting from a number of mechanisms like strong electron–lattice coupling or magnetic bipolarons [14].

In this paper we discuss a recent progress in understanding the crossover behavior in the case of s–wave attraction channel, mostly derived from analytic and numerical studies of the attractive EHM model. The highly interesting and for cuprates experimentally relevant cases of anisotropic extended s–wave or d–wave pairing, require however further studies. Nevertheless, we think that some of the presented results can be of general importance for the short coherence length superconductors.

It is well known that in the ground state, but not for the finite temperatures, the BCS ansatz for superconducting state allows to reach the case of strong coupling limit of small preformed local pairs. It is of interest to ask if a similar continuation holds for certain response functions and electromagnetic properties of superconductors. Other questions are related to the problem of phase separation and effects of Coulomb repulsion on the crossover.

We shall briefly discuss the density-density response function and collective excitation spectrum of the extended Hubbard model over the superconducting ground state for arbitrary electron concentration [19]. Next the evolution of the Meissner kernel is studied in the ground state. The electromagnetic properties of the local pair regime, where the problem reduces to that of hard core bosons on a lattice in the magnetic field are summarized. One principal feature which can distinguish the Cooper pair limit from the preformed local pair regime is the normal state behavior. In this respect we present some recent results for the normal state of 2D attractive Hubbard model by going beyond the Hartree–Fock approximation (HFA). Finally, we summarize briefly the main physical ideas involved in the crossover problem.

II. EVOLUTION OF COLLECTIVE EXCITATION SPECTRUM

In Ref. [19] the density response function was determined by using the diagrammatic perturbation theory in the Random Phase Approximation (RPA). The collective excitations at $T = 0K$ were analyzed in both the weak coupling and strong coupling limits and it was shown that they smoothly interpolate between the limits, for short- and long-range intersite Coulomb interaction. From numerical analysis of the collective modes in 2D square lattice, we also determined the ground state phase diagram of the system for nearest-neighbours Coulomb interaction. The effects of long-range Coulomb interaction and general frequency behavior of the density-density response function were also discussed. These collective modes are related to fluctuations of electron density and the phase of the superconducting order parameter and they can play a role in thermodynamic and spectral properties of the system. Experimentally, spectroscopies such as electron-energy loss experiments can give a direct measure of these modes.

Energies of collective modes related to electron density oscillations are given by the poles of density-density response function [20]:

$$\tilde{\Pi}(\mathbf{q},t) = -i\Theta(t) \sum_{\sigma\sigma'} \langle [\rho_{\mathbf{q}}^{\sigma}(t), \rho_{-\mathbf{q}}^{\sigma'}] \rangle, \tag{3}$$

where $\rho_{\mathbf{q}}^{\sigma}$ denotes Fourier component of electron density operator $n_{i\sigma} = c_{i\sigma}^{\dagger} c_{i\sigma}$ with spin σ.

The response function have been evaluated by means of diagrammatic method in the Nambu formalism [19,21]. For the density-density response function one obtains formally exact result:

$$\tilde{\Pi}(q) = \frac{\Pi(q)}{1 - (U/2 + W_{\mathbf{q}})\Pi(q)}, \tag{4}$$

where $q = (\mathbf{q}, \omega)$ and $W_\mathbf{q}$ is the spatial Fourier transform of W_{ij}. The irreducible polarization part $\Pi(q)$ evaluated in the generalized RPA has the following form [19]:

$$\Pi(q)/2 = A(q) - D(q) + 2\,U\,\frac{\Gamma(-q)B(q) + \Gamma(q)B(-q)}{(1 - UC(q))(1 - UC(-q)) - U^2 D^2(q)}, \quad (5)$$

where $\Gamma(q) = 1 - UC(q)B(-q) + UD(q)B(q)$ is a vertex correction. The unperturbed susceptibilities $A(q)$, $B(q)$, $C(q)$ and $D(q)$ are given in terms of one particle Matsubara-Green function in the HFA:

$$\mathcal{G}_{\alpha\alpha}(\mathbf{k}, i\omega_n) = \frac{1}{2} \sum_{\sigma=\pm} \frac{1 + \sigma(-1)^\alpha \lambda_\mathbf{k}/E_\mathbf{k}}{i\omega_n + \sigma E_\mathbf{k}}, \quad \mathcal{G}_{-\alpha\alpha}(\mathbf{k}, i\omega_n) = \frac{\Delta}{2} \sum_{\sigma=\pm} \frac{\sigma}{i\omega_n - \sigma E_\mathbf{k}}$$

$$(6)$$

where $E_\mathbf{k} = \sqrt{\lambda_\mathbf{k}^2 + \Delta^2}$ is the quasiparticle excitation energy, $\lambda_\mathbf{k} = \epsilon_\mathbf{k} - \bar\mu$, and $\epsilon_\mathbf{k}$ is the spatial Fourier transform of t_{ij}, as the following sums:

$$A(\mathbf{q}, i\nu_m) = \frac{1}{\beta N} \sum_{\mathbf{k} i\omega_n} \mathcal{G}_{11}(\mathbf{k} + \mathbf{q}, i\omega_n + i\nu_m)\mathcal{G}_{11}(\mathbf{k}, i\omega_n), \quad (7)$$

$$B(\mathbf{q}, i\nu_m) = \frac{1}{\beta N} \sum_{\mathbf{k} i\omega_n} \mathcal{G}_{11}(\mathbf{k} + \mathbf{q}, i\omega_n + i\nu_m)\mathcal{G}_{21}(\mathbf{k}, i\omega_n), \quad (8)$$

$$C(\mathbf{q}, i\nu_m) = \frac{1}{\beta N} \sum_{\mathbf{k} i\omega_n} \mathcal{G}_{11}(\mathbf{k} + \mathbf{q}, i\omega_n + i\nu_m)\mathcal{G}_{22}(\mathbf{k}, i\omega_n), \quad (9)$$

$$D(\mathbf{q}, i\nu_m) = \frac{1}{\beta N} \sum_{\mathbf{k} i\omega_n} \mathcal{G}_{12}(\mathbf{k} + \mathbf{q}, i\omega_n + i\nu_m)\mathcal{G}_{21}(\mathbf{k}, i\omega_n), \quad (10)$$

and after subsequent analytic continuation to the real frequency axis: $i\nu_m \rightarrow \omega + i0^+$. $\beta = (k_B T)^{-1}$.

The gap parameter Δ and the chemical potential $\mu = \bar\mu + Un/2 + W_0 n$ are determined from the set of HFA self–consistent equations [8]:

$$1 = -\frac{U}{2N} \sum_\mathbf{k} \frac{1}{E_\mathbf{k}} \tanh(\beta E_\mathbf{k}/2), \quad 1 - n = \frac{1}{N} \sum_\mathbf{k} \frac{\lambda_\mathbf{k}}{E_\mathbf{k}} \tanh(\beta E_\mathbf{k}/2)$$

At $T = 0K$ the polarization part $\Pi(\mathbf{q}, \omega)$ is a real quantity provided that $\omega < \min_\mathbf{k}(E_\mathbf{k} + E_{\mathbf{k}+\mathbf{q}})$, where the last quantity determines a boundary of a quasiparticle continuum. Below this boundary and for small $\mathbf{q}, \omega$, we obtain:

$$\Pi(\mathbf{q}, \omega) = q^2 \frac{L(\hat{\mathbf{q}})}{\omega^2 M_1 - q^2 M_2(\hat{\mathbf{q}})}. \quad (11)$$

In above: $L(\hat{\mathbf{q}})$, M_1, $M_2(\hat{\mathbf{q}})$ are lattice sums depending on the one–electron dispersion $\epsilon_\mathbf{k}$ only, and they are given by

$$L(\hat{\mathbf{q}}) = \frac{\Delta^2}{U} S_2(\hat{\mathbf{q}})(\Delta^2 S_0^2 + U^2 S_1^2), \quad M_1 = \Delta^2 S_0^2 + U^2 S_1^2, \quad M_2(\hat{\mathbf{q}}) = \Delta^2 U^2 S_2(\hat{\mathbf{q}}) S_0,$$

$$(12)$$

where $\mathbf{q} = \hat{\mathbf{q}}|\mathbf{q}|$ and

$$S_0 = \frac{U^3}{N} \sum_{\mathbf{k}} \frac{1}{E_{\mathbf{k}}^3}, \quad S_1 = \frac{U^2}{N} \sum_{\mathbf{k}} \frac{\lambda_{\mathbf{k}}}{E_{\mathbf{k}}^3}, \quad S_2(\hat{\mathbf{q}}) = \frac{U}{N} \sum_{\mathbf{k}} \frac{(\hat{\mathbf{q}} \cdot \nabla \epsilon_{\mathbf{k}})^2}{E_{\mathbf{k}}^3}. \quad (13)$$

A. Long wave collective modes for short-range intersite interaction

For $W_{\mathbf{q}} \to W_0 < \infty$, $\tilde{\Pi}$ has a pole for:

$$\omega = \Omega_{\mathbf{q}} = |\mathbf{q}|\sqrt{\frac{(U/2 + W_0)L + M_2}{M_1}} = qv \ .$$

The velocity, v, in the weak–U limit ($|t/U| \gg 1$, k^2–dispersion of the one-electron energy) is given by:

$$v^2 = \frac{v_F^2}{d} [1 + (U + 2W_0)\mathcal{N}(\varepsilon_F)] \ ,$$

d – dimensionality of the lattice (see Ref. [22]). The velocity in the strong–U limit ($|t/U| \ll 1$) is given by:

$$v^2 = \frac{8t^4 Z}{U^2} \left(1 + \frac{W|U|}{2t^2}\right) n(2 - n) \ ,$$

which is the result obtained using the effective Heisenberg model [8] (Bogoliubov sound mode for weakly interacting bosons). These results show an evolution of the mode from weak coupling limit to strong coupling bosonic limit. The velocity has been determined numerically for cubic lattices and the analysis confirms continuous evolution [19,23].

B. Collective modes for $q \to 0$ and long-range intersite interaction

In the 3D system with long range Coulomb interaction of the form $W_{\mathbf{q}} = 4\pi e^2/q^2$ the acoustic branch is transformed into a plasma branch and in the weak–U case it reads:

$$\Omega_{\mathbf{q} \to 0}^2 = \frac{4\pi \bar{e}^2 \bar{n}}{m^\star} \ , \qquad (14)$$

with $\bar{e} = e$, $m^\star = \hbar^2/2t$ for small concentration and $\bar{n} = n$. The same equation holds in the strong–U region as well where we have the system of local pairs with charge $\bar{e} = 2e$, density $\bar{n} = n/2$ and the effective mass $m^\star = \hbar^2/2J = \hbar^2|U|/4t^2$.

In the 2D system due to a different form of the Coulomb interaction $W_{\mathbf{q}} = 2\pi e^2/q$, one has an 'acoustic' plasmon branch with a square root dispersion:

$$\Omega_{\mathbf{q}}^2 = q\, 2\pi e^2 \Delta S_2/U \ , \tag{15}$$

which for the weak-U and small electron concentration reproduce a familiar result for 2D interacting electron gas:

$$\Omega_{\mathbf{q}}^2 = q\, 2\pi e^2 n/m^\star \ . \tag{16}$$

In the opposite limit of the large U we again rederive a result obtained with a help of effective pseudospin model:

$$\Omega_{\mathbf{q}}^2 = q\, 2\pi e^2 n(2-n)4t^2/|U| \ . \tag{17}$$

C. Collective modes for finite momentum

For general $\mathbf{q}$ the dispersion relations have been determined numerically for 2D square lattice. In the case of $W = 0$, collective modes along high symmetry lines of the Brillouin zone (BZ) and $n = 1$ are soft at $\mathbf{q} = (\pi, \pi) = \mathbf{M}$, what reflects degeneracy of the charge density wave (CDW) and the singlet superconductivity (SS) states. With increase of $|U|$, $\Omega_{\mathbf{q}}$ gradually approaches the strong coupling theory result:

$$\Omega_{\mathbf{q}} \; = \; \frac{2Zt^2}{|U|}\sqrt{(1-\gamma_{\mathbf{q}})[1-\gamma_{\mathbf{q}}(2n^2-4n+1)]} \, , \quad \gamma_{\mathbf{q}} \; = \; \frac{1}{Z}\sum_{\delta} e^{i\mathbf{q}\cdot\delta} \, , \tag{18}$$

(δ–elementary unit lattice vector). Away from half-filling there are also roton like minima near or at the zone boundary, depending on $|U/t|$.

Fig.1 shows the evolution of the excitation spectrum with increasing U in the absence of intersite Coulomb interaction [19,24]. For weak coupling there exists some critical value of q, $q_c \approx \frac{\pi\Delta}{\hbar v_F}$, for which the collective mode reaches the two–particle continuum, but for larger $q(> 2k_F)$ there is a splitting of the mode (Fig.1a). In this case the pair breaking excitations will dominate thermodynamics. With increasing $|U|$, in the crossover regime, there is a qualitative change in the spectrum and the collective mode mode splits from the continuum for all q (Fig.1b). In this crossover regime we expect mode coupling effects and low temperature thermodynamics will be determined by both pair breaking excitations and collective modes. Finally, for large $|U|$, the collective mode approaches the strong coupling theory result Eq.(18) in the pair Bose condensation regime, and it will drive superfluid transition (Fig.1c). We also notice that our results for the mode at $\mathbf{q} = (\pi, \pi)$ point agree with the exact ones thus confirming the $SO(4)$ symmetry of the Hubbard model [25]. The boundary between the BCS and Pair Bose Condensation regimes can be located within the HFA from the requirement that the chemical potential reaches the bottom of the bare fermionic band.

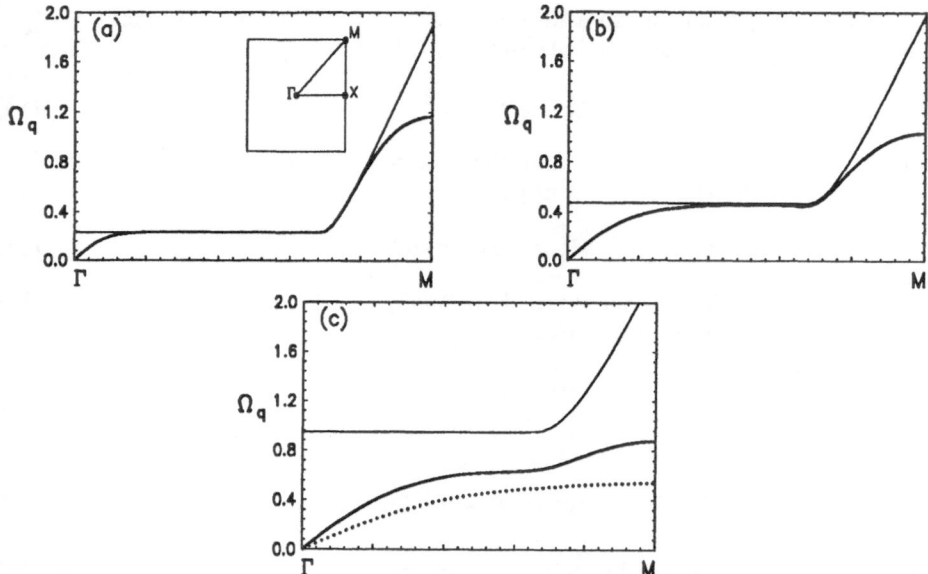

FIG. 1. Energy of the collective modes (Ω_q) as a function of the wave vector (q) along the direction ΓM for the 2D square lattice for $W_0 = 0, n = 0.4$ and four values of U. (a) $U = -1.2$, (b) $U = -1.5$, (c) $U = -2$ and $U = -4$ (dotted line). The inset shows the BZ for 2D square lattice. The boundary of the two–particle continuum is shown by a thin solid line. Here and in Fig.2, $2|t|$ is an energy unit.

For a critical value of W_q depending upon the $|U/t|$ ratio and the electron density n, the collective mode energy can soften completely for some critical wave vector q_{min}. This effect is the most clearly seen in the strong$-U$ limit where Ω_M goes to zero first for:

$$W_0^c = \frac{2Zt^2}{|U|} \frac{(1-n)^2}{1-(1-n)^2} .$$ (19)

Softening at $q_{min} = M$ means that the simple s-wave state becomes unstable with respect to formation of the commensurate CDW. In a resulting ground state the electron charge density alternates along lattice directions, and the superconducting order parameter is still non-zero [8]. The situation is however different in the small$-U$ limit where the critical value of the wave vector depends almost linearly on electron density. From Eq.(4) the W_0^c can be determined as an absolute minimum in the BZ of a function:

$$W_q^c = \frac{1}{\gamma_q} \left(\frac{1}{\Pi(q,0)} - \frac{U}{2} \right) .$$ (20)

We thus find an instability of singlet superconducting state toward a new super-conducting state accompanied by the incommensurate CDW for $W_0 > W_0^c$. For the *negative* W_0 the system is unstable toward a phase separation or electron 'droplet' formation. Our RPA analysis of the energy of Ω_q allows us to determine boundaries separating the simple s-wave superconducting phase and the other discussed phases in the whole range of n and $|t/U|$. The resulting phase diagram for the $NNSQ$ lattice with $W_q = \frac{1}{2}W_0[\cos(k_x) + \cos(k_y)]$ is shown in Fig.2.

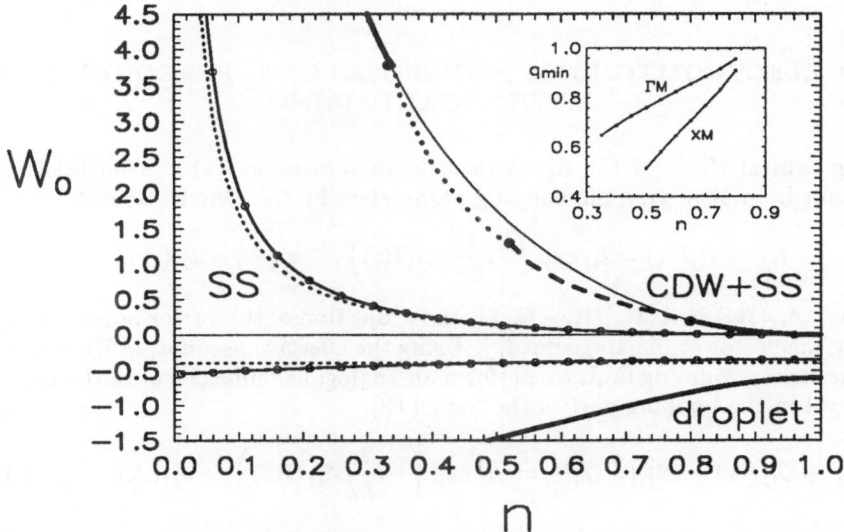

FIG. 2. Ground state phase diagram of the extended Hubbard model for the 2D SQ lattice. $U = -1$ (solid lines), $U = -5$ (empty circles). Tiny–dashed line shows the limit of the strong-coupling theory. The SS state is stable between a line of 'droplet' instability (for $W_0 < 0$) and a line of CDW instability (for $W_0 > 0$) for each U. Softening of Ω_q along the XM (ΓM) direction of the BZ takes place along the heavy dashed (heavy dotted) curve. In the rest of these curves Ω_q softens exactly at the M point of the BZ first for $W > 0$. On the thin solid line Ω_M goes to zero for $U = -1$. The inset shows the mimimal values of q (in units of π) vs n, for which the Ω_q softens along the XM and ΓM, directions of the BZ.

In conclusion:

- RPA provides a reasonable interpolation scheme for $T = 0K$ evolution from BCS to Bose superfluid cases and description of the crossover regime. Evolution is smooth for the ground state characteristics and collective modes. Collective modes evolve continuously from BCS to Local Pair limit for both short range (neutral) and long–range (charged) Coulomb interaction.

- A new prediction for the ground state of the EHM is a possibility of incommensurate CDW+SS state, which can be supersolid phase (with both DLRO i ODLRO present) or phase separation, in a weak coupling regime. We note, that such a phase with incommensurate CDW does not occur in the bosonic limit if only near-neighbour repulsion is considered.

- Due to the attraction–repulsion transformation [1] the presented results can be directly applied to the Hubbard model with $U > 0$.

- A case of quasi 2D lattice with long–range Coulomb repulsion shows quite interesting mode coupling effects, which can be relevant for superconducting superlattices [26,19].

III. ELECTROMAGNETIC PROPERTIES OF SUPERCONDUCTOR WITH LOCAL PAIRING

An orbital effect of the magnetic field in the model (1) can be taken into account by multiplying the hopping parameters by the Peierls factors:

$$t_{ij} \to t_{ij}(\mathbf{A}) = t_{ij} \exp\left(-i\frac{e}{\hbar c}A_{ij}(\mathbf{R}_i)\right), \quad t_{ji}(\mathbf{A}) = t_{ij}^*(\mathbf{A}), \tag{21}$$

where: $A_{ij}(\mathbf{R}_i) = \mathbf{A}(\mathbf{R}_i)(\mathbf{R}_i - \mathbf{R}_j)$ is the projection of the vector potential at $\mathbf{R}_i$ along the vector connecting sites i, j. Using the effective pseudospin Hamiltonian in the strong coupling limit we perform an analogous replacement of the transfer integral in the hopping part of the model (2):

$$J_{ij} \to J_{ij}(\mathbf{A}) = 2t_{ij}^2(\mathbf{A})/|U| = J_{ij} \exp\left(-i\frac{2e}{\hbar c}A_{ij}(\mathbf{R}_i)\right), \quad J_{ji}(\mathbf{A}) = J_{ij}^*(\mathbf{A}). \tag{22}$$

The current operator can be obtained as a functional derivative of the Hamiltonian with respect to the vector potential [27]: $j_{ij}(\mathbf{R}_i) = -c\delta\mathcal{H}/\delta A_{ij}(\mathbf{R}_i)$. For the hypercubic lattices with NN hopping with we have $j_{ij} = j_\alpha$ and $A_{ij} = A_\alpha$ with $\alpha = x, y, z$. In the case of the extended Hubbard model and the linear approximation with respect to $\mathbf{A}$ we get:

$$j_\alpha(\mathbf{R}_i) = \frac{e^2}{\hbar^2 c}A_\alpha(\mathbf{R}_i)t\sum_\sigma\left(c_{i\sigma}^\dagger c_{j\sigma} + c_{j\sigma}^\dagger c_{i\sigma}\right) + \frac{ie}{\hbar}t\sum_\sigma\left(c_{i\sigma}^\dagger c_{j\sigma} - c_{j\sigma}^\dagger c_{i\sigma}\right)$$

$$= j_\alpha^{dia}(\mathbf{R}_i) + j_\alpha^{para}(\mathbf{R}_i),$$

with $\mathbf{R}_i - \mathbf{R}_j$ along direction α. In the same way the current operator describing hopping of the composite bosons in the system modelled by the pseudospin effective Hamiltonian takes the form:

$$j_\alpha(\mathbf{R}_i) = \frac{\bar{e}^2}{\hbar^2 c}A_\alpha(\mathbf{R}_i)J\left(\rho_i^+\rho_j^- + \text{H.c.}\right) + \frac{i\bar{e}}{\hbar}J\left(\rho_i^+\rho_j^- - \text{H.c.}\right) = j_\alpha^{dia}(\mathbf{R}_i) + j_\alpha^{para}(\mathbf{R}_i).$$

Within the framework the linear response theory the current response to a vector potential $A_\alpha(\mathbf{R}_i, t)$ is given by the expectation value of the F. T. of the current-density operator $J_\alpha(\mathbf{q}, \omega)$:

$$J_\alpha(\mathbf{q}, \omega) = \frac{c}{4\pi}\sum_\beta\left[\delta_{\alpha\beta}K^{dia} + K_{\alpha\beta}^{para}(\mathbf{q}, \omega)\right]A_\beta(\mathbf{q}, \omega).$$

The diamagnetic part of the response kernel is determined by the expectation value of the hopping part of the Hamiltonian whereas the paramagnetic part reads:

$$K_{\alpha\beta}^{para}(\mathbf{q},\omega) = \frac{i}{N} \int_{-\infty}^{+\infty} dt e^{-i\omega t} \theta(t) \langle \left[j_\alpha^{para}(\mathbf{q},t), j_\beta^{para}(-\mathbf{q}) \right] \rangle . \qquad (23)$$

In above $j_\alpha^{para}(\mathbf{q},t)$ is the space–Fourier transform of the paramagnetic part of the current operator in the Heisenberg representation.

A. Meissner effect for the attractive Hubbard model

The Meissner kernel for the negative-U Hubbard model in the 2D square lattice was analyzed in Ref. [28] using the diagrammatic Nambu formalism. The kernel calculated with a help of RPA satisfies a 'lattice gauge invariance' condition:

$$\sum_\beta K_{\alpha\beta}(\mathbf{q})(e^{iq_\beta} - 1) = 0 \qquad (24)$$

which implies the usual gauge invariant form of the kernel matrix in the limit of continuous medium [20]: $K_{\alpha\beta}(\mathbf{q}) = K(\mathbf{q})(\delta_{\alpha\beta} - q_\alpha q_\beta/q^2)$. For $\mathbf{q} \to 0$ the kernel eigenvalue, $K(\mathbf{q})$, is given at $T = 0K$, by the diamagnetic part of the response:

$$K(\mathbf{q} \to 0) = K^{dia} = -\frac{1}{\lambda_L^2}, \quad \lambda_L^{-2} = 4\pi e^2(n_s/m^*c^2) = \frac{8\pi e^2}{\hbar^2 c^2 ZN} \sum_\mathbf{k} \frac{\varepsilon_\mathbf{k}\lambda_\mathbf{k}}{E_\mathbf{k}}, \quad (25)$$

where n_s is the superfluid number density, and λ_L stands for the London penetration depth. In the strong attraction limit ($|U| \gg t$) λ_L goes over to the result of order $t^2/|U|$ in agreement with a corresponding result for the composite bosons given in the next Section.

$K(\mathbf{q})$ shows significant anisotropy even in the small $\mathbf{q}$ limit

$$K(\mathbf{q}) = K^{dia} + q^2 S_{iso} + \frac{q_x^2 q_y^2}{q^2} S_{ani} + \mathcal{O}(q^4), \quad S_{iso} = \frac{4\pi e^2}{\hbar^2 c^2} \frac{(2t)^4 \Delta^2}{4N} \sum \frac{\sin^2 k_x \sin^2 k_y}{E_\mathbf{k}},$$

$$S_{ani} = -6S_{iso} + \frac{4\pi e^2}{\hbar^2 c^2} \frac{(2t)^4 \Delta^2}{2N} \sum \frac{\sin^4 k_x}{E_\mathbf{k}}.$$

In the continuous medium with $\sin k_\alpha \sim k_\alpha$, the anisotropic part vanishes and we return to the usual form of the kernel eigenvalue. The ratio: $r_0 = -S_{iso}/K^{dia}$ can be treated then as a useful estimation of the electromagnetic correlation length, by comparison with an analogous expansion for the Pippard's kernel. In the weak U limit we rederive in this way a result of the BCS theory, $r_0 \sim \hbar v_F/\pi\Delta$. For strong U, $r_0 \sim 2t/|U|$ and may become less than lattice spacing which shows that the RPA-HFA calculation becomes less reliable for the description of space dependence of the kernel in this limit.

In the weak U region the kernel q-dependence is very pronounced, which is characteristic for the behaviour of the Pippard superconductor. In the strong U region the dispersion of the kernel is small which is consistent with a strongly local electromagnetic properties of the system in this limit. Thus, the evolution from a Pippard to London type of behaviour is smooth. The numerical investigation of the wavelength dependence of the Meissner kernel troughout the whole Brillouin zone of the 2D lattice shows a gradual evolution of the kernel with a change of U (Fig.3).

In Fig.3 we also compare the RPA result with a one obtained from the HFA calculation after projecting out a spurious longitudinal part present in the HFA kernel. The comparison shows that the collective effects included in RPA vanish along the high symmetry directions of the BZ. For general q they are relatively small in the weak U limit but become relatively more significant with increase of coupling strength. We conclude that the collective effects, being less important for the calculation of the London penetration depth for $|U| \gg Zt$, may be substantial in evaluation of the spatial dependence of the kernel in this limit.

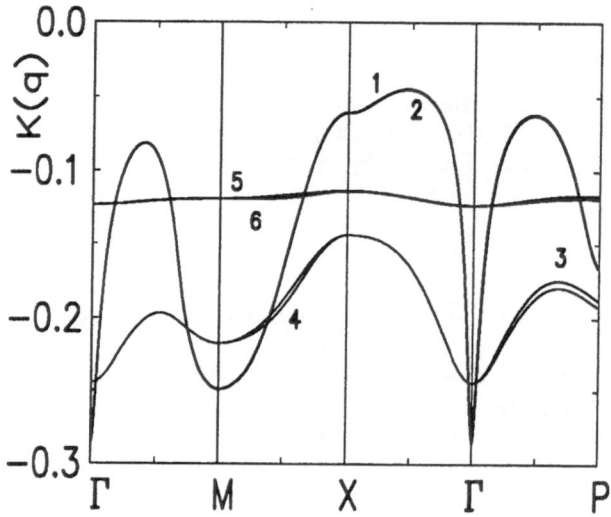

FIG. 3. Eigenvalue of the kernel matrix (in units of $8\pi|t|e^2/\hbar^2c^2$) in the 2D BZ for the electron concentration $n=0.4$. $P=(\pi/2,\pi)$. Curves 1, 3 and 5 are for the *corrected* HFA kernel, and curves 2, 4 and 6 are for the RPA kernel. $|U/(2t)|$ is 1 for curves 1 and 2, 2 for curves 3 and 4, and 5 for curves 5 and 6. Within the resolution of plotting, curves 1 and 2, and curves 5 and 6 are indistinguishable.

B. Hard–core charged bosons on a lattice

Let us now consider the system of local electron pairs on a lattice in the external magnetic field. The diamagnetic part of the kernel is given by:

$$K^{dia} = \frac{4\pi}{\hbar^2c^2}\frac{2\bar{e}^2}{ZN}\left\langle -\sum_{\mathbf{k}} J_{\mathbf{k}}\rho_{\mathbf{k}}^+\rho_{\mathbf{k}}^- \right\rangle , \qquad (26)$$

where: $\rho_{\mathbf{k}}^+$, $\rho_{\mathbf{k}}^-$ and $J_{\mathbf{k}}$ denote the space–Fourier transform of the charge operators and the transfer integral, respectively. The paramagnetic part is given by the time–Fourier transform of the retarded current–current Green function (Eq.(23)) with $j_\alpha^{para}(\mathbf{q}, t)$ being the space–Fourier transform of the paramagnetic part of the bosonic current operator in the Heisenberg representation. In evaluation of quantities given by Eqs.(26,23) we have employed the self–consistent RPA method [29,31]. We will concentrate on the homogenous SS phase characterized by the order parameter $\langle \rho^\pm \rangle = N^{-1} \sum_l \langle \rho_l^\pm \rangle \neq 0$. The collective excitation energy spectrum has the form:

$$E_{\mathbf{k}} = R\sqrt{(\varepsilon_{\mathbf{k}}^0)^2 + V_{\mathbf{k}} \varepsilon_{\mathbf{k}}^0 \sin^2 \theta}. \tag{27}$$

$K_{\mathbf{q}}$ is the Fourier transform of K_{jm}, $V_{\mathbf{k}} = J_{\mathbf{k}} + K_{\mathbf{k}}$, $\cos^2 \theta = (n-1)^2/R^2 = 4\langle \rho^z \rangle^2/R^2$, $R \equiv 2\sqrt{\langle \rho^x \rangle^2 + \langle \rho^z \rangle^2}$. The length of pseudospin R is given as a solution of the equation:

$$1 = \frac{R}{N} \sum_{\mathbf{k}} \frac{A_{\mathbf{k}}}{E_{\mathbf{k}}} \coth\left(\frac{E_{\mathbf{k}}}{2k_B T}\right), \tag{28}$$

where: $A_{\mathbf{k}} = R\varepsilon_{\mathbf{k}}^0 + B_{\mathbf{k}}$, $\varepsilon_{\mathbf{k}}^0 = J_0 - J_{\mathbf{k}}$, $B_{\mathbf{k}} = \frac{1}{2}RV_{\mathbf{k}} \sin^2 \theta$. In the case of near-neighbor interactions and in the long-wave limit $E_{\mathbf{k}} = \hbar s|\mathbf{k}|$, and $s = 2\langle \rho^x \rangle J \sqrt{1 + K/J} \sqrt{Z}/\hbar$ is the sound velocity. The diamagnetic part of the kernel calculated in RPA yields the following result:

$$K^{dia} = \frac{-8\pi \bar{e}^2}{Z\hbar^2 c^2} \left[\sin^2 \theta \frac{R^2 J_0}{4} + \frac{R}{4N} \sum_{\mathbf{k}} J_{\mathbf{k}} \frac{B_{\mathbf{k}} \sin^2 \theta + (1 + \cos^2 \theta)A_{\mathbf{k}}}{E_{\mathbf{k}}} \coth\left(\frac{\beta E_{\mathbf{k}}}{2}\right) \right] \tag{29}$$

In evaluation of the paramagnetic part of the kernel we used the corresponding current–current Matsubara Green's function and symmetric RPA-like factorization of higher order Green's functions. In the continuous medium approximation, upon neglecting the quantum corrections to the diamagnetic part, the total kernel, in the static limit can be written as

$$K_{\alpha\alpha'}(\mathbf{q}, 0) = 4\pi \frac{\bar{e}^2 n_0}{m^* c^2} \left(\frac{q_\alpha q_{\alpha'}}{q^2} - \delta_{\alpha\alpha'}\right) + K_{\alpha\alpha'}^{para,nc}(\mathbf{q}, 0), \tag{30}$$

with the first part again being explicitly gauge invariant and the second part representing the noncondensate contribution.

It is also of interest to point out that at $T = 0K$, $n_0 = \langle \rho^x \rangle^2 \approx \frac{1}{4}n(2-n)$, and the first part of Eq.(30) is the large $|U|$ limit of the gauge invariant RPA–Hartree Fock theory of the response kernel for the negative U Hubbard model [28].

In general case in the limit: $\mathbf{q} \to 0$ we obtain:

$$K_{\alpha\alpha}^{para}(0, 0) = -4\pi \left(\frac{2\bar{e}JR \cos \theta}{\hbar c}\right)^2 \frac{1}{N} \sum_{\mathbf{k}} \sin^2 k_\alpha \frac{\partial f(E_{\mathbf{k}})}{\partial E_{\mathbf{k}}}, \tag{31}$$

where $f(x) = 1/(\exp(\beta x) - 1)$ is the Bose distribution function. In the low density limit Eqs.(29,31) can be evaluated in continuous medium approximation, which yields the following simple form for the total kernel:

$$K_{\alpha\alpha} = -\frac{4\pi\bar{e}^2\bar{n}}{m^\star c^2}\left[1 - \frac{\hbar^2}{\bar{n}m^\star}\int\frac{d^dp}{(2\pi)^d}p_\alpha^2\left(-\frac{\partial f(E_{\mathbf{p}})}{\partial E_{\mathbf{p}}}\right)\right] = -\frac{4\pi\bar{e}^2 n_s(T)}{m^\star c^2}, \qquad (32)$$

(where $E_{\mathbf{q}}$ is given by: $E_{\mathbf{q}} = \sqrt{(\varepsilon_{\mathbf{q}}^0)^2 + 4V_{\mathbf{q}}\varepsilon_{\mathbf{q}}^0 n_0}$, $\varepsilon_{\mathbf{q}}^0 = \frac{\hbar^2 q^2}{2m^\star}$ and $n_0 = \langle\rho^x\rangle^2 \approx$ $\frac{1}{4}\sin^2\theta$ is the condensate density being equal to $\approx n/2$ at $T = 0K$) which is in agreement with the result of Fetter [32]. The 'superfluid number density' $n_s(T)$ is given by

$$\frac{n_s(T)}{\bar{n}} = 1 - \frac{\rho_n(T)}{\rho} = 1 - \frac{\hbar^2}{\bar{n}m^\star d}\int\frac{d^dp}{(2\pi)^d}p^2\left(-\frac{\partial f(E_{\mathbf{p}})}{\partial E_{\mathbf{p}}}\right), \qquad (33)$$

with $\rho_n = m^\star\bar{n}$ being the normal fluid density.

Eqs.(32,33) provide microscopic justification for the Landau two-fluid model. The corresponding current induced by the static transverse vector potential $\mathbf{A}(\mathbf{q})$ (in the transverse gauge $\mathbf{q}\cdot\mathbf{A}(\mathbf{q}) = 0$) satisfies the usual London equation:

$$\mathbf{J}(\mathbf{q}) = -\frac{\bar{e}^2 n_s(T)}{m^\star c}\mathbf{A}(\mathbf{q}), \quad \mathbf{q}\to 0. \qquad (34)$$

In particular, an applied magnetic field will be shielded with the London penetration depth:

$$\lambda_L^{-2}(T) = \frac{4\pi\bar{e}^2 n_s(T)}{m^\star c^2}. \qquad (35)$$

The theory of electromagnetic response presented in this Section is applicable to the systems with arbitrary form of the density–density interaction K_{ij}. In relation to real materials the two cases are most relevant: i) short–ranged K_{ij}, corresponding to a screened Coulomb interaction (such a screening can be due to the presence of the other electronic subsystem of wide band electrons) ii) unscreened long–range Coulomb repulsion ($K(\mathbf{r}) \sim 1/|\mathbf{r}|$, $|\mathbf{r}| \to \infty$). A detailed analysis of the excitation spectra, thermodynamic and electromagnetic properties of these cases has been given elsewhere [29,30,33]. The analysis of the spatial dependence of the response kernel indicates that the coherence length ξ is indeed of order of the lattice constant a multiplied by a function of boson concentration. For short–range repulsion one can go beyond RPA method and in the low density limit a consistent description of superfluid characteristics was obtained, based on the knowledge of the exact scattering amplitude [33].

The results obtained for hard core bosons have been also applied to some recent experiments on HTS. In particular the temperature dependence of the penetration depth, and the universal plots of T_c versus $\lambda_L^{-2}(0)$ first reported by Uemura et al [2] for various families of superconducting cuprates, were analyzed [30]. The theory predicts an almost universal $\bar{T}_c$ versus $\bar{\lambda}^{-2}(0)$ behaviour (being only a very weakly dependent on K/J) in the underdoped regime (low concentration) and possible deviations from the universality for the systems in the overdoped regime [30].

In Table I we have summarized superfluid characteristics of the model for the short–range intersite interaction and low temperature characteristics of the model for the case of long–range intersite interaction.

Table 1. Superfluid characteristics of the hard–core boson model.

Quantity	Short–range repulsion				Long–range repulsion
	$0 < T < T_0$		$T_0 < T < T_c$ *		$0 < T \ll T_c$
	$d = 3$	$d = 2 + \varepsilon$	$d = 3$	$d = 2 + \varepsilon$	$d = 3, n \ll 1$
$\langle \rho^x(0) \rangle - \langle \rho^x(T) \rangle$	T^2	T	$T^{3/2}$	T	$T^{1/2} \exp(-\bar{\Delta}/k_B T)$
$E(T) - E(0)$	T^4	T^3	$T^{5/2}$	T^2	$T^{3/2} \exp(-\bar{\Delta}/k_B T)$
$C(T)$	T^3	T^2	$T^{3/2}$	T	$T^{-1/2} \exp(-\bar{\Delta}/k_B T)$
$\rho_s - \rho_s(T) \sim$ $1 - (\lambda_L(0)/\lambda_L(T))^2$	T^4	T^3	$T^{3/2}$	T	$T^{3/2} \exp(-\bar{\Delta}/k_B T)$

* only for low concentration and beyond the critical regime

For a system with short range repulsion, the temperature dependences are power law, e.g. in $3D$ $\langle \rho^x(0) - \rho^x(T) \rangle \sim T^2$ while $\rho_n/\rho = (\rho_s - \rho_s(T))/\rho \sim T^4$. However, situation is quite different for long–range, unscreened repulsion, where collective mode is pushed to a plasma frequency Ω_0^*, $(\Omega_0^*)^2 = 4\pi\bar{e}^2 \langle \rho^x(0) \rangle^2/m^*$, and the leading exponential temperature dependence of ρ_n/ρ and $\langle \rho^x(0) - \rho^x(T) \rangle$ occur. In general, these exponential T dependences of the thermodynamic characteristics will occur in a restricted temperature range. With increasing T, a crossover to power–law characteristics, $c_v(T) \sim T^{d/2}$, $\langle \rho^x - \rho^x(T) \rangle \sim \rho_s - \rho_s(T) \sim T^{d/2}$, can take place at higher T, if $k_B T_c > k_B T > \bar{\Delta} = \min_{\mathbf{k}} E_{\mathbf{k}} \leq \hbar\Omega_0^*$, with $E_{\mathbf{k}}$ given by Eq.(27).

We notice that the preexponential factors found for our model and quoted in Table 1 are different from the corresponding prefactors found for a standard $3D$ charged Bose gas in continuum [32]. It is due to the hard–core effects in the lattice boson model [33]. We should also add that in the presence of a long-range Coulomb interaction, Wigner crystallization can compete with the superconducting state especially in the low density limit of the model [34].

IV. NORMAL STATE PROPERTIES OF THE ATTRACTIVE HUBBARD MODEL FOR $T \neq 0$K

Recently, there have been several studies of the attractive Hubbard model going beyond the Hartree-Fock theory(see [35] and Refs.therein). In Ref. [35] the excitation spectrum and normal state properties of the 2D attractive Hubbard model have been studied, using the conserving, self-consistent T-matrix formalism, in the intermediate coupling regime and for low electron concentration. In particular, the competition between the bound state formation in 2D and the Fermi sea was examined.

The T-matrix for the Hubbard model is given by the sum of particle-particle ladder diagrams, in the perturbation expansion in terms of U:

$$T(\mathbf{q}, i\nu_m) = \frac{U}{1 + U\chi(\mathbf{q}, i\nu_m)}, \tag{36}$$

where $\chi(\mathbf{q}, i\nu_m)$ is the independent pairing susceptibility:

$$\chi(\mathbf{q}, i\nu_m) = \frac{1}{\beta N} \sum_{\mathbf{k}, \omega_n} G(\mathbf{k}, i\omega_n) G(\mathbf{q} - \mathbf{k}, i\nu_m - i\omega_n), \tag{37}$$

$\omega_n = 2\pi(n + 1/2)/\beta$ and $\nu_m = 2\pi m/\beta$, $n, m = 0, \pm 1, \pm 2, ...$, are the Fermionic (odd) and Bosonic (even) Matsubara frequencies, respectively. The one-particle Green's function obeys the Dyson equation $G^{-1}(\mathbf{k}, i\omega_n) = i\omega_n - \varepsilon(\mathbf{k}) + \mu - \Sigma(\mathbf{k}, i\omega_n)$, with the self energy given by

$$\Sigma(\mathbf{k}, i\omega_n) = \frac{1}{\beta N} \sum_{\mathbf{q}, \nu_m} T(\mathbf{q}, i\nu_m) G(\mathbf{q} - \mathbf{k}, i\nu_m - i\omega_n), \tag{38}$$

which closes the system of self–consistent equations. The results for real frequencies are obtained upon the analytic continuation. The above system of equations (together with the number equation determining μ), valid in the normal state, is conserving [37] and known to include Gaussian fluctuations of the pairing field in a selfconsistent manner. This approximation is supposed to be accurate in the low density limit. The superconducting instability of the normal state at T_c can be signalized by the divergence of the T-matrix at zero energy and zero momentum (the Thouless criterion): $1 + U\chi(0,0) = 0$. Let us point out that the above T-matrix scheme is more general than special cases treated in Refs. [9,12,38,40].

A. Numerical Results

The 2D attractive Hubbard model is believed to undergo a Kosterlitz–Thouless phase transition away from the half-filling [41,42]. In general there are two relevant temperature scales in the problem, the HFA and the Kosterlitz-Thouless (KT) transition temperatures. The KT transition temperature can be estimated from the universal jump in the superfluid density [36] or from Quantum Monte Carlo (QMC) studies [42]. This allows us to identify the regime where the T-matrix formalism without the ODLRO is expected to work. Two other energy scales are provided by the Fermi energy E_F and the binding energy E_b for a pair of electrons in the empty lattice [1]. One gets for $U/t = -4.0$, which is the case studied below, $E_b/t = 0.144$. The binding energy also provides a crude estimate of T_p, the pair formation temperature, yielding $k_B T_p/t = 0.144$, which is in between the estimates of HFA(0.36) and KT(0.11) transition temperatures [36]. We notice that T_c and T_p can be different in contrast to the conventional BCS picture. For small $|U|$, T_p is expected to the same as the HF transition temperature, and the regime $T_c < T < T_{HF}$, can be described in terms of the superconducting fluctuations. For large $|U|$, however, this two scales are widely separated, $T_c \sim t^2/|U|$ while $T_p \sim |U|$. In such a case well defined pairs exist above T_c and they will drive the superfluid transition [1].

If the binding energy is much larger that the Fermi energy, one expects existence of a state of stable tightly bound pairs with a negligible fraction of dissociated fermions. In the opposite case ($E_b \ll E_F$), however, due to interaction of bound pairs with medium, it is not clear if well defined bound pairs can exist for a time scale exceeding the quasiparticle life time.

The numerical results presented below have been obtained by the solution of the self–consistent T-matrix equations by the Fast Fourier Transform technique for a square lattice of 32×32 lattice sites and 1024 Matsubara frequencies subjected to periodic boundary conditions. The cases of $|U/t| = 2, 3.5, 4, 4.5$ and $n/2 = 0.1$ were studied [35]. (For these values of $|U/t|$ the binding energy is less than the Fermi energy.) The one particle and two particle excitation spectra, the scattering phase shift, the one-particle momentum distribution, the chemical potential, the density of states and the static spin susceptibility were determined [35,45].

For small $|U/t|$ ($|U/t| \leq 2$) and sufficiently large densities the two–particle states, determined from the poles of the T–matrix, are located practically in the continuum. Their life–time is short and for most of the q values, they do not separate from the continuum of the two-particle excitations. These states resemble the large q bound states found in the study of the continuum Fermi gas with s-wave attraction [12]. We note that in the 2D case these states always occur for $q > 2k_F$, since any point–like attraction will produce bound states, which is not the case in 3D. However, in contrast to the non–selfconsistent calculations, where these states always split from the continuum for $q > 2k_F$, fully selfconsistent T–matrix calculations reveal that the bound states are not well separated from the two-particle continuum [35,39].

The effect of bound states in the small $|U/t|$ case is rather weak as far as the one–particle properties are concerned. The low frequency behavior is a Fermi liquid type with $Im\Sigma(k_F, \omega) \sim \omega^2$(apart from a logarithmic correction) [43].

With the increase of $|U|$ the bound states begin to separate from the continuum and they can split off for large q. Effects of these bound states (fluctuating pairs) are quite remarkable for $U/t = -4$. They produce a satellite band in $A(k, \omega)$, because they are almost separated from the continuum for $q > \pi/2$ and their lifetime is larger. In this intermediate coupling we have a sort of mixture of fermionic quasiparticles and fluctuating pairs with the chemical potential inside the quasiparticle band.

In Fig.4 we show the one-electron spectral function, $A(k, \omega) = -ImG(k, \omega + i0^+)/\pi$, along the diagonal of the BZ for $U/t=-4$, $n/2 = 0.1$, and $T/t = 0.17$, a temperature which is above the critical one. The spectral function shows a main peak and, depending on k, a secondary maximum or a shoulder. The main peak follows a dispersion which is similar to the tight-binding band structure of the non-interacting system having a bandwidth of $8t$. This main "quasi particle" band crosses a secondary band which is an effect of the attractive electron–electron interaction. This secondary band has a much smaller weight and less dispersion, its width being of the order of t, rather than $8t$.

The evaluation of the momentum distribution function, $n(k)$, shows reduction for the small k values and a tail for large momenta, which is a signature of correlated system. Within the T-matrix scheme the static spin susceptibility, $\chi_s(T)$ has also been evaluated and the results are given in Fig.5, together with $\chi_s(T)$ calculated from the RPA, and the moments of spectral density based on the two pole ansatz for $A(k, \omega)$. The pairing correlations have significant effects on $\chi_s(T)$ yielding $\partial\chi_s/\partial T > 0$ in low T regime. T-matrix and moment approaches, both including pairing correlations exhibit a temperature bending. Similar behavior has been observed in recent QMC calculations at higher densities [44]. RPA and HFA, which do not include such effects, show monotonic temperature behavior.

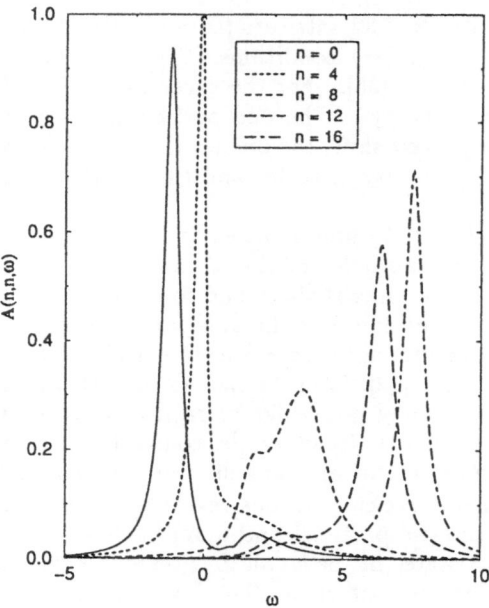

FIG. 4. One–particle spectral function, $A(n, n, \omega)$ vs ω for $n/2 = 0.1$, $U/t = -4$ and $T/t = 0.17$. The momenta are given in units of $\pi/16$. After Ref. [35]

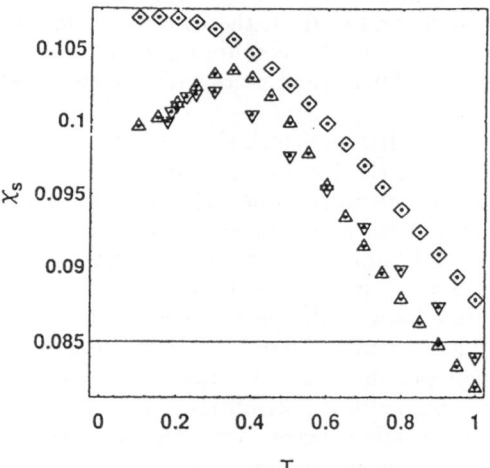

FIG. 5. Static spin susceptibility, $\chi_s(T)$ vs T, for $U = -4.0$ and $n/2 = 0.1$ in the T–matrix approximation. RPA–calculation ($\Diamond$); moment approach ($\triangle$). (In units of $|t|$). After Ref. [35]

Finally, from the picture obtained in the regime of intermediate $|U/t|$ values, we judge that increase in $|U|$ will cause the pair states to get closer to the chemical potential, and then they will move below the bottom of the quasiparticle band.

For large $|U|$ one will approach the situation of the two Hubbard bands with the chemical potential pinned around $-|U|/2$. In this limit, the low energy physics $(k_B T \ll |U|)$ can be described by interacting bound pairs on a lattice [1].

B. Discussion

The T–matrix studies show the occurrence of two types of excitations in the one-particle spectral function. One peak can be associated with pair formation while the other corresponds to renormalized quasiparticle excitation. The presence of quasibosonic pair states affects the self–energy (Eq.(38)) and this produces the secondary band in the one-particle spectral function. It turns out that the two band feature is reasonably well described by a two–pole ansatz for $A(\mathbf{k}, \omega)$, which satisfies the first four frequency moments [35].

For the electron concentration and the coupling strength $|U/t|$ considered in Ref [35], the chemical potential is located in the lower branch of the spectrum, which implies the existence of a Fermi surface. Thus, we expect below T_c the opening of a gap signaling the occurrence of superconductivity [45].

The existence of pair states, despite of a rather well defined Fermi surface, affects the response functions, like the magnetic susceptibility. This explains the behaviour of static spin susceptibility, which has been observed both experimentally and in QMC studies [44] of the attractive Hubbard model. In particular, we notice that $\chi_s(T)$ has a Curie type behavior for high temperatures but it does not go over to the Pauli susceptibility at low temperatures. Instead a bending of $\chi_s(T)$ at a particular temperature T^* occurs, in agreement with the moment approach. This temperature bending is due to the singlet pairing correlations and it is not found in the HFA and RPA calculations. Similar features have been observed in recent QMC calculations of the static and dynamic spin susceptibility at higher densities. [44]

Other effect of pairing correlations manifests in occurrence of a pseudogap around μ above T_c, obtained with T–matrix approach [45] and by QMC [46,47] for higher densities.

V. SUMMARY

In this paper we have analyzed thermodynamic and electromagnetic properties of the simple systems with short–range pairing interaction, focussing on the crossover problem from BCS to local electron pair limit, where the system is equivalent to that of hard-core charged bosons on a lattice.

Let us now briefly summarize main physical ideas regarding the crossover from cooperative Cooper Pairing(BCS) to Bose condensation (see also Refs[1,4,5]).

1. There is only one phase transition at T_c, as long as no other broken symmetry phases intervene or the system does not phase separate. The transition is from normal state to superfluid state with ODLRO (or algebraic order in 2D).

2. The nature of phase transition is in the both limits quite different. In the BCS limit a formation of Cooper Pairs and condensation at T_c takes place simultaneously and $T_c \sim \exp(-1/V)$. A first deviation from this scenario can be described in terms of superconducting fluctuations.

   In the preformed pair (BC) regime, however the pair formation (T_p) and their condensation (T_c) are independent processes. T_p and T_c are widely separated and T_p is a characteristic energy scale not a phase transition temperature. T_c will decrease with the increase of coupling constant V. For $T > T_p$ the local pairs are thermally dissociated.

3. In the weak coupling limit, below T_c, we have BCS condensate of large number overlapping Cooper pairs $(\xi \gg a)$. Thermodynamics and T_c are determined by single particle excitations (broken Cooper pairs) with exponentially small gap. In the opposite, strong coupling regime one has the Bose condensate of tightly bound local pairs $(\xi \sim a)$, and thermodynamics and T_c are governed by the collective modes.

   The collective modes evolve smoothly between the two regimes. In the BCS we have the Anderson–Bogoliubov modes for neutral case and plasmons for charged case, respectively. In the BC limit there are either sound wave Bogoliubov modes for screened Coulomb repulsion or plasma modes for charged boson superfluid. As concerns the electromagnetic properties, we demonstrated the smooth evolution of the Meissner kernel from a Pippard type to London type behavior at $T = 0K$. Also, λ_L evolves smoothly from the BCS to BC regime. In the BC case: $\kappa = \lambda_L/\xi \gg 1$, superconductivity is clearly of extreme type II, H_{c2} is very large, $H_{c2}(T)$ can exhibit a positive curvature vs T and $H_{c1}/H_{c2} \ll 1$.

4. For the BCS superconductor the normal state is a Fermi liquid. In the BC regime one has for $T_c < T < T_p$ a normal Bose liquid (of bound pairs). The evolution of normal state from Fermi liquid to Bose liquid is quite unusual, especially in 2D systems. From recent studies, it appears that even for moderate attraction (degenerate) Fermi liquid regime shows anomalous features due to pairing correlations such as pseudogap, anomalous behaviour of spin susceptibility and a spin gap in the normal state. It is of interest that these anomalies (superfluid precursor effects) are similar to experimentally observed in NMR and optical conductivity on underdoped high T_c cuprates. From these studies it follows that deviations from conventional Fermi liquid behaviour are generic to the normal state of short–coherence length superconductors.

Finally, let us stress the importance of a mixed model of coexisting bosons and fermions [1,48,49]. This model can be naturally considered as an extension of the extreme bosonic limit and a candidate to describe an intermediate coupling crossover regime. Such a generalization can be also of interest for many non-standard superconductors.

   Regarding open problems we point out the question of crossover for anisotropic pairings of s or d–wave symmetry in systems of reduced dimensionality. Enhanced thermodynamic fluctuations and short coherence length effects are clearly challenging issues for successful theory of the cuprates.

ACKNOWLEDGMENTS

R.M. would like to thank T. Schneider, M. H. Pedersen, J.J. Rodríguez-Núñez, S. Schafroth and H. Beck for invaluable input, the results presented in Section IV were obtained in collaboration with them. R.M. and T.K. thank S. Robaszkiewicz for discussions and comments. R.M. acknowledges hospitality of the Institute for Scientific Interchange, Turin, Italy, where part of this work was discussed. This work has financial support from the Committee of Scientific Research (K.B.N. Poland, project numbers 2 P03 B 057 09 and 2 P302 038 07).

---

[1] R. Micnas, J. Ranninger and S. Robaszkiewicz, Rev. Mod. Phys. **62**, 113 (1990).

[2] Y.J. Uemura et al., Phys. Rev. Lett. **66**, 2665 (1991); Y. J. Uemura et al., Nature **352**, 605 (1991).

[3] T. Schneider and H. Keller, Phys. Rev. Lett. **69**, 3374 (1992); T. Schneider and H. Keller, Int. J. Mod. Phys.**B 8**, 487 (1994).

[4] R. Micnas and S. Robaszkiewicz, in *"Ordering Phenomena in Condensed Matter Physics"* (Eds. Z. M. Galasiewicz and A. Pękalski, World Scientific, 1991), p. 127.

[5] M. Randeria, in *"Bose-Einstein Condensation"* edited by A. Griffin, D. Snoke and S. Stringari (Cambridge University Press, 1994).

[6] D. M. Eagles, Phys. Rev. **186**, 456 (1969).

[7] A. J. Leggett, in "Modern Trends in the Theory of Condensed Matter", edited by A. Pękalski and J. Przystawa (Springer Verlag, Berlin 1980).

[8] S. Robaszkiewicz, R. Micnas, and K.A. Chao, Phys. Rev. **B 23**, 1447 (1981); ibid.**B 24**, 1579 (1981); **B 24**, 4018 (1981); **B 26**, 3915 (1982).

[9] P. Nozieres and S. Schmitt-Rink, J.Low Temp. Phys.,**59**, 195 (1985).

[10] K. Miyake, Prog. Theor. Phys. **69**, 195 (1983).

[11] M. Randeria, J.-M. Duan and L.-Y. Shieh, Phys. Rev. Lett. **62**, 981 (1989); Phys. Rev. **B 41**, 327 (1990).

[12] S. Schmitt-Rink, C.M. Varma and A. E. Ruckenstein, Phys. Rev. Lett. **63**, 445 (1989).

[13] R. Micnas, J. Ranninger, S. Robaszkiewicz and S.Tabor, Phys. Rev. **B37**, 9410 (1988); R. Micnas, J. Ranninger and S. Robaszkiewicz Phys. Rev. **B39**, 11653 (1989).

[14] N.F. Mott, Adv. Phys. **39** 55 (1990); A.S. Alexandrov and N.F. Mott, Rep. Prog. Phys., **57**, 1197 (1994); L.J. de Jongh, Physica **C 152**, 171 (1988); ibid. **C 161**, 631 (1989).

[15] A.R. Bishop, P.S. Lomdahl, J.R. Schrieffer, and S.A. Trugman, Phys. Rev. Lett. **61**, 2709 (1988).

[16] C. M. Varma, Phys. Rev. Lett. **61**, 2713 (1988).

[17] S. Chakravarty and S. Kivelson, Europhys. Lett. **16**, 751 (1991).

[18] F. C. Zhang, M. Ogata and T.M. Rice, Phys. Rev. Lett. **67**, 3452 (1991); J.A. Wilson, Physica **C 182** 1, (1991).

[19] T. Kostyrko and R. Micnas, Phys. Rev. **B 46**, 11025 (1992) and Acta Physica Polonica **A 83**, 837 (1993); T. Kostyrko, Phys. Stat. Sol.(b) **143**, 149 (1987).

[20] J.R. Schrieffer, *"Theory of Superconductivity"*, (Benjamin, New York 1964).

242

[21] The response functions can also be obtained by the functional differentiation technique of Kadanoff and Baym, see: R. Côte and A. Griffin, Phys.Rev. **B 48**, 10404 (1993)

[22] P.W. Anderson, Phys. Rev. **112**, 1900 (1958).

[23] L. Belkhir and M. Randeria, Phys. Rev.**B 45**, 5087 (1992); K. Nasu, Physica B **143**, 229 (1986).

[24] J.O. Sofo, C.A. Balseiro and H.E. Castillo, Phys. Rev. **B 45**, 9860 (1992); D. van der Marel, Phys. Rev. **B 51**, 1147 (1995).

[25] S. Zhang, Phys. Rev. Lett. **65**, 120 (1990).

[26] H.A. Fertig and S. Das Sarma, Phys. Rev. Lett. **65**, 1482 (1990); Phys. Rev. **B 44**, 4480 (1991).

[27] D.J. Scalapino, S. R. White and S. C. Zhang, Phys. Rev. Lett. **68** 2830 (1992).

[28] T. Kostyrko, R. Micnas and K.A. Chao, Phys. Rev. **B 49**, 6158 (1994).

[29] R. Micnas and S. Robaszkiewicz, Phys. Rev. **B 45**, 9900 (1992).

[30] S. Robaszkiewicz, R. Micnas and T. Kostyrko, in *"Polarons and Bipolarons in High T_c Superconductors and Related Materials"*, (W.Y. Liang, E.S. Salje and A.S. Alexandrov, eds., Cambridge University Press 1995), p. 244.

[31] R.T. Whitlock and P.R. Zilsel, Phys. Rev. **131**, 2409 (1963).

[32] A.L. Fetter, Ann. Phys.(N.Y.) **60**, 464 (1970).

[33] R. Micnas, S. Robaszkiewicz and T. Kostyrko, Phys. Rev. **B 52**, 6863 (1995).

[34] B.J. Alder amd D.S. Peters, Europhys. Lett. **10**,1 (1989); R. Friedberg, T.D. Lee and H.C. Ren, Ann. Phys.(N.Y.) **208**, 149 (1991).

[35] R. Micnas, M.H. Pedersen, S. Schafroth, T. Schneider,J.J. Rodríguez-Núñez and H. Beck, Phys. Rev.**B 52**, 16223 (1995).

[36] P.J.H. Denteneer, Guozhong An and J.M.J. van Leeuwen, Europhys. Lett. **16**, 5 (1991).

[37] G. Baym, Phys.Rev. **127**, 1391 (1962).

[38] J.W. Serene, Phys.Rev. **B 40**, 10 873 (1989).

[39] R. Frésard, B. Glasser and P. Wölfle, J. Phys.: Condens. Matter **4**, 8565 (1992).

[40] For a critical discussion of the nonconserving T-matrix approach see also: J.O. Sofo and C.A.Balseiro, Phys. Rev. **B 45**, 8197 (1992); A. Tokumitu, K. Miyake and K. Yamada, Phys. Rev. **B 47**, 11 988 (1993).

[41] R.T. Scalettar, E. Loh, J. Gubernatis, A. Moreo, S. White, D. Scalapino, R. Sugar and E. Dagotto, Phys. Rev. Lett. **62**, 1407 (1989).

[42] A. Moreo and D.J. Scalapino, Phys. Rev. Lett. **66**, 946 (1991).

[43] C. Hodges, H. Smith and J.W. Wilkins, Phys. Rev. **B 4**, 302 (1971); P. Bloom, Phys. Rev. **B 12**, 125 (1975).

[44] M. Randeria, N.Trivedi, A. Moreo and R. T. Scalettar, Phys. Rev. Lett.**49**, 2001 (1992).

[45] J.J. Rodríguez-Núñez et al, Physica **B 206 & 207**, 654 (1995); J.J. Rodríguez-Núñez et al, J. Low Temp. Phys. **99**, 315 (1995).

[46] A. Moreo, D.J. Scalapino and S.R. White, Phys. Rev. **B 45**, 7544 (1992).

[47] N. Trivedi and M. Randeria (Preprint cond-mat/9411121, SISSA 1994).

[48] S. Robaszkiewicz, R. Micnas and J. Ranninger, Phys. Rev. **B 36**, 180 (1987).

[49] R. Friedberg and T. D. Lee, Phys. Rev. **B 40**, 6745 (1989).

Electron–Phonon Interactions in Correlated Systems

Karol I. Wysokiński

Institute of Physics, M.Curie-Sklodowska University, ul.Radziszewskiego 10A, PL–20-031 Lublin, Poland

> The consensus is that there is absolutely no consensus on the theory of high-T_c superconductivity.
> *P.W.Anderson*
>
> One hopes that ideas that are ultimately found to be inappropriate for high T_c will be important elswhere. (...) Finally, as for phonons, they must certainly play some role in setting T_c.
> *J.R.Schrieffer*

1 Introduction

Shortly after the discovery of high T_c superconducting (HTS) oxides [1] with transition temperatures T_c around 100 K *much* exceeding the most optimistic estimate of T_c resulting from electron–phonon interactions [2] the hot discussion started about the applicability of BCS theory [3] and the electron–phonon (EP) mechanism [4]. At the beginning of the HTS era the electron–phonon mechanism was completely dismissed mainly on theoretical grounds. Subsequent experimental discoveries have shown that the electron–phonon interaction, though perhaps not singularly responsible for the superconductivity in these materials, does play an important role and should not be abandoned completely.

The issue of competing retarded electron–phonon and usually assumed non-retarded Coulomb interactions is valid also for materials with low (and even more with elevated) transition temperatures. The serious studies of the interplay between them started early on with work of Eliashberg [5], Anderson and Morel [6] and others [7].

The parent compounds of superconducting oxides are antiferromagnetic insulators with Neél ordering temperature T_N as high as 250 K in LaCuO$_4$. By doping with Sr or Ba, T_N decreases to zero. The neutron scattering studies show the persistence of sizeable antiferromagnetic correlations deep in the metallic state [8]. This points out that short range Coulomb interactions are important in the description of (at least) normal state properties of HTS, making the materials strongly correlated systems.

There exist attempts to describe the superconducting mechanism operating in HTS as based on antiferromagnetic fluctuations [9].

It is not our intention to dwell on the superconducting mechanism, even though this is very a important issue. The main aim is to discuss the problem of interplay between electron-phonon and electron–electron interactions in correlated systems. We believe such analysis can be of importance for various materials and not only HTS'S. We shall however mainly refer to experiments on this last class of superconductors.

Severe complications are to be expected by studying the problem. As is well known electron correlations are very important in narrow band systems, where the relevant electronic scale E_F is quite small. In those circumstances, the phonon energy scale ω_D is of comparable magnitude, with the ratio ω_D/E_F of order 1 signalling a possible break down of the Migdal [10] – Eliashberg description of the electron–phonon interaction in metals. Here we shall assume the validity of the Migdal–Eliashberg approximation and concentrate on the mutual influence of electron and phonon subsystems.

In the next Section we shall discuss experimental motivation for and theoretical work related to the present problem. Section 3 contains a brief discussion of our theory. It is a self–consistent theory à la Migdal with strong correlations treated with an auxiliary boson technique. We conclude with results and their discussion.

2　Experimental Motivation and Previous Theories

In this Section we shall briefly mention those experiments which indicate that electron–phonon coupling is the relevant interaction in HTS. We will also discuss recent theoretical work dealing with the electron–phonon interaction in correlated systems.

There have been found Raman–active and infrared–active modes with frquencies and linewidths strongly temperature dependent [11]. This finding is most easily explained by involving electron–phonon coupling [12]. The absence or very small values of the isotope shift exponent ($T_c \propto M^{-\alpha}$) has in the early days of the high T_c era, been taken as an argument against the electron–phonon interaction acting in the HTS, even though it is well known that small or even negative values of isotope shift exponents are possible for electron–phonon mediated superconductivity. This is due to the retarded nature of this interaction. In La–based oxides the values of $\alpha \approx 0.16$ have been reported [13]. Smaller but clearly nonzero values of α have been found in $YBa_2Cu_3O_{7-\delta}$ [14]. In the BCS model one gets $\alpha = 0.5$. It has been found that in HTS α strongly depends on the material studied, and for a given class of HTS, on the doping level with smallest values of α found for optimally doped systems.

Clearly the electron–phonon interaction is of central importance to the superconducting pairing in $Bi_{1-x}K_xBiO_3$ and $BaPb_{1-x}Bi_xO_3$ as is evident from the isotope shift, tunneling and photoemission measurements [15].

The theoretical approach used to study systems with competing interactions depends on their strength in relation to the bandwidth W and each other.

Recently there appeared a number of theoretical papers dealing with various aspects of the physics of the normal and superconducting states in correlated systems with electron–phonon interactions [16–28]. Solution of the Eliashberg equations [16] for a "hole superconductivity" model supplemented by electron–phonon coupling has shown a striking feature that at some parameter range this coupling has diminished T_c. The effect takes place for sufficiently strong attraction from the Coulomb channel and was due to self energy corrections.

Essentially exact calculations [17] of the spectral properties of quasiparticles described by the $t - J$ model and coupled to optical phonons show phonon-induced mass renormalisation of carriers that is higher than in the uncorrelated case. Sufficiently strong local EP coupling may outweigh the (local) Coulomb repulsion and lead to local attraction and local pairing superconductivity usually described by the negative U Hubbard model [2]. The models relevant to the description of HTS have been studied systematically in Refs. [22–25]. It has been found *inter alia* that: 1) the isotope coefficient is inversly correlated with superconducting T_c [23] in accordance with experimental findings, 2) the electron correlations have a relatively small influence on the Eliashberg function $a^2 F(\omega)$ but strongly suppress the corresponding transport function $a_{tr}^2 F(\omega)$. This explains the small values of EP coupling found in transport measurements on superconducting oxides. 3) sufficiently large electron–phonon coupling may produce a phase separation [24] instability in the correlated system. Analysis of the three–band model showed that near the phase separation region, there appear superconducting instabilities in both s– and d–wave channels. The authors [24] found that the electron–phonon parameter λ_{e-ph} is not strongly modified by the Coulomb correlations.

The increase of λ_{e-ph} and thus superconducting T_c due to correlations has been found in [25]. These authors argue that phonon mediated superconductivity should be characterised by the gap of d–wave symmetry. Rigorous studies of one–dimensional models with strong repulsive and weak retarded attractive interactions show the existence of superconducting fluctuations near half–filling [26]. The models without the attractive part of the interaction do not show superconductivity. The most "radical" approach to the HTS has been recently reviewed in [28], where the very existence of strong correlations in these materials has been questioned and electron–phonon coupling invoked as the only mechanism of superconductivity and the reason for "noncanonical" Fermi–liquid behaviour of the materials.

Summarising this short review of various approaches for treating the EP interaction in correlated materials, we emphasize that there is increasing evidence coming both from experiments and calculations that the electron–phonon interaction plays an important and previously unanticipated role in HTS's.

3 Self-consistent Theory of the Correlated Electron–Phonon System

We start with the one band Hubbard Hamiltonian with electron–phonon interaction term derived from modulation of the hopping integral [30]

$$\mathcal{H}_o = \sum_{ij\,\sigma}(t_{ij} - \mu\delta_{ij})\tilde{c}^+_{i\sigma}\,\tilde{c}_{i\sigma} + \sum_{ij\,s\alpha\sigma}T^\alpha_{ij\,s}u^\alpha_i\,\tilde{c}^+_{i\sigma}\,\tilde{c}_{i\sigma} + U\sum_i \tilde{n}_{i\uparrow}\tilde{n}_{i\downarrow}\,. \quad (1)$$

Here μ denotes the chemical potential, t_{ij} is hopping integral assumed to take on nonzero value $-t$ for i,j being nearest neighbour sites, U is the Hubbard on–site repulsion of carriers, $T^\alpha_{ij\,s} = \frac{\partial t_{ij}}{\partial R^{o\alpha}_{ij}}\delta_{ij} - \frac{\partial t_{ij}}{\partial R^{o\alpha}_{ij}}\delta_{j\,s}$, is the electron–lattice interaction parameter and u^α_i is the a-th component of the displacement vector of i-th ion, $\tilde{n}_{i\sigma} = \tilde{c}^+_{i\sigma}\tilde{c}_{i\sigma}$. In the large U limit ($U \gg W$) double occupancy of sites is prohibited. It has been shown [31] that to project out doubly occupied sites it is convenient to rewrite the electron operators in (1) in terms of new fermion operators $c^+_{i\sigma}$ and boson operators b^+_i. The term U (in the $U = \infty$ limit) may be dropped at the expense of introducing a constraint

$$\sum_\sigma c^+_{i\sigma}\,\tilde{c}_{i\sigma} + b^+_i b_i = 1 \quad (2)$$

via Lagrange multiplier λ. The constraint allows for at most single occupation of each site. In mean field theory one assumes $\langle b^+_i \rangle = \langle b_i \rangle = r$ and $\lambda_i = \lambda$ at each site.

Equations (1) and (2) predict the renormalised band structure of fermions $\zeta_k = r^2\varepsilon_k - \mu + \lambda$, where $\varepsilon = \frac{1}{N}\sum_{ij}e^{ik(R_i - R_j)}$ is the underlying lattice structure function. In square lattice $\varepsilon_k = -2t(\cos k_x a + \cos k_y a)$, a is the lattice constant. The parameters λ and r are choosen in such a way as to minimize the ground state energy $E_{GS} = \langle H \rangle$. One gets [32]

$$r^2 = 1 - n,$$

$$-\lambda = \frac{1}{N}\sum_{k\sigma}\varepsilon_k\langle c^+_{k\sigma}\,c_{k\sigma}\rangle + \frac{1}{N^2}\sum_{kq\nu\sigma}|M^\nu_{kq}|^2\langle(a_{q\nu} + a^+_{-q\nu})c^+_{kq\nu\sigma}c_{k\sigma}\rangle\,. \quad (3)$$

Here $a^+_{q\nu}$ ($a_{q\nu}$) are the creation (annihilation) operators for phonons with wave vector q and branch index ν, $M^\nu_{kq} = i\hbar\frac{Q}{a}\sum_{\alpha=x,y}g_\nu(q)e^\alpha_\nu(q)(v^\alpha_k - v^\alpha_{k-q})$ is the electron–phonon matrix element, $Q = \frac{1}{t}\frac{\partial t}{\partial R_{ij}}$ measures changes of the hopping integral [30], $g_\nu(q) = (\hbar/2M\omega_{q\nu})^{1/2}$, $\omega_{q\nu}$ is the dispersion of bare harmonic phonons. $v^\alpha_k = \frac{1}{\hbar}\frac{\partial\varepsilon_k}{\partial k_\alpha}$ and $e_\nu(q)$ is the polarisation vector.

In the following we shall neglect the phonon contribution to the Lagrange multiplier λ, i.e., the second term in (3). To go beyond mean field theory for auxiliary bosons we define the boson fluctuating fields via: $b^+_i = r + \delta b^+_i$.

Because we shall work out our theory with helpt of two–time thermodynamic Green's functions, we shall not treat the Langrage multiplier as an additional field. Instead we choose again λ and r with the help of a variational method applied to the total Hamiltonian with δb_i^+ (δb_i) fields. The total Hamiltonian of electron–phonon–slave boson system then becomes

$$\mathcal{H} = \mathcal{H}_o + \mathcal{H}_1 + \mathcal{H}_2 + \mathcal{H}_{ph} , \qquad (4)$$

where we have denoted

$$\mathcal{H}_o = \sum_{k\sigma} (r^2 \varepsilon_k - \mu) c_{k\sigma}^+ c_{k\sigma} + \lambda \Big(\frac{1}{M} \sum_{k\sigma} c_{k\sigma}^+ c_{k\sigma} + r^2 - 1 \Big) , \qquad (4a)$$

which is the Hamiltonian of our system obtained in mean field approximation for slave bosons,

$$\mathcal{H}_1 = \frac{r^2}{\sqrt{N}} \sum_{kq\sigma\nu} M_{k+q,\sigma} c_{k+q\,\sigma}^+ c_{k\sigma} \phi_{q\nu}$$
$$+ \frac{r}{\sqrt{N}} \sum_{k} \varepsilon_k (c_{k\sigma}^+ c_{k+q\,\sigma} \delta b_q^+ + c_{k+q\,\sigma}^+ c_{k\sigma} \delta b_q) , \qquad (4b)$$

the first term of which describes the fermion–phonon interaction and the second term, the fermion–boson interaction. Note the presence of factors r^2 and r. Here $\phi_{q\nu} = a_{q\nu} + a_{-q\nu}^+$ is the phonon field operator, and δb_q^+ (δb_q) are Fourier transforms of corresponding real space fields δb_i^+ (δb_i).

More complicated interactions together with a part describing fluctuating boson fields are contained in H_2 which reads

$$\mathcal{H}_2 = \lambda \sum_{k} \delta b_k^+ \delta b_k + \frac{1}{N} \sum_{kk'q\sigma} \varepsilon_{k-q} c_{k\sigma}^+ c_{k'\sigma} \delta b_q \delta b_{k'+q-k}^+$$
$$+ \frac{1}{\sqrt{N}} \sum r M_{kq}^\nu (c_{k\sigma}^+ c_{k'\sigma} \delta b_{k'+q-k}^+ c_{k+k'\sigma}^+ c_{k-q\sigma} \delta b_{k'}) \phi_{q\nu}$$
$$+ \frac{1}{N} \sum_{kk'qp\sigma\nu} M_{kq}^\nu c_{k+p\sigma}^+ c_{k'\sigma} \phi_{q\nu} \delta b_p \delta b_{k'+q-k}^+ . \qquad (4c)$$

The last term describes the phonons in the system which we take in the harmonic approximation

$$\mathcal{H}_{ph} = \sum_{q\nu} \hbar\omega_{q\nu} \Big(a_{q\nu}^+ a_{q\nu} + \frac{1}{2} \Big) . \qquad (4d)$$

As already mentioned we use the equation of motion method for the double time thermodynamic Green's function (GF). For arbitrary operators A and B the equation of motion reads

$$\omega \ll A|B \gg_\omega = \langle [A, B]_\pm \rangle + \ll [A, H]_- | B \gg_\omega$$
$$= \langle [A, B]_\pm \rangle - \ll A | [B, H]_- \gg_\omega , \qquad (5)$$

where $\ll A|B \gg_\omega$ is the Fourier transformed frequency dependent GF in Zubariev's [33] notation; $[A, B]_\pm$ denotes anticommutator $(+)$ or commutator $(-)$.

To solve for the spectrum of the coupled fermion–phonon–boson Hamiltonian (4) we define the Green's functions of electrons $G_{k\sigma}(\omega) = \ll c_{k\sigma}|c^+_{k\sigma} \gg_\omega$, phonons $D_{q\nu}(\omega) = \ll \phi_{q\nu}|\phi_{-q\nu} \gg_\omega$ and bosons $B_q(\omega) = \ll \delta b_q|\delta b^+_q \gg_\omega$. The general expressions are as follows [34]

$$G_{k\sigma}(\omega) = \frac{1}{\omega - r^2\varepsilon_k + \mu - \lambda - \Sigma_k(\omega)}, \tag{6a}$$

$$D_{q\nu}(\omega) = \frac{1}{\omega^2 - \omega^2_{q\nu} - 2\omega_{q\nu}\Pi_\nu(q,\omega)}, \tag{6b}$$

$$B_q\omega = \frac{1}{\omega - \lambda - \Pi_B(q\omega)}. \tag{6c}$$

The expressions for mass operator $\Sigma_k(\omega)$ and polarisation operators $\Pi_\nu(q,\omega)$ and $\Pi_B(q,\omega)$ depend in turn on the full GF (6). This fact introduces an important self–consistency aspect into the theory, a full account of which will be given elsewhere [34].

To be specific let us write down the first few terms of the electron self–energy. They read

$$\Sigma_k(\omega) = \tfrac{1}{N}\sum_q \varepsilon_{k-q}\langle \delta b_q \delta b^+_q \rangle$$

$$+\tfrac{r^2}{N}\sum_q \left(\varepsilon_k^2 \ll c_{k+q\sigma}\delta b^+_q|c^+_{k+q\sigma}\delta b_q \gg_\omega +\varepsilon_{k-q}^2 \ll c_{k-q\sigma}\delta b_q|c^+_{k-q\sigma}\delta b^+_q \gg_\omega \right)$$

$$+\tfrac{r^4}{N}\sum_{q\nu} |M^\nu_{kq}|^2 \ll \phi_{q\nu}c_{k-q\sigma}|c^+_{k-q\sigma}\phi_{-q\sigma} \gg_\omega + \cdots \tag{7}$$

The first term is real and represents a wave vector dependent shift of the fermion spectrum due to a fluctuating boson field, the second stems from the fermion–boson interaction in the Hamiltonian, while the last in (7) comes from the electron–phonon interaction. Note, however, that the last term contains an additional, as compared to the usual electron–phonon contribution, factor $r^4 \approx (1 - n)^2$ which makes it look relatively unimportant for low dopings $x = 1 - n$. In writing down (7) we have neglected a number of higher order terms coming from more complicated vertices present in the Hamiltonian (cf. 4c).

The remaining task is to express the higher order GF's appearing in (7) through the known propagators. We shall achieve this goal by using the spectral representation for the GF and then decoupling the resulting correlation functions. Evaluating the electron–phonon part of $\Sigma_k(\omega)$ to be denoted by $\Sigma^{e-p}_k(\omega)$ we make the decoupling

$$\langle \phi_{-q\nu}(t)c^+_{k-q}(t)c_{k-q\sigma}(0)\phi_{q\nu}(0)\rangle \approx \langle \phi_{-q\nu}(t)\phi_{q\nu}(0)\rangle \langle c^+_{k-q\sigma}(t)c_{k-q\sigma}(0)\rangle. \tag{8}$$

The fluctuation–dissipation theorem enables us to express the correlation functions on the right hand side of the last expression via the GF's (6), depending in turn on $\Sigma_k(\omega)$ and other self–energies. Solution of the total selfconsistent problem is a rather complicated numerical task. Here we shall present the results of nonselfconsistent calculations. This means that the GF appearing in formulas for $\Sigma_k(\omega)$ be approximated by the bare GF; *i.e.*, those resulting from (6) when self–energies are neglected. In this approximation we get

$$\Sigma_k^{e-P}(\omega) = \frac{r^4}{2N} \sum_{q\nu} |M_{kq}^\nu|^2 \left\{ \frac{\tanh\frac{\beta}{2}(r^2\varepsilon_{k-q} + \lambda - \mu) + \coth\frac{\beta}{2}\omega_{q\nu}}{\omega - r^2\varepsilon_{k-q} + \lambda - \mu - \omega_{q\nu}} \right.$$
$$\left. - \frac{\tanh\frac{\beta}{2}(r^2\varepsilon_{k-q} + \lambda - \mu) - \coth\frac{\beta}{2}\omega_{q\nu}}{\omega - r^2\varepsilon_{k-q} + \lambda + \mu + \omega_{q\nu}} \right\}. \tag{9}$$

4 Results and Discussion

The complete solution of the self–consistent set of equations for the Green's functions (6) and the self–energies [34] is extremaly difficult, if not impossible, and has not been achieved yet. In this review we stick to the mean field approximation for slave bosons. It amounts to complete neglect of the GF $B_q(\omega)$ and calculation of the parameters r and λ from (3). For calculation of the coupling constant M_{kq} we take $\omega_q = \omega_0 \simeq 0.04t \approx 0.02$ eV and ion mass corresponding to the oxygen mass. We shall illustrate the results of our calculations by discussing the carrier concentration dependence of the total electron–phonon coupling parameter λ_{e-ph} and its symmetry components. To be precise, we have defined the weak coupling (BCS like nonretarded) interactions

$$V_{kk'} = -r^4 |M_{kk'}|^2 D_{k-k'}(0) \,, \tag{10}$$

where $D_q(0)$ is the phonon G_F at zero frequency. The factor r^4 in front of it is due to the nonlocal nature of our electron–phonon interaction in (1). For local on–site Holstein type of coupling this factor would be replaced by 1.

To obtain λ_{e-ph} we have averaged $V_{kk'}$ over the Fermi energy. Its symmetry components λ_i are defined as

$$\lambda_i = \sum_{kk'} \frac{\delta_F(\zeta_k)\delta_F(\zeta_{k'})V_{kk'}g_i(k)g_i(k')}{A_i^2} \,, \tag{11}$$

where $\delta_F(\zeta_k)$ denotes that the sum goes over wave vectors lying on the Fermi surface. We have taken $g_1 = \cos k_x a + \cos k_y a$, $g_2 = \cos k_x a - \cos k_y a$, $g_3 = \sin k_x a \cdot \sin k_y a$ and $g_4 = \sin k_x a$; $A_i = \Sigma_k \delta_F(\zeta_k)g_i^2(k)$. Contrary to some recent findings [25], our result shows that the s-wave contribution λ_1 is the largest one. The couplings λ_2 and λ_4 are negative and thus do not lead to superconducting instability, while λ_3 is positive but two orders of magnitude smaller than $\lambda_1 \approx \lambda_{e-ph}$.

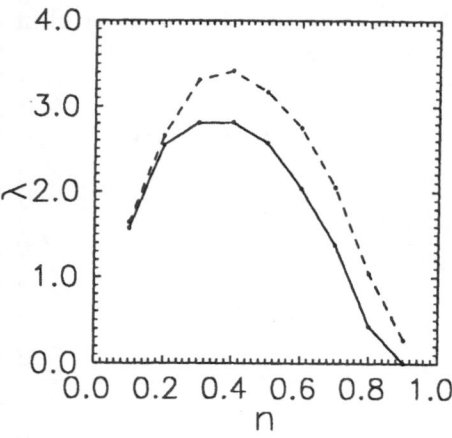

Fig.1. The *s*-wave component of the coupling constant λ_1 (solid curve) is compared with total λ_{e-ph} (dashed line)

Fig.2. λ_{e-ph} of uncorrelated (dashed) and of correlated (solid) system as a function of electron concentration

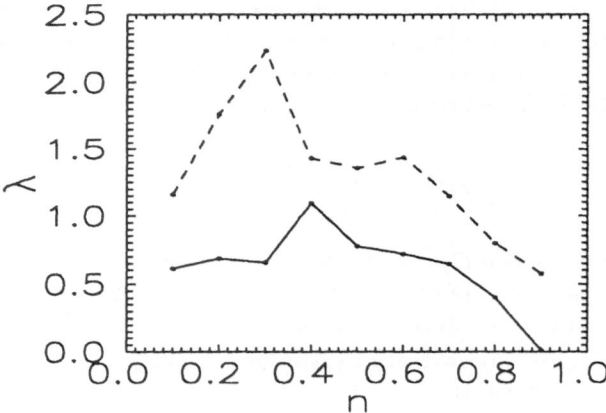

Fig.3. The same as in Fig.(1) but with phonon renormalisation taken into account

In figure (1) we show the frequency dependence of both parameters $\lambda_1 A$, and λ_{e-ph}. Figure (2) shows the effect of correlations on A,-Ph. The dashed curve was calculated for a system without correlations, and the solid one with strong $(U = \infty)$ correlations. It is seen that the concentration dependence of r in a correlated metal near half–filling $(n = 1)$ is dominated by the term r^4 in (10). Two powers of it are roughly cancelled by the density of states which contains the factor $1/r^2$. As already mentioned, the factor r^4 is absent in a Holstein model and this leads to the result $\lambda \propto 1/r^2$ which gives strong enhancement of EP coupling by correlations. Models with both local and non–local EP couplings may thus exhibit complex doping dependences. The results shown in figures (1) and (2) were obtained with the phonon polarisation operator neglected, $\Pi_{q\nu}(\omega) = 0$. The effect of phonon renormalisation and thus self–consistency of the theory is shown in figure (3), where again we plot λ against concentration. The comparison of Figs.(1) and (3) shows that, even for large dopings, phonon renormalisation tends to diminish the coupling constants. At dopings $\delta = 1 - n \approx 0.1$ the renormalisation of phonons leads to an increase of λ_{e-ph} .

We wish to emphasize the internal consistency of our approach. We have started with the Hamiltonian of an electronic system and derived consistently the electron–phonon interaction. The correlations treated by a mean field approach to slave bosons lead to strong downward renormalisation of the electron–phonon coupling. The symmetry analysis shows that the s-wave component is the dominating one. There are, however, possible d-wave fluctuations but with coupling which is two orders of magnitude smaller. The importance the of d-wave component increases a little when phonon self–energy is taken into account. The calculations of thermodynamic properties of superconductors require the full solution of the Eliashberg equation which takes the retarded nature of electron–phonon interactions into account. Such an approach has been recently presented by Zieliński and his coworkers [25] for a slightly different model.

Acknowledgements: This work has been supported by KBN through grant No. 2P 302 070 06. Part of it has been done during author's stay at ISI Torino supported by CEC, Brussels.

References

1. J.G.Bednorz and K.A.Müller, *Z.Phys.*B **64**, 189 (1986); *Rev.Mod.Phys.* **60**, 585 (1988).

2. see *e.g.* C.M.Varma, in: *Superconductivity in d- and f-Band Metals*, ed.by W.Buckel and W.Weber (KfK, Karlsruhe 1982) p.603.

3. J.Bardeen, L.N.Cooper, and J.R.Schrieffer, *Phys.Rev.* **108**, 1175 (1957).

4. P.W.Anderson, *Science* **235**, 1196 (1987).

5. G.M.Eliashberg, *Zh.Eksp.Teor.Fiz.* **25**, 966 (1960) (in Russian).

6. P.Morel and P.W.Anderson, *Phys.Rev.* **125**, 1263 (1962).

7. for review see *e.g.*: P.B.Allen and B.Mitrović, in: *Solid State Physics*, ed.by H.Ehrenreich, F.Seitz, and D.Turnbull (Academic, New York, 1982) p.1;
see also: *Problems of High Temperature Superconductivity*, ed.by V.L.Ginzburg and D.A.Kirzhnits (Nauka, Moscow 1977) (in Russian).

8. J.M.Tranquada, *J.Appl.Phys.* **64**, 6071 (1988).

9. D.Pines, *Physica* C **235–240**, 113 (1994); see also these proceedings.

10. A.B.Migdal, *Sov.Phys.- JETP* **7**, 996 (1958).

11. J.Chrzanowski, S.Gygax, J.C.Irvin, and W.N.Hardy, *Solid State Commun.* **65**, 139 (1988);   S.L.Cooper, M.V.Klein, B.G.Pazol, J.P.Rice, and D.M.Ginsberg, *Phys.Rev.*B **37**, 5817 (1988).

12. R.Zeyher and G.Zwicknagl, *Solid State Commun.* **66**, 617 (1988); *Z.Phys.*B **78**, 175 (1990).

13. K.Alex Müller, *Z.Phys.*B **80**, 193 (1990).

14. H.J.Bornemann and D.E.Morris, *Phys.Rev.*B **44**, 5322 (1991).

15. D.G.Hinks *et al.*, *Physica* C **162-164**, 1405 (1989); J.F.Zasadzinski *et al.*, *Physica* C **162-164**, 1053 (1989); H.Namatame *et al.*, *Phys.Rev.*B **50**, 13 674 (1994).

16. F.Marsiglio, *Physica* C **160**, 305 (1989); P.Marsiglio and J.G.Hirsch, *Phys.Rev.*B **49**, 1366 (1994).

17. A.Ramsak, P.Horsch, and P.Rulde, *Phys.Rev.*B **46**, 14305 (1992).

18. U.Trapper *et al.*, *Z.Phys.*B **93**, 465 (1994).

19. H.Feshke *et al.*, *Phys.Rev.*B **61**, 16 582 (1995).

20. D.Manske, C.T.Rieck, and D.Fay, preprint 1994.

21. J.K.Freericks and Mark Jarrel, preprint 1995 (SISSA Cond.Matt.9502098).

22. J.H.Kim, K.Levin, R.Wentzovitch, and A.Auerbach, *Phys.Rev.*B **44**, 5148 (1991); M.L.Kulić and R.Zeyher, *Phys.Rev.*B **49**, 4398 (1994).

23. Ju H.Kim and Zlatko Tesanović, *Phys.Rev.Lett.* **71**, 4218 (1993).

24. M.Grilli and C.Castellani, *Phys.Rev.*B **50**, 16 880 (1994).

25. J.Zieliński, M.Matlak, and P.Entel, *Phys.Lett.*A **166**, 285 (1992); M.Mierzejewski and J.Zieliński, *Phys.Rev.*B **52**, 3079 (1995); M.Mierzejewski, J.Zieliński, and P.Entel, *Physica* C **235-240**, 2143 (1994); J.Zieliński *et al.*, *J.Supercond.* **8**, 135 (1995).

26. G.T.Zimanyi, S.A.Kivelson, and A.Luther, *Phys.Rev.Lett.* **60**, 2089 (1988).

27. N.M.Plakida and R.Hayn, *Z.Phys.*B **93**, 313 (1994).

28. V.L.Ginzburg and E.G.Maximov, *Physica* C **236-240**, 193 (1994).

29. R.Micnas, J.Ranninger, and S.Robaszkiewicz, *Rev.Mod.Phys.* **62**, 13 (1990).

30. S.Barisic, J.Labbé, J.Frieder, *Phys.Rev.Lett.* **25**, 919 (1970); K.I.Wysokiński, A.L.Kuzemsky, *J.Low Temp.Phys.* **52**, 81 (1983), — , *phys.stat.sol.*(b) **113**, 409 (1982); K.I.Wysokiński, in: *Ordering Phenomena in Condensed Matter Physics*, ed.by Z.M.Galasiewicz and A.Pękalski (World Sci., Singapore 1991) p.187.

31. Piers Coleman, *Phys.Rev.*B **29**, 3035 (1984); G.Kotliar and A.Ruckenstein, *Phys.Rev.Lett.* **57**, 1362 (1986).

32. D.M.Newns and N.Read, *Adv.Phys.* **36**, 799 (1987).

33. D.N.Zubarev, *Sov.Phys.- Usp.* **3**, 320 (1960).

34. K.I.Wysokiński, *to be published.*

Voltage–Probe–Position Dependence and Magnetic–Flux Contribution to the Measured Voltage in ac Tansport Measurements: Which Measuring Circuit Determines the Real Losses?

John R. Clem, Thomas Pe, and Jason McDonald

Ames Laboratory–USDOE and Department of Physics and Astronomy, Iowa State University, Ames, IA

Abstract. The voltage V_{ab} measured between two voltage taps a and b during magnetic flux transport in a type–II superconductor carrying current I is the sum of two contributions, the line integral from a to b of the electric field along an arbitrary path C_s through the superconductor and a term proportional to the time rate of change of magnetic flux thrugh the area bounded by the path C_s and the measuring circuit leads. When the current $I(t)$ is oscillating with time t, the apparent ac loss (the time average of the product IV_{ab}) depends upon the measuring circuit used. Only when the measuring–circuit leads are brought out far from the surface does the apparent power dissipation approach the real (or true) ac loss associated with the length of sample probed. Calculations showing comparisons between the apparent and real ac losses in a flat strip of rectangular cross section will be presented, showing the behavior as a function of the measuring–circuit, dmensions. Corresponding calculations also are presented for a sample of elliptical cross section.

1 Introduction

As high–temperature superconducting materials move closer to large–scale electric–power applications, it is increasingly important to understand the magnitude and origin of the ac losses in these materials. Ideally, measurements of such losses should be carried out under expermiental conditions close to those of the proposed applications. For example, for testing materials intended for use in ac power transmission cables, it is preferable that the ac losses be measured while the conductor is carrying an applied ac transport current.

In measuring the low-frequency ac losses of normal metals, it is safe to use the simple procedure of (a) applying an ac current $I(t) = I_0 \cos \omega t$, (b) using a high-impedance voltmeter to measure the corresponding voltage $V(t)$ between a pair of voltage taps across a representative segment of the conductor, and (c) obtaining the average rate of power dissipation from $P = \langle I(t)V(t) \rangle$, where the brackets denote the time average. It is at first surprising to learn that when the sample is a type-II superconducting tape or strip, this simple method runs into serious difficulties: The apparent rate of power dissipation P_{app} so obtained depends on where the voltage taps are placed on the tape (along an edge or along the centerline) and on how far away from the tape the voltage leads are extended before they are brought together, twisted, and led out to the voltmeter ([1],[2]).

As emphasized by Campbell ([3]), to explain these results it is important to account for the fact that when low-resistance leads are attached at contact points a and b on the conductor, the time-dependent voltage V_{ab} measured by a high-impedance voltmeter is the sum of an electric-field integral term and a magnetic-flux term. As shown in Refs. ([4][5][6]), this voltage is

$$V_{\text{ab}} = \int_a^b \boldsymbol{E} \cdot d\boldsymbol{l} - \frac{d}{dt} \Phi_{\text{sm}} \quad , \tag{1}$$

where the line integral is to be carried out from a to b along a path C_{s} through the conductor, and Φ_{sm} is the magnetic flux up through the loop bounded by the path C_{s} and the measuring circuit leads (which define the contour C_{m}). It can be shown with the help of Faraday's law that the voltage V_{ab} is independent of the contour C_{s}, because any change in the first term on the right-hand side of Eq. (1) is compensated by a canceling change in the second term. It often is convenient to chose the contour C_{s} so that the first term is zero.

Because of magnetic hysteresis, the flux term, i.e., the second term on the right-hand side of Eq. (1), has terms in strip geometry that are both in phase and out of phase with the current $I(t)$. Only for normal-metal wires in which the current density is uniform and for superconducting wires that have circular cross section does the contribution to the flux term from magnetic fields outside the sample have a vanishing in-phase component. As pointed out by Campbell ([3]), in general only when the leads are brought out to a large distance before bringing them together does the measurement give the true loss, i.e., the dissipated power delivered by the power supply to the segment between a and b. Since the true loss involves the voltage measured across the terminals of the power supply, the flux Φ_{sm} involved in Eq. (1) is the total flux through the area bounded by the contour C_{s} along the the sample of interest and the contour C_{m} along the leads that connect the sample to the power supply.

In Sec. 2, we present the details of how to calculate the measured time-dependent voltage $V(t)$ generated by hysteretic losses in a flat strip of rectangular cross section carrying an alternating current $I(t)$. We show how the apparent loss depends upon the measuring circuit geometry and examine the conditions under which the apparent loss is a good approximation to the true loss. We also present the corresponding results for a sample of elliptical cross section. In Sec. 3, we briefly summarize our results.

2 Theoretical Approach and Results

Consider a type-II superconducting strip of width $2W$ and thickness $d \ll 2W$ in the xy plane, centered on the y axis, as shown in Fig. 1, such that the edges are at $x = \pm W$. Assume that the London penetration depth λ is less than the sample thickness. Suppose that an alternating current $I_y(t) = I_0 \cos \omega t$ is applied. A corresponding self-field, which wraps around the sample, will be produced. If the current amplitude is very tiny, the magnitude of the field at the sample edges will be less than the lower critical field H_{c1} and thus will be too small to cause any vortices to penetrate into the sample. However, we are most interested in the case for which the current amplitude I_0 is substantial, such that vortices or antivortices are nucleated at the edges and driven into the sample during each half cycle.

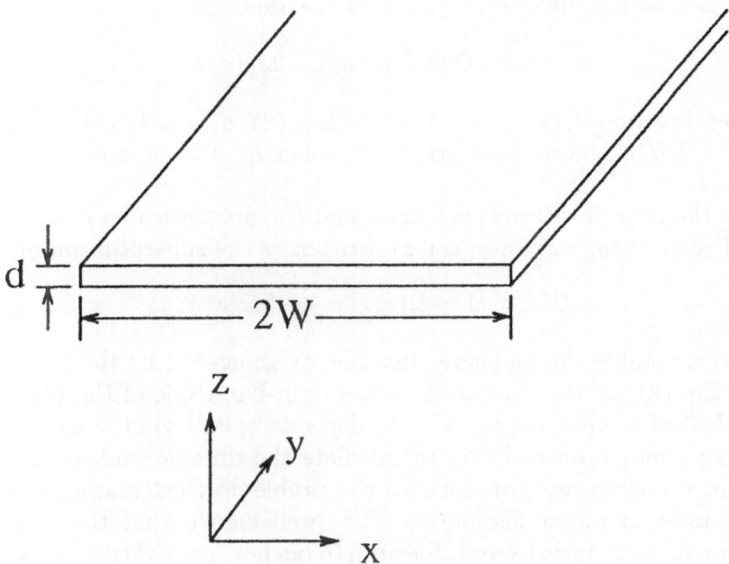

Fig. 1. Sketch of strip geometry considered in this paper. The properties of long current-carrying type-II superconductors of width $2W$ and thickness $d \ll 2W$ are examined.

To calculate the hysteretic losses under such circumstances, we use the critical state model, which is characterized by a critical depinning current density J_c.

We assume for simplicity in this paper that J_c is independent of the flux density $\boldsymbol{B}$. We also assume that over most of the vortex-filled region, the local field is sufficiently large by comparison with H_{c1} that we may take $\boldsymbol{B} = \mu_0 \boldsymbol{H}$ to good approximation. At each value of the current $I_y(t)$ (assuming that the current amplitude I_0 is less than the critical current $I_c = 2WdJ_c$), the quasistatic profiles of the current density $J_y(x,t)$ (averaged across the sample thickness) and the magnetic flux density $\boldsymbol{B}(x,z,t)$ may be calculated using the Norris method [7] as described in detail in Refs. [8] and [9].

When the current $I_y(t)$ is equal to I_0, magnetic flux penetrates from both edges to the coordinates $x = \pm a_0$, where $a_0 = W\sqrt{1-(I_0/I_c)^2}$, but $B_z(x,0,t)$ remains zero in the unpenetrated region $|x| < a_0$. The current density J_y in the strip is given by

$$J_y(x;a_0) = \begin{cases} \frac{2}{\pi} J_c \arctan \sqrt{\frac{W^2-a_0^2}{a_0^2-x^2}}, & |x| < a_0, \\ J_c, & a_0 < |x| < W. \end{cases} \tag{2}$$

As $I_y(t)$ decreases, however, new flux fronts move in from both edges. The current density in each of these new regions has magnitude J_c, but it is reversed in direction. Let $x = \pm a(t)$ denote the time-dependent coordinates of these incoming flux fronts. As shown in Refs. [8] and [9], the current density in the strip can be expressed as the superposition of two distributions,

$$J_y(x,t) = J_y(x;a_0) - 2J_y(x;a), \tag{3}$$

where the function $J_y(x;a)$ is defined in Eq. (2), $a(t) = W\sqrt{1-(I_a/I_c)^2}$, and $I_a = [I_0 - I_y(t)]/2$. Note that $a(t) = W$ when $I_y(t) = I_0$, and $a(t) = a_0$ when $I_y(t) = -I_0$.

From the Biot-Savart law, it follows that the magnetic flux density $\boldsymbol{B}(x,z,t)$ in the vicinity of the strip also can be written as the superposition of two fields,

$$\boldsymbol{B}(x,z,t) = \boldsymbol{B}(x,z;a_0) - 2\boldsymbol{B}(x,z;a), \tag{4}$$

where $\boldsymbol{B}(x,z;a_0)$ is the magnetic flux density generated by the current density given in Eq. (2). Of the two terms on the right-hand side of Eq. (4), the first is independent of t, while the second one depends upon t via the coordinate $a(t)$.

Having briefly discussed how to calculate the time-dependent magnetic flux density in the strip, we now turn to the problem of calculating the apparent and real rates of power dissipation. It is well known that the rate of power dissipation for hysteretic losses is linear in frequency, since there is a fixed amount of energy dissipated each cycle. We thus calculate the apparent and real loss per cycle per unit length of the strip. For simplicity, we consider two special circuit configurations:

Perpendicular case (Fig. 2). Here the voltage taps are placed along the centerline of the strip ($x = 0, z = d/2$), and the leads extend perpendicular to the surface to a height z before they are brought together, twisted, and led out to the voltmeter.

Parallel case (Fig. 3). Here the voltage taps are placed along the edge of the strip ($x = W, z = 0$), and the leads lie in the plane of the strip but extend perpendicular to the edge to the coordinate $x > W$ before they are brought together, twisted, and led out to the voltmeter.

The measured voltage for both cases is determined by Eq. (1), where the contours C_s extend along the line $x = z = 0$ except for segments that extend perpendicular to this line out to the voltage taps, as shown by the dotted lines in Figs. 2 and 3. It can be shown that the line integrals of the electric field [the first term of Eq. (1)] vanish for these choices of C_s. To calculate the voltage $V_{ab}(t)$ for these cases, we therefore need only to calculate the time derivative of the magnetic flux through the shaded areas in the measuring circuits sketched in Figs. 2 and 3. The time derivative of the magnetic flux is obtained by first using the Biot-Savart law to express the magnetic field in terms of $J_y(x,t)$, as sketched in Eq. (4), and by taking the time derivative of the resulting expression using Eq. (3) and the connection between $a(t)$ and $I_y(t)$. The apparent loss per cycle per unit length is then obtained by evaluating $\int I_y(t)V_{ab}(t)\,dt$ over one period ($2\pi/\omega$), where $V_{ab}(t)$ is the voltage across unit length of the sample.

The apparent loss per cycle per unit length for the strip in the perpendicular case is

$$L_\perp(F) = \left(\frac{\mu_0 I_c^2}{\pi}\right) \times$$

$$\left[2\left(\frac{z}{W}\right)\left(\sqrt{\left(\frac{z}{W}\right)^2 + 1} - \sqrt{\left(\frac{z}{W}\right)^2 + 1 - F^2}\right.\right.$$

$$\left.- F\arctan\left(\frac{F}{\sqrt{\left(\frac{z}{W}\right)^2 + 1 - F^2}}\right)\right)$$

$$-\frac{F^2}{2}\log\left(\frac{\sqrt{1-F^2}\left(\left(\frac{z}{W}\right) + \sqrt{\left(\frac{z}{W}\right)^2 + 1}\right)}{\left(\frac{z}{W}\right) + \sqrt{\left(\frac{z}{W}\right)^2 + 1 - F^2}}\right)$$

$$-\left(1 - \frac{F}{2}\right)^2\log\left(\frac{\left(\frac{z}{W}\right)^2 + 1 + \left(\frac{z}{W}\right)\sqrt{\left(\frac{z}{W}\right)^2 + 1 - F^2} - F}{(1-F)\sqrt{\left(\frac{z}{W}\right)^2 + 1}\left(\left(\frac{z}{W}\right) + \sqrt{\left(\frac{z}{W}\right)^2 + 1}\right)}\right)$$

$$\left.-\left(1 + \frac{F}{2}\right)^2\log\left(\frac{\left(\frac{z}{W}\right)^2 + 1 + \left(\frac{z}{W}\right)\sqrt{\left(\frac{z}{W}\right)^2 + 1 - F^2} + F}{(1+F)\sqrt{\left(\frac{z}{W}\right)^2 + 1}\left(\left(\frac{z}{W}\right) + \sqrt{\left(\frac{z}{W}\right)^2 + 1}\right)}\right)\right],$$

(5)

where $F = I_0/I_c$, $I_c = 2WdJ_c$, and the thickness d of the strip is ignored relative to the width $2W$. Note that the apparent loss is zero when the leads are brought together at the height $z = 0$. The reason for this is that both terms on the right-hand side of Eq. (1) are then zero.

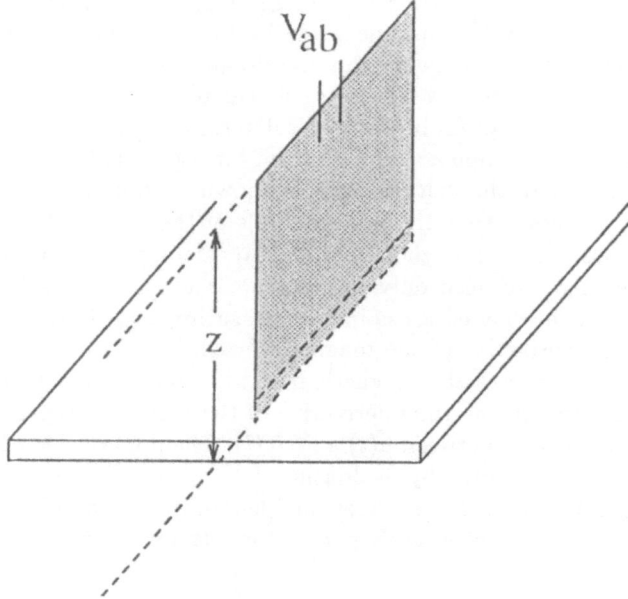

Fig. 2. The perpendicular case: Measuring-circuit leads extend to a height z above the sample.

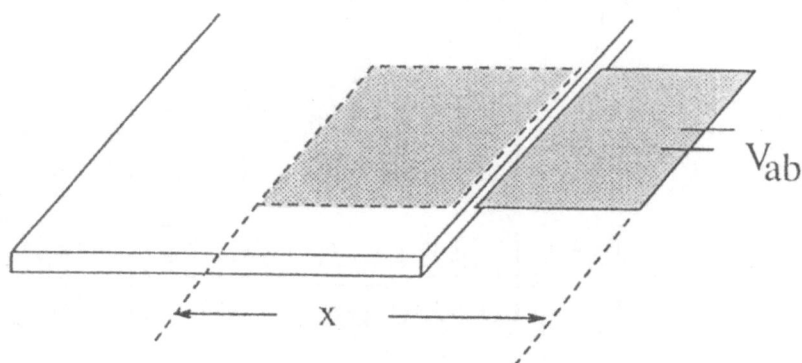

Fig. 3. The parallel case: Measuring-circuit leads extend to the coordinate $x > W$.

Similarly, the apparent loss per cycle per unit length for the strip in the parallel case is

$$L_{\parallel}(F) = \left(\frac{\mu_0 I_c^2}{\pi}\right) \times$$

$$\left[2\left(\frac{x}{W}\right)\left(\sqrt{\left(\frac{x}{W}\right)^2 - 1 + F^2} - \sqrt{\left(\frac{x}{W}\right)^2 - 1}\right.\right.$$

$$\left.- F\log\left(\frac{\sqrt{\left(\frac{x}{W}\right)^2 - 1}}{\sqrt{\left(\frac{x}{W}\right)^2 - 1 + F^2} - F}\right)\right)$$

$$- \frac{F^2}{2}\log\left(\frac{\sqrt{1 - F^2}\left(\left(\frac{x}{W}\right) + \sqrt{\left(\frac{x}{W}\right)^2 - 1}\right)}{\left(\frac{x}{W}\right) + \sqrt{\left(\frac{x}{W}\right)^2 - 1 + F^2}}\right)$$

$$- \left(1 - \frac{F}{2}\right)^2\log\left(\frac{\left(\frac{x}{W}\right)^2 - 1 + \left(\frac{x}{W}\right)\sqrt{\left(\frac{x}{W}\right)^2 - 1 + F^2} + F}{(1 - F)\sqrt{\left(\frac{x}{W}\right)^2 - 1}\left(\left(\frac{x}{W}\right) + \sqrt{\left(\frac{x}{W}\right)^2 - 1}\right)}\right)$$

$$- \left(1 + \frac{F}{2}\right)^2\log\left(\frac{\left(\frac{x}{W}\right)^2 - 1 - \left(\frac{x}{W}\right)\sqrt{\left(\frac{z}{W}\right)^2 - 1 + F^2} - F}{(1 + F)\sqrt{\left(\frac{x}{W}\right)^2 - 1}\left(\left(\frac{x}{W}\right) + \sqrt{\left(\frac{x}{W}\right)^2 - 1}\right)}\right)\right]. \tag{6}$$

Figure 4 shows plots of the apparent loss per cycle per unit length in both the perpendicular and parallel cases for strip geometry versus $F = I_0/I_c$ on a semilogarithmic scale. Shown for comparison is the real (or true loss) per cycle per unit length for strip geometry (solid curve) [7],

$$L_0(F) = \left(\frac{\mu_0 I_c^2}{\pi}\right)\left[(1 - F)\log(1 - F) + (1 + F)\log(1 + F) - F^2\right]. \tag{7}$$

Note that $L_0 \propto F^4$ for small $F \ll 1$. We see that the apparent losses $L_\perp$ and $L_{\parallel}$ agree with L_0 within about 6% when $z/W > 3$ or $x/W > 3$. For large values of z/W and x/W, the following expansions are useful:

$$L_\perp(F) = L_0(F) +$$
$$\left(\frac{\mu_0 I_c^2}{\pi}\right)\left[-\frac{F^4}{12\left(\frac{z}{W}\right)^2} + \frac{F^4(\frac{1}{2} - \frac{F^2}{5})}{8\left(\frac{z}{W}\right)^4} + O\left(\frac{1}{\left(\frac{z}{W}\right)^6}\right)\right], \tag{8}$$

$$L_{\parallel}(F) = L_0(F) +$$
$$\left(\frac{\mu_0 I_c^2}{\pi}\right)\left[\frac{F^4}{12\left(\frac{x}{W}\right)^2} + \frac{F^4(\frac{1}{2} - \frac{F^2}{5})}{8\left(\frac{x}{W}\right)^4} + O\left(\frac{1}{\left(\frac{x}{W}\right)^6}\right)\right]. \tag{9}$$

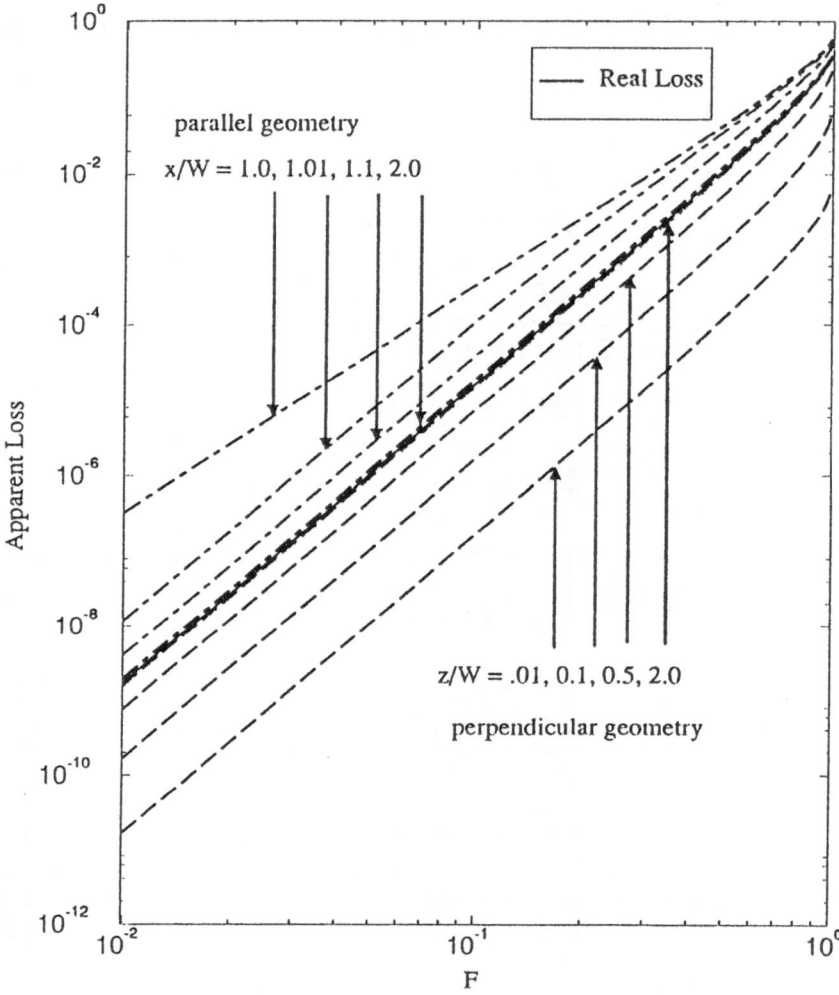

Fig. 4. Apparent loss per cycle per unit length (in units of $\mu_0 I_c^2/\pi$)for a strip of width $2W$ and thickness d, calculated as described in the text for parallel (dot-dashed curves) and perpendicular geometry (dashed curves). The real loss, representing the actual power dissipation made up by the power supply is shown by the solid curve [7].

For samples of elliptical cross section, the apparent loss per cycle per unit length can be calculated by a procedure very similar to that described above, except that the current density and magnetic fields are those found in Ref. [7]. We consider the elliptical cross section to be characterized by a semimajor axis W and semiminor axis $d/2$, such that the width of the sample is $2W$ and the thickness at the thickest point is d. The critical current is thus $I_c = (\pi/2)WdJ_c$, and the ratio of the semiminor axis to the semimajor axis is $\alpha = d/2W$. The results are, for the perpendicular case,

$$L_\perp(F) = \left(\frac{\mu_0 I_c^2}{\pi}\right) \times$$

$$\left[\frac{2\left(\frac{z}{a}\right)}{3(1-\alpha^2)^2}\left(\left(2\left(\frac{z}{W}\right)^2 - (3F-5)(1-\alpha^2)\right)\sqrt{\left(\frac{z}{W}\right)^2 + (1-\alpha^2)}\right.\right.$$

$$\left. - \left(2\left(\frac{z}{W}\right)^2 - (2F-5)(1-\alpha^2)\right)\sqrt{\left(\frac{z}{W}\right)^2 + (1-F)(1-\alpha^2)}\right)$$

$$- \frac{F^2}{2}\log\left(\frac{\sqrt{1-F}\left(\left(\frac{z}{W}\right) + \sqrt{\left(\frac{z}{W}\right)^2 + (1-\alpha^2)}\right)}{\left(\frac{z}{W}\right) + \sqrt{\left(\frac{z}{W}\right)^2 + (1-F)(1-\alpha^2)}}\right)$$

$$- \left(1-\frac{F}{2}\right)^2\log\left(\frac{\left(\frac{z}{W}\right) - \sqrt{\left(\frac{z}{W}\right)^2 + (1-\alpha^2)}}{\left(\frac{z}{W}\right) + \sqrt{\left(\frac{z}{W}\right)^2 + (1-\alpha^2)}}\right)$$

$$\left. - \left(1-\frac{F}{2}\right)^2\log\left(\frac{\left(\frac{z}{W}\right) + \sqrt{\left(\frac{z}{W}\right)^2 + (1-F)(1-\alpha^2)}}{\left(\frac{z}{W}\right) - \sqrt{\left(\frac{z}{W}\right)^2 + (1-F)(1-\alpha^2)}}\right)\right]$$

$$(10)$$

and, for the parallel case,

$$L_\parallel(F) = \left(\frac{\mu_0 I_c^2}{\pi}\right) \times$$

$$\left[\frac{2\left(\frac{x}{W}\right)}{3(1-\alpha^2)^2}\left(\left(2\left(\frac{x}{W}\right)^2 + (3F-5)(1-\alpha^2)\right)\sqrt{\left(\frac{x}{W}\right)^2 - (1-\alpha^2)}\right.\right.$$

$$\left. - \left(2\left(\frac{x}{W}\right)^2 + (2F-5)(1-\alpha^2)\right)\sqrt{\left(\frac{x}{W}\right)^2 - (1-F)(1-\alpha^2)}\right)$$

$$- \frac{F^2}{2}\log\left(\frac{\sqrt{1-F}\left(\left(\frac{x}{W}\right) + \sqrt{\left(\frac{x}{W}\right)^2 - (1-\alpha^2)}\right)}{\left(\frac{x}{W}\right) + \sqrt{\left(\frac{x}{W}\right)^2 - (1-F)(1-\alpha^2)}}\right)$$

$$- \left(1-\frac{F}{2}\right)^2\log\left(\frac{\left(\frac{x}{W}\right) - \sqrt{\left(\frac{x}{W}\right)^2 - (1-\alpha^2)}}{\left(\frac{x}{W}\right) + \sqrt{\left(\frac{x}{W}\right)^2 - (1-\alpha^2)}}\right)$$

$$\left. - \left(1-\frac{F}{2}\right)^2\log\left(\frac{\left(\frac{x}{W}\right) + \sqrt{\left(\frac{x}{W}\right)^2 - (1-F)(1-\alpha^2)}}{\left(\frac{x}{W}\right) - \sqrt{\left(\frac{x}{W}\right)^2 - (1-F)(1-\alpha^2)}}\right)\right].$$

$$(11)$$

Figure 5 shows plots of the apparent loss per cycle per unit length in both the perpendicular and parallel cases for elliptical cross sections versus $F = I_0/I_c$ on a semilogarithmic scale. Shown for comparison is the real (or true loss) per cycle per unit length for elliptical cross sections (solid curve) [7],

$$L_0(F) = \left(\frac{\mu_0 I_c^2}{\pi}\right)\left[(1-F)\log(1-F) + (2-F)\frac{F}{2}\right]. \tag{12}$$

Note that $L_0 \propto F^3$ for small $F \ll 1$. We find that the apparent losses $L_\perp$ and $L_\parallel$ agree with L_0 within about 6% when $z/W > 3$ or $x/W > 3$. For large values of z/W and x/W, the following expansions for elliptical cross sections are useful:

$$L_\perp(F) = L_0(F)+$$
$$\left(\frac{\mu_0 I_c^2}{\pi}\right)\left[-\frac{(1-\alpha^2)F^3}{12\left(\frac{z}{W}\right)^2} + \frac{(1-\alpha^2)^2(2-F)F^3}{32\left(\frac{z}{W}\right)^4} + O\left(\frac{1}{\left(\frac{z}{W}\right)^6}\right)\right], \tag{13}$$

$$L_\parallel(F) = L_0(F)+$$
$$\left(\frac{\mu_0 I_c^2}{\pi}\right)\left[\frac{(1-\alpha^2)F^3}{12\left(\frac{x}{W}\right)^2} + \frac{(1-\alpha^2)^2(2-F)F^3}{32\left(\frac{x}{W}\right)^4} + O\left(\frac{1}{\left(\frac{x}{W}\right)^6}\right)\right]. \tag{14}$$

Note that for both the strip geometry and elliptical cross sections the apparent loss in the parallel geometry is an overestimate of the true loss, but the apparent loss in the perpendicular geometry is an underestimate. The predicted behavior has been confirmed, at least qualitatively by Fleshler et al. [10][11].

The above calculations have been carried out for monolithic type-II superconducting strips of rectangular or elliptical cross section, in which the critical depinning current density is J_c. Our calculations also should apply, with minor modifications, to strip-like composite conductors containing a uniform density (volume fraction f_s) of untwisted superconducting filaments (each filament characterized by J_c) embedded in a normal-metal matrix. It can be shown that application of a current to such a composite conductor induces a current that flows initially with highest density in the filaments near the edge. Only when the current density in the outermost filaments exceeds J_c does current transfer to filaments farther from the edge, thereby permitting magnetic flux to penetrate more deeply into the composite conductor. Under the application of an ac current, the penetration of magnetic flux into composites is thus very similar to that into monolithic superconductors. The maximum supercurrent density, averaged over the composite's total cross section, is the engineering critical current density, $J_e = f_s J_c$. To describe losses of multifilamentary composite conductors using Eqs. (5)-(14), one must therefore replace J_c by $J_e = f_s J_c$ and I_c by $I_c = 2Wd J_e$ for rectangular cross section or $I_c = (\pi/2)Wd J_e$ for elliptical cross section.

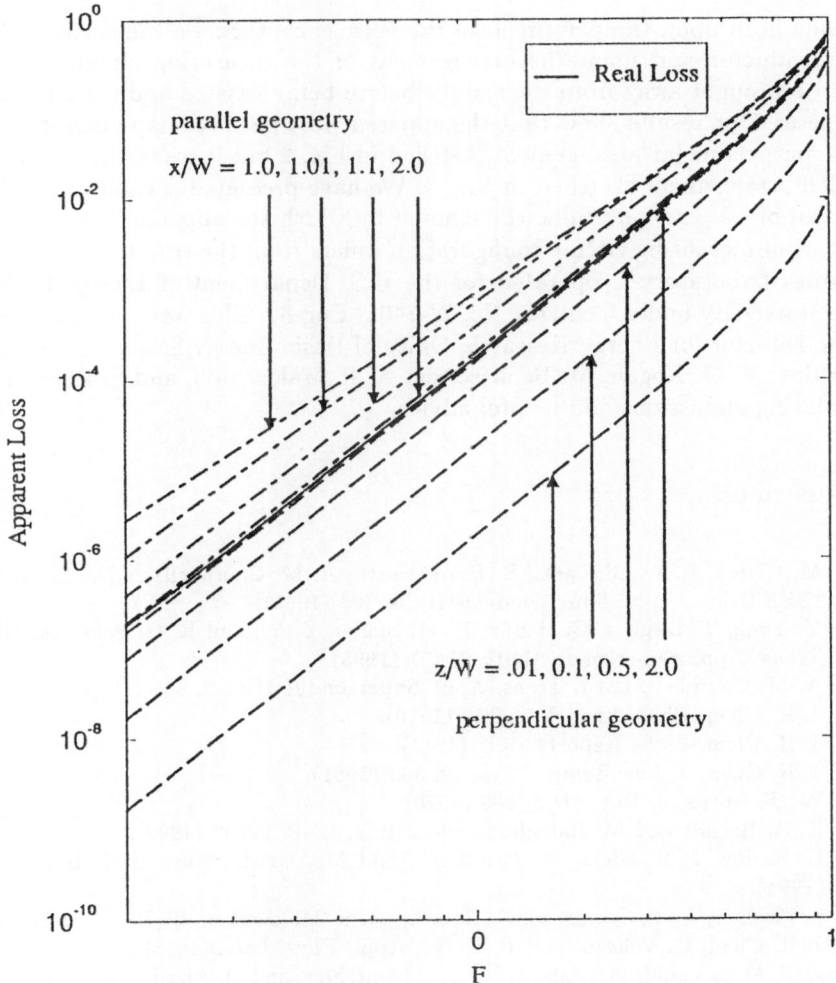

Fig. 5. Apparent loss per cycle per unit length (in units of $\mu_0 I_c^2/\pi$) for a sample of elliptical cross section ($\alpha \to 0$), calculated as described in the text for parallel (dot-dashed curves) and perpendicular geometry (dashed curves). The real loss, representing the actual power dissipation made up by the power supply is shown by the solid curve [7].

3 Summary and Acknowledgments

In this paper we have theoretically studied the hysteretic ac transport losses of type-II superconducting strips of rectangular and elliptical cross section carrying an ac current. Our theory shows that the apparent loss per cycle per unit length depends both upon the placement of the voltage contacts on the surface of the superconductor and upon the arrangement of the measuring circuit leads as they are brought away from the sample before being twisted and led out to the voltmeter. Our results show that the apparent loss per cycle is underestimated in the perpendicular arrangement sketched in Fig. 2 but is overestimated in the parallel arrangement sketched in Fig. 3. We have presented several expressions that can be used to determine the amount by which the apparent loss per cycle for a given measuring circuit configuration differs from the true loss per cycle.

Ames Laboratory is operated for the U.S. Department of Energy by Iowa State University under Contract No. W-7405- Eng-82. This work was supported by the Director for Energy Research, Office of Basic Energy Sciences. We thank E. Zeldov, V. G. Kogan, M. Benkraouda, A. P. Malozemoff, and S. Fleshler for stimulating discussions and helpful advice.

References

1. M. Ciszek, B. A. Glowacki, S. P. Ashworth, A. M. Campbell, and J. E. Evetts, IEEE Trans. Appl. Superconductivity 5, 709 (1995).
2. Y. Yang, T. Hughes, C. Beduz, D. M. Spiller, Z. Yi, and R. G. Scurlock, IEEE Trans. Appl. Superconductivity 5, 701 (1995).
3. A. M. Campbell, IEEE Trans. Appl. Superconductivity 5, 687 (1995).
4. J. R. Clem, Phys. Rev. B 1, 2140 (1970).
5. J. R. Clem, Phys. Reports 75, 1 (1981).
6. J. R. Clem, J. Low Temp. Phys. 42, 363 (1981).
7. W. T. Norris, J. Phys. D 3, 489 (1970).
8. E. H. Brandt and M. Indenbom, Phys. Rev. B 48, 12893 (1993).
9. E. Zeldov, J. R. Clem, M. McElfresh, and M. Darwin, Phys. Rev. B 49, 9802 (1994).
10. S. Fleshler, L. T. Cronis, G. E. Conway, A. P. Malozemoff, T. Pe, J. McDonald, J. R. Clem, G. Vellego, and P. Metra, Appl. Phys. Lett. (1995).
11. A. P. Malozemoff, Q. Li, S. Fleshler, G. Snitchler, and D. Aized, Taiwan International Conference on Superconductivity, Hualien, Taiwan, R.O.C., August 8-11, 1995.

Superconductivity in Doped Insulators

V. J. Emery[1] and S. A. Kivelson[2]

[1] Dept. of Physics, Brookhaven National Laboratory, Upton, NY 11973, USA
[2] Dept. of Physics, UCLA, Los Angeles, CA 90095, USA

Abstract. It is shown that many synthetic metals, including high temperature super-conductors are "bad metals", with such a poor conductivity that the usual mean–field theory of superconductivity breaks down because of anomalously large classical and quantum fluctuations of the phase of the superconducting order parameter. It is argued that the supression of a first order phase transition (phase separation) by the long–range Coulomb interaction leads to high temperature superconductivity accompanied by static or dynamical charge inhomogeneity. Evidence in support of this picture for high temperature superconductors is described.

1 Introduction

The past few years have seen a resurgence of interest in the behavior of correlated electron systems. Much of this activity has been directed towards an understanding of high temperature superconductivity, but it also has led to a renewed interest in novel materials in general, especially oxides, heavy–fermion superconductors, Kondo insulators, and synthetic metals such as organic conductors and alkali–doped C_{60}. As a whole, these materials display a wide variety of magnetic and charge–ordering behavior, together with transport properties, ranging from high temperature superconductivity to a giant magnetoresistance, which challenge the foundations of the theory of metals.

The conventional theory of metals is based on the ideas of Fermi liquid theory, which is a semi–phenomenological description of the behavior of a system of interacting fermions in terms of its low–energy excitations: collective modes and quasiparticles. Fermi liquid theory has provided a remarkably successful description of simple metals, liquid ^{3}He, and nucleons in an atomic nucleus, despite the presence of rather strong interactions between the fundamental constituents. The concept of a propagating quasiparticle is valid provided its mean free path is longer than its de Broglie wavelength: $l > \lambda_F = 2\pi/k_F$ (Ioffe and Regel 1960). This condition may not be violated in typical solids, until l is roughly equal to the lattice spacing a. But many novel materials, especially those which are close to a metal–insulator transition, have a rather low carrier concentration and small Fermi wave vector, and the quasiparticle concept may break down even though l/a is relatively large. In conventional metals with strong electron–phonon coupling, the resistivity saturates at a value ρ_s before the Ioffe–Regel condition is violated; in the A15's, for example, ($\rho_s \approx 150\mu\Omega - cm$) corresponds to $l \approx 2\lambda_F$.

Many synthetic metals are, in fact, "bad metals" for which a theory based on conventional quasiparticles with reasonably well–defined crystal momenta suffering occasional scattering events does not apply, as a number of examples will show. The optical conductivity of a typical organic conductor does not display a Drude peak, the hallmark of a quasiparticle. In the case of Rb_3C_{60}, Boltzmann transport theory would imply that the mean free path of a quasiparticle is less than the size of a C_{60} molecule, and much less than the lattice spacing. The in-plane resistivity of a high temperature superconductors such as $La_{1.85}Sr_{.15}CuO_4$ increases linearly with temperature from T_c up to $900°K$, where its magnitude is about 0.7 mΩcm, with no sign of saturation. Thus the quasiparticle picture breaks down at high temperature and indeed at all temperatures above T_c, since the simple temperature dependence of the resistivity indicates a single mechanism of charge transport. The resistivity of the $La_{1-x}Ca_xMnO_3$ is metallic but again it violates the Ioffe–Regel condition, attaining an even larger value than in the high temperature superconductors. All in all, a large but metallic resistivity is typical of oxides and synthetic metals in general, and an intrinsic linear temperature dependence is frequently, but not universally, observed. This behavior should be contrasted with that of materials with strong electron–phonon coupling, such as the A15's, which are not bad metals as the term is used here.

A major consequence of the poor conductivity of bad metals is that the long–range part of the Coulomb interaction is poorly screened. This is a characteristic of another group of materials, granular conductors and ultra–thin two-dimensional films, which are of considerable interest in their own right, and serve as an interesting counterpoint for bad metals. Granular conductors are close to a metal–insulator transition and they share with the high temperature superconductors the properties of charge inhomogeneity, poor screening, and an enhanced role for superconducting phase fluctuations. In granular conductors, ultra–thin films, and radiation–damaged high temperature superconductors, the effects of poor screening are particularly evident, and the conductivity at T_c must exceed a material–dependent critical value σ_c in order to overcome the disordering effects of the long–range Coulomb interaction. However, granular conductors and ultra–thin films differ from bad metals in that their resistivity is activated at high temperature.

The anomalous normal state of a bad metal requires that the physics of the superconducting transition be reconsidered. Here we give a summary of the main ideas, and argue that the usual mean field theory breaks down because quantum and classical phase fluctuations may not be neglected. detailed accounts are given BCS–Eliashberg mean–field theory (Schrieffer 1964), which is an extremely good approximation for conventional superconductors, electron pairing and long–range phase coherence occur at the same temperature T_c^{MF}. There are essentially three types of fluctuations about the BCS ground state which can ultimately destroy the superconducting order: classical phase fluctuations, quantum phase fluctuations, and the effects of all other degrees of freedom which affect the local magnitude of the order parameter.

As described in Sec. 2, the stiffness of the system to classical phase fluctuations is determined by the superfluid density, $\rho_s(T)$; the smaller the superfluid density, the more significant classical phase fluctuations. In bad metals, especially the high temperature superconductors, the value of the bare superfluid density n_{s0} is quite low at zero temperature, so the *classical* phase ordering temperature, which is proportional to n_{s0}/m^*, can be substantially lower than T_c^{MF}.

Quantum phase fluctuations, which will be considered in Sec. 3, are associated with the number–phase uncertainty relation (Anderson 1966) according to which phase coherence between neighboring regions implies large relative number fluctuations and correspondingly large Coulomb energies unless there is adequate screening. Thus, the smaller the dielectric function, or equivalently the frequency dependent conductivity, the more significant the effects of quantum phase fluctuations.

Together, these effects determine the superconducting transition temperature T_c in a bad metal. But they have a number of other consequences for the physical properties of the materials. Local superconducting fluctuations are important for a much larger range of temperatures above T_c than in good metals, and phase fluctuations control the temperature–dependence of the penetration depth and other parameters of the superconducting state. In particular, the unusual linear temperature–dependence of the penetration depth of high temperature superconductors (Hardy et al. 1993) may be explained in this way (Emery and Kivelson 1995b).

Another important feature of doped insulators is static or dynamical charge inhomogeneity. The theoretical reasons to expect this behavior are given in Sec. 4, and some of the experimental evidence to show that it actually occurs in high temperature superconductors is described in Sec. 5. Implications for the mechanism of high temperature superconductivity are summarised in Sec. 6.

2   Classical Phase Fluctuations

The importance of phase fluctuations may be assessed by using experimentally–determined quantities to evaluate the temperature T_θ^{max} at which phase order would disappear if the disordering effects of all other degrees of freedom were ignored. If $T_c \ll T_\theta^{max}$, phase fluctuations are relatively unimportant, and T_c will be close to the mean–field transition temperature, T^{MF}, predicted by BCS–Eliashberg theory (Schrieffer 1964). On the other hand, if $T_\theta^{max} \sim T_c$, then the value of T_c is determined primarily by phase ordering, and T^{MF} is simply the characteristic temperature below which local pairing becomes significant.

In order to evaluate T_θ^{max}, the system must be divided into regions of linear dimension a which are large enough for the order parameter to be well–defined locally. A region j is characterised by a phase angle θ_j and its dynamically conjugate variable, the number of electrons N_j. The Hamiltonian governing the thermodynamic effects of long wavelength phase fluctuations at low temperature is the kinetic energy of the superfluid:

$$\mathcal{H} = \frac{1}{2}\rho_s(0) \int d\mathbf{r}\, \mathbf{v}_s^2 \,, \tag{1}$$

where $\mathbf{v}_s = \hbar \nabla \theta / 2m^*$ is the superfluid velocity, and m^* is the effective mass of an electron. The system described by $\mathcal{H}$ undergoes a phase–ordering transition, since θ is an angle variable (defined modulo 2π), and there is a short–distance spatial cutoff, *i.e.* the variables are defined in regions of size a, and the integral and derivative in Eq.(1) should be regarded as a sum and a finite difference respectively.

The characteristic energy scale for phase fluctuations is the zero–temperature "phase stiffness" $V_0 = (\hbar/2m^*)^2 \rho_s(0) a$, which may be expressed in terms of the length scale a and a measurable quantity, the penetration depth $\lambda(T)$, via the relation $4\pi \rho_s(0) = (m^* c / e\lambda(0))^2$ to obtain:

$$V_0 = \frac{(\hbar c)^2 a}{16\pi e^2 \lambda^2(0)} \,. \tag{2}$$

Since V_o gives the energy scale of the model, it follows that the transition temperature $T_\theta^{max} = AV_0$ where A is a dimensionless number of order 1 which depends on the details of the short–distance physics. In three dimensions, our prescription for the short–distance cutoff is equivalent to defining the model on a simple cubic lattice, for which $A = 2.2$ (Adler et al. 1993). In order to complete the specification of V_0, we identify a^2 with the area $\pi\xi^2$ defined by the superconducting coherence length ξ; the precise numerical relation between a and ξ will not be important for the subsequent discussion.

Quasi two–dimensional systems, such as oxide superconductors consist of weakly coupled planes in which the phase variables may be defined on a square lattice. The value of A lies between 0.9 (for a two–dimensional system (Olsson and Minnhagen 1991)) and 2.2 (for the isotropic three–dimensional system). For definiteness, we shall use $A = 0.9$ for all quasi–two dimensional materials, but the true phase ordering temperature may typically be 50% larger. The in–plane cutoff does not enter the expression for V_0, so that a in Eq. (2) is the larger of the average spacing between layers d or $\sqrt{\pi}\xi_\perp$, where $\xi_\perp$ is the coherence length perpendicular to the layers. Since d is known very accurately, there is much less uncertainty in the calculated value of V_0 for materials, such as organic and high temperature superconductors, for which $d > \xi_\perp$.

Other ways of introducing the short–distance cutoff into the Hamiltonian will give somewhat different values of A but the one we have chosen is physically natural and, as we shall see, it leads to a very suggestive interpretation of the phase diagram of high–temperature superconductors.

T_θ^{max} is an upper bound on the true superconducting transition temperature T_c because of the neglect of quantum phase fluctuations, as well as the tempera-ture dependent effects of the other degrees of freedom. Similarly, T^{MF} is also an upper bound on T_c, since phase fluctuations will always depress T_c somewhat. Thus, the ratio T_θ^{max}/T_c provides a very useful criterion for the importance of phase fluctuations; if it is large, it is clear that phase fluctuations have a minor effect on T_c, whereas if it is close to 1, then phase fluctuations depress T_c sub-stantially below T^{MF} or, as we shall see, they may even be the major factor in determining T_c itself. In fact V_0 is important in its own right since, as a measure of the rigidity of the superconducting state to variations of the phase, it characterises the energetics of vortex lattices.

The values of T_θ^{max}/T_c for a wide variety of materials have been tabulated (Emery and Kivelson 1995a). For elemental metals such as lead, $T_\theta^{max}/T_c \sim 10^5$ is extraordinarily high; phase fluctuations have a negligible effect on T_c and the superconducting state has a substantial phase rigidity at all temperatures below T_c. In other words, pairing and long–range phase coherence occur essentially simultaneously. With the exceptions to be described below, other superconducting materials behave in the same way, although the values of T_θ^{max}/T_c are somewhat smaller.

On the other hand, the organic superconductor $(BEDT-TTF)_2Cu(NCS)_2$ and the hole–doped oxide superconductors are in an entirely different range of parameters inasmuch as T_θ^{max}/T_c is of order unity.

In particular, the analysis suggests a new interpretation of the usual phenomenological classification of high temperature superconductors into three, more or less distinct categories: "underdoped", "optimally doped", and "overdoped." The value of T_c is predominantly determined by phase fluctuations in underdoped high temperature superconductors such as $YBa_2Cu_4O_8$, ($T_\theta^{max}/T_c \approx 1$), and by the mean–field transition temperature T^{MF} in overdoped materials such as Tl 2201 ($T_\theta^{max}/T_c \geq 2$). Optimally doped materials, such as $YBa_2Cu_3O_{7-\delta}$ and $La_{2-x}Sr_xCuO_4$, with δ and x in the neighbourhood of 0.05 and 0.15 respectively, are in the crossover region between the two. Of course there is no precise dividing line between the two kinds of superconductor but rather a more or less gradual crossover in behavior as the value of T_θ^{max}/T_c changes. Furthermore, the values of this ratio are subject to the uncertainties in the experimental values of $\lambda(0)$ and the precise form of the short–distance cutoff. But the systematic variation of the properties of high temperature superconductors from one material to another clearly supports the importance of phase fluctuations.

When T_c is much smaller than T^{MF}, the effects of pairing manifest themselves as a pseudogap in the the temperature range $T_c < T < T^{MF}$. This observation provides a natural explanation of a variety of measurements, including NMR [1] and optical conductivity (Basov et al. 1994), (Wachter et al. 1994), on underdoped high temperature superconductors, such as the stoichiometric material $YBa_2Cu_4O_8$ (T_c=80K), which show a pseudogap opening below a temperature of order 160–180K. Similar behavior is seen (Mehring 1992) in underdoped $YBa_2Cu_3O_{7-\delta}$, extending to the large δ end of the the 90K plateau, beyond which there is a rather rapid change in behavior as the oxygen content is increased.

This analysis is macroscopic and independent of the underlying "mechanism" of high temperature superconductivity. However, some additional constraints on the appropriate microscopic theory emerge: 1) The theory must account for the

---

[1] For a review, see Mehring (1992).

very high values of $T^{MF} > T_c$ and be able to survive any loss of low–energy spectral weight due to the opening of a pseudogap. 2) The fundamental reason why phase fluctuations are so important in the oxide superconductors is that they are doped insulators with a very low Drude weight which, in turn, implies a low superfluid density. This requires a significant departure from the BCS–Eliashberg theory. 3) When phase fluctuations are important, the characteristic consequences of the BCS mean–field theory, such as the existence of NMR coherence peaks, the jump in the specific heat at T_c, the value of the gap ratio, $2\Delta/k_B T_c$, the isotope effect, and the temperature dependence of physical properties must be modified.

These ideas give a new perspective on the relation between T_c and $\lambda^{-2}(0)$ suggested by muon spin relaxation experiments. From our point of view, the "universal relation" between T_c and the muon depolarization rate in high temperature superconductors proposed by Uemura and coworkers (Uemura 1991) should in fact be reinterpreted as an *upper bound* on T_c given by the ordering temperature for phase fluctuations, as shown above. Indeed the picture that underdoped materials are close to the bound (but depressed below it by an amount that depends on the conductivity) and that overdoped materials are further from the bound because the mean field transition temperature takes control, gives a very good description of the systematics of the relation between T_c and the muon depolarisation rate. Other explanations of the relationship between $\lambda^{-2}(0)$ and T_c also make use of the low superfluid density, but they also assume that the size of a pair is much smaller than the average spacing between the charge carriers [2]. Our analysis shows that the second assumption is not necessary in this context, and must be justified on other grounds.

3   Quantum Phase Fluctuations

In elemental superconductors, $\sigma(T_c) \gg \sigma_Q$, so quantum phase fluctuations are entirely insignificant. However, as indicated above, poor conductivity implies poor screening, which supresses the charge fluctuations implied by phase order. A method of calculating these effects has been presented elsewhere (Emery and Kivelson 1995b); two consequences will be mentioned here.

In the case of pristine stoichiometric high temperature superconductors, $\sigma(T_c)/\sigma_Q \sim 10$, and there is a measurable suppression of T_c below T_θ^{max}. According to our estimates, quantum fluctuations produce a 5%–7% suppression of T_c in optimally–doped $YBa_2Cu_3O_{7-\delta}$. Damaged or relatively more disordered samples tend to have a substantially smaller conductivity, and quantum effects are correspondingly more important. Indeed it is possible to increase the resistance so as to destroy superconductivity altogether. The idea that the superconducting transition is influenced by the value of the conductivity at T_c correctly predicts many trends in the transition temperature of high temperature superconductors. It has been considered as an explanation of the variation of T_c upon purposely

---

[2] See the contribution by R.Micnas in this volume.

reducing the conductivity of $YBa_2Cu_3O_{7-\delta}$ by radiation damage (Sun et al. 1994).

A second important implication for high temperature superconductors is the existence of a substantial range of temperatures in the neighborhood of T_c in which critical phase fluctuations dominate the low frequency electromagnetic response, and a low temperature regime in which $n_s(T)$ will have an anomalous temperature dependence as a result of phase fluctuations: $n_s(T)/n_s(0) = 1 - \alpha \mathcal{F} k_B T/[2V_o(\alpha - 1)]$, to first order in T. Here, it is assumed that the Coulomb interaction is screened by the finite residual far IR conductivity observed in $YBa_2Cu_3O_{7-\delta}$ (Basov et al. 1995). If we use $\sigma(\omega) = 500(\Omega cm)^{-1}$, (which is roughly consistent with experiment) together with $R_Q \approx 40k\Omega$ ($\mathcal{F} \approx 1/6$), as deduced from the data of Sun et al. (1994), the linear temperature dependence of the superfluid density agrees quantitatively with the observations (Hardy et al. 1993).

4 Charge Inhomogeneity

High temperature superconductors provide the most important examples of doped insulators. They are obtained by chemically adding charge carriers to a highly–correlated antiferromagnetic insulating state, and by now there is a good deal of theoretical evidence that, in the absence of long–range Coulomb interactions (i.e. for neutral holes), a low concentration of holes is unstable to phase separation into a hole–rich "metallic" phase and a hole–deficient antiferromagnetic phase (Emery, Kivelson, and Lin 1990).

It has always been clear that macroscopic phase separation is suppressed by the long–range Coulomb interaction (Emery, Kivelson, and Lin 1990). However this does *not* mean that the Coulomb interaction merely stablises a state of uniform density in the neutral system. Indeed, for jellium, the Coulomb interaction favors *local* charge inhomogeneity (a Wigner crystal) whereas it is the *kinetic energy* that forces the system to be uniform. As we have shown, the situation is entirely different for a correlated electron systems for which minimization of the zero–point kinetic energy is achieved by separation into hole–rich and hole–free regions: all energies conspire to produce a state that is inhomogeneous on some length scale and time scale, although of course macroscopically it must be uniform.

A simple argument illustrates how charge inhomogeneity emerges from the competition between phase separation and the long–range part of the Coulomb interaction. In linear response theory, the Debye screening length λ_D is given by

$$\lambda_D^2 = \frac{1}{4\pi e^2} \frac{\partial \mu}{\partial n}, \tag{3}$$

where μ and n are the chemical potential and number density of the neutral system, and e is the charge. Now $\frac{\partial n}{\partial \mu} = n^2 \kappa$, where κ is the compressibility. Thus, between the pseudospinodals of the neutral system, $\kappa < 0$ and λ_D is imaginary.

This implies that the uniform state is unstable to charge inhomogeneity, with a length scale determined by the value of $|\lambda_D|$. The inhomogeneity may be static, as in a charge density wave or a "cluster spin glass" phase, or fluctuating in time. The dynamical character of frustrated phase separation stems from the *quantum* nature of the problem and is not easily displayed by solving a microscopic model. Consequently we consider two versions of a coarse–grained representation of the problem; a classical Ising pseudospin model (Löw et al. 1994) and a quantum version of the corresponding spherical model (Chayes et al. 1996). In particular, it will be shown that the Coulomb interactions do *not* generally favor a uniform density phase but rather produce charge–modulated structures, with periods that are unrelated to nesting wave vectors of the Fermi surface.

4.1 Coarse–Grained Models

The Hamiltonian for the Ising pseudospin model is given by:

$$H = K \sum_j S_j^2 - L \sum_{<ij>} S_i S_j + \frac{Q}{2} \sum_{i \neq j} \frac{S_i S_j}{r_{ij}} . \tag{4}$$

Here $S_j = \pm 1, 0$ is a coarse–grained variable representing the local density of mobile holes (Emery and Kivelson 1993). Each site j lies on a two–dimensional square lattice and represents a small region of space in which $S_j = +1$ and $S_j = -1$ correspond to hole–rich and hole–poor phases respectively, whereas $S_j = 0$ indicates that the local density is equal to the average value. The fully phase separated state has $S_j^2 = 1$ and is ferromagnetically ordered, with $S_j = +1$ in one half of the volume, and $S_j = -1$ in the other, so as to maintain overall charge neutrality.

The zero temperature phase diagram was determined for the complete range of parameters by using a combination of numerical and analytical techniques (Löw et al. 1994). It was found that the pure Coulomb interaction favors a Néel state (equivalent to a Wigner crystal) but, as Q decreases, the system crosses over to a ferromagnetic (phase–separated) state via a rich structure of highly symmetric striped and checkerboard phases. Regions with uniform charge density, corresponding to sites with $S_j = 0$, do not occur unless K is positive and sufficiently large.

In the spherical version of the model (Chayes et al. 1996), the S_j are real numbers in the range $[-\infty, \infty]$, and quantum conjugate "momenta" P_j are introduced. Momentum order corresponds to superconductivity. The Hamiltonian includes a term proportional to $\sum (P_i - P_j)^2$ and a constraint in which the mean value of $\sum [S_i^2 + P_i^2]$ is equal to a constant. Thus the model is Gaussian, so it may be solved exactly and correlation functions and other properties may be evaluated at finite temperature. The constraint guarantees a non–trivial phase diagram, in which superconductivity competes with charge density wave order. The disordered region displays crossovers to fluctuating hole–free droplets and to orientationally–ordered stripes, as the temperature is lowered (Chayes et al. 1996).

The solution of these simple models confirms our intuition that local inhomogeneity is the expected consequence of frustrated phase separation and that it should be a characteristic behavior of metallic correlated electron systems. But it also indicates that *ordered* charge–modulated states are a likely outcome unless they are destroyed by quantum effects and/or frustration. A specific example is $La_{2-x}Sr_xNiO_{4+\delta}$, which is identical in structure to $La_{2-x}Sr_xCuO_{4+\delta}$ with Ni replacing the Cu. When undoped, this system is a spin–one antiferromagnet, so it is expected that quantum fluctuations will be considerably less important than in the cuprates. When doped, La_2NiO_4 is known to form a variety of modulated phases (Tranquada et al. 1993), (Sachan et al. 1995), (Tranquada et al. 1995b).

The hole concentration in a stripe is governed by the energetics of phase separation. Thus, in general, the stripes are partially–filled and *metallic*, and the associated wave vectors do not nest the Fermi surface. The intervening hole–free regions are antiferromagnetic antiphase spin domains, with a wave vector of π parallel to the stripes, and a period equal to twice the stripe period in the perpendicular direction. In this sense, the charge order drives the spin order.

4.2 Evidence for Stripes

Recent neutron scattering experiments (Tranquada et al. 1995a) have shown that the suppression of superconductivity in $La_{1.6-x}Nd_{0.4}Sr_xCuO_4$, for x near to 0.125, is caused by the formation of ordered charge and spin–density waves in the CuO_2 planes. The ordered state consists of an array of charged stripes which form antiphase domain walls between antiferromagnetically ordered spin domains. This observation explains the peculiar behavior of the La_2CuO_4 family of compounds near to $\frac{1}{8}$ doping (Moodenbaugh et al. 1988), and strongly supports the idea that disordered, or fluctuating stripe phases are of central importance for the physics of high temperature superconductors (Emery and Kivelson 1993).

In the neutron scattering experiments the principal signature of the antiphase spin domains in $La_{1.6-x}Nd_{0.4}Sr_xCuO_4$ is a set of resolution–limited peaks in the spin structure factor at wave vectors $(\frac{1}{2} \pm \epsilon, \frac{1}{2})$ and $(\frac{1}{2}, \frac{1}{2} \pm \epsilon)$ in units of $2\pi/a$. The associated charge stripes are indicated by peaks in the nuclear structure factor at wave vectors $(0, \pm 2\epsilon)$ and $(\pm 2\epsilon, 0)$. Thus it is natural to interpret the *inelastic* peaks in the magnetic structure factor previously observed in similar locations in reciprocal space for superconducting samples of $La_{2-x}Sr_xCuO_{4-\delta}$ (Cheong et al. 1991), Mason at al. 1992), Thurston et al. 1992) as evidence of stripe *fluctuations* in which the stripes are oriented along vertical or horizontal Cu—O bond directions respectively.

Another conceivable mechanism for the stripe phases in $La_{2-x}Sr_xCuO_{4-\delta}$ is Fermi–surface nesting (Schulz 1989), (Zaanen et al. 1989), (Poilblanc and Rice 1989), (Schulz 1990), (Giamarchi and Lhuillier 1990), (Verges et al. 1991), (Inui and Littlewood 1991), (An and van Leeuwen 1991), which produces an insulating state, with a reduced density of states at the Fermi energy (an energy gap). On the other hand, in the frustrated phase separation picture, the period of the ordered density wave is generally unrelated to any nesting vector of the Fermi surface. The charge forms a periodic array of *metallic* stripes, with a hole density determined by the energetics of phase separation. The spin order has twice the period of the charge order, and consists of undoped antiferromagnetic regions, which are weakly antiferromagnetically coupled across the charge stripes. The experiments clearly favor the latter point of view. The relevant wave vectors do not nest the Fermi surface, the stripes are partially filled, and the ordered system is not an insulator. The peaks associated with magnetic ordering develop below the charge–ordering temperature (Tranquada et al. 1995a), which shows that the transition is driven by the charge, rather than the spin.

5 Some Consequences of Charge Ordering

In this section we describe some of the consequences of charge ordering and the low effective carrier concentration of high temperature superconductors.

5.1 Angle–Resolved Photemission Spectroscopy

The single–particle properties of a disordered striped phase also account for the peculiar features of the electronic structure of high temperature super-conductors observed by angle–resolved photoemission spectroscopy (ARPES) in $Bi_2SrCaCu_2O_{8+x}$, the best studied of the hole–doped high temperature superconductors (Salkola, Emery, and Kivelson, unpublished). In particular, the spectral function of holes moving in a disordered striped background reproduces the experimentally–observed shape of the Fermi surface, the existence of nearly dispersionless states at the Fermi energy ("flat bands") (Dessau et al. 1993), and the weak additional states ("shadow bands") (Aebi et al. 1994), features which have no natural explanation within conventional band theory. In our picture, the "flat bands" arise as follows: The ordered system has energy gaps at specific points on the Fermi surface that are spanned by the wave vectors of the charge and spin structures. An energy gap serves to flatten the energy bands in its vicinity. In the case of disordered but slowly–fluctuating stripes, the energy gaps are smeared, leaving a region of dispersionless states, which give the appearance of flat bands in the ARPES experiments, although they do not correspond to quasiparticle states.

5.2 Magnetic Resonance

Since the stripes are charged, they are easily pinned by disorder. Thus, if the temperature is not too high, we can think of the system as a quenched disordered array of stripes, which divides the Cu—O plane into long thin regions, with weak antiphase coupling between the intervening hole–deficient regions. This picture rationalises a) the observation (Hammel et al., unpublished) by NQR of two distinct species of Cu nuclei (one in a pinned stripe, the other between the stripes), and b) the existence of a "cluster–spin–glass" phase in samples with $x < 15\,\%$ (Cho et al. 1992) (since the antiphase coupling between regions is frustrating). Moreover, there is evidence that the creation of dilute meandering stripes can account for the rapid supression of the Néel temperature for $x < 2\,\%$ (Borsa et al. 1995), (Castro–Nieto and Hone, unpublished).

6 Mechanism of High Temperature Superconductivity

In our view, high temperature superconductivity is associated with the inhibition of a first order phase transition (phase separation) by the long–range Coulomb interaction, which converts a relatively large condensation energy into a strong pairing force. As we have seen in Sec. 4, this mechanism of superconductivity may compete with or coexist with charge inhomogeneity in the form of droplets or ordered structures. Here we consider some aspects of this behavior in the cuprates.

The phase diagram which follows from the role of classical phase fluctuations in underdoped high temperature superconductors (Emery and Kivelson 1995a) strongly suggests that there is a very high energy scale for pairing in lightly-doped but metallic high temperature superconductors. In other words, a single stripe in an undoped antiferromagnetic environment should manifest the mechanism of pairing, although full phase coherence and long–range order could not be established. As a model for this problem, we have analysed the behavior of a one–dimensional electron gas (the stripe) in an active environment (the undoped antiferromagnet). This is a generalization of the theory of the one–dimensional electron gas. We have found several processes that involve the coupling between the mobile holes and the environment and lead to pairing, *even though the basic Hamiltonian contains only repusive interactions* (Emery, Kivelson, and Zachar, unpublished). Here we mention one which involves a pair of holes hopping from the stripe into a bound state of the environment. This process has a number of advantages for high temperature superconductivity. A pair of holes in the medium may have a large binding energy, but such a tightly bound pair is typically immobile, since it cannot easily move without breaking up. Thus it does not, by itself, lead to high temperature superconductivity. However, the holes in the stripe are able to utilise this large binding energy to form pairs, without losing their own mobility, and in this way they achieve a high superconducting transition temperature. Secondly, a stripe phase already has incorporated the long range part of the Coulomb interaction, and a pair may hop into the close neighborhood of a stripe without too much cost in energy. Thus the poorly-screened Coulomb force, which is especially damaging to pairing in systems with a small coherence length (such as the high temperature superconductors), is not a severe problem.

It is important to note that this model provides a counterexample to the argument that antiferromagnetism cannot be relevant for true high temperature superconductivity because it is so difficult to detect by neutron scattering experiments on optimally–doped $YBa_2Cu_3O_{7-\delta}$. If the "environment" consisted of two coupled spin chains (a spin ladder), then the magnetic excitations would have a spin gap of about $0.5J$ (White et al. 1994). Here J is the exchange integral which is about 100 meV in the high temperature superconductors, so there would be no spin excitations in the range of energies that have been used in most neutron scattering experiments. Nevertheless two holes on a ladder form a bound state (Tsunetsugu, Troyer, and Rice 1995) of just the kind required to account for high temperature superconductivity.

Acknowledgements: SK was supported in part by NSF grant #DMR-90-11803. This work also was supported by the Division of Materials Sciences, U.S.Department of Energy, under contract DE-AC02-76CH00016.

References

Adler, J., Holm, C., Janke, W. (1993): *Physica* A **201**, 581–592.

Aebi, P., Osterwalder, J., Schwaller, P., Schlapbach, L., Shimoda, M., Mochiku, T., Kadowaki, K. (1994): *Phys.Rev.Lett.* **72**, 2757–2760.

Allen, P.B. (1992): *Comm.on Cond.Matt.Phys.* **15** 327–331.

An, G, van Leeuwen, G.M.J., (1991): *Phys.Rev.*B **44**, 9410–9417.

Anderson, P.W. (1966): *Quantum Fluids*, ed.by Brewer, D. F. (North–Holland. Amsterdam) pp.146–171.

Basov, D.N., Liang, R., Bonn, D.A., Hardy, W.N., Dabrowski, B., Quijada, M., Tanner, D.B., Rice, J.P., Ginsberg, D.M., Timusk, T. (1995): *Phys.Rev.Lett.* **74**, 598–601.

Basov, D.N., Timusk, T., Dabrowski, B., Jorgensen, J.D. (1994): *Phys.Rev.*B **50**, 3511–3514.

Borsa, F., Carretta, P., Cho, J.H., Chou, F.C., Hu, Q., Johnston, D.C., Lascialfari, A., Torgensen, D.R., Gooding, R.J., Salem, N.M., Vos, K.J.E. (1995): *Phys.Rev.*B **52**, 7334–7345.

Castro–Nieto A., Hone D., unpublished.

Chayes, L., Emery, V.J., Kivelson, S.A., Nussinov, Z., Tarjus, J. (1996): *Physica* A, in press.

Cheong, S-W., Aeppli, G., Mason, T.E., Mook, H., Hayden, S.M., Canfield, P.C., Fisk, Z., Clausen, K.N., Martinez, J.L. (1991): *Phys.Rev.Lett.* **67**, 1791–1794.

Cho, J.H., Borsa, F., Johnston, D.C., Torgensen, D.R. (1992): *Phys.Rev.*B **46**, 3179–3182.

Dessau, D.S., Shen, Z.-X., King, D.M., Marshall, D.S., Lombardo, L.W., Dickinson, P.H., Loeser, A.G., Di Carlo, J., Park, C.-H., Kapitulnik, A., Spicer, W.E. (1993): *Phys.Rev.Lett.* **71**, 2781–2784.

Emery, V.J., Kivelson, S.A., (1993); *Physica* C **209** 597–621.

Emery, V.J., Kivelson, S.A. (1995): *Nature* **374**, 434–437.

Emery, V.J., Kivelson, S.A. (1995): *Phys.Rev.Lett.* **74**, 3253–3256.

Emery, V.J., Kivelson, S.A., Lin, H-Q. (1990): *Phys.Rev.Lett.* **64**, 475–478.

Emery, V.J., Kivelson, S.A., Zachar, O., unpublished.

Giamarchi, T., Lhuillier, C. (1990): *Phys.Rev.*B **42**, 10641–10647.

Hardy, W.N., Bonn, D.A., Morgan, D.C., Liang, R., Zhang, K. (1993): *Phys.Rev.Lett.* **70**, 3999–4003.

Inui, M., Littlewood, P. (1991): *Phys.Rev.*B **44**, 4415–4422.

Ioffe, A.F., Regel, A.R. (1960): *Semicond.* **4**, 237–251.

Löw, U., Emery, V.J., Fabricius, K., Kivelson, S.A. (1994): *Phys.Rev.Lett.* **72**, 1918–1921.

Mason, T.E., Aeppli, G., Mook, H.A. (1992): *Phys.Rev.Lett.* **68**, 1414–1417.

Mehring, M. (1992): *Appl.Magn.Reson.* **3**, 383–421.

Moodenbaugh, A.R., Xu, Y., Suenaga, M., Folkerts, T.J., Shelton, R.N. (1988): *Phys.Rev.*B **38**, 4596–4600.

Olsson, P., Minnhagen, P. (1991): *Phys.Scr.* **43**, 203–209.

Poilblanc, D., Rice, T.M. (1989): *Phys.Rev.*B **39**, 9749–9752.

Sachan, V., Buttrey, D.J., Tranquada, J.M., Lorenzo, J.E., Shirane, G. (1995): *Phys.Rev.*B **51**, 12742–12746.

Salkola, M., Emery, V.J., Kivelson, S.A., unpublished.

Schrieffer, J.R. (1964): *Theory of Superconductivity,* (Benjamin, New York).

Schultz, H. (1989): *J.Phys.(Paris)* **50** 2833–2841.

Schultz, H. (1990): *Phys.Rev.Lett.* **64**, 1445–1448.

Hammel, P.C., Statt, B., Martin, R.L., Cheorg, S-W., Chou, F.C., Johnston, D.C., unpublished.

Sun, A.G., Paulius, L.M., Gajewski, D.A., Maple, M.B., Dynes, R.C. (1994): *Phys.Rev.*B **50**, 3266–3270.

Thurston, T.R., Gehring, P.M., Shirane, G., Eirgeneau, R.J., Kastner, M.A., Endoh, Y., Matsuda, M., Yamada, K., Kojima, H., Tanaka, I., (1992): *Phys.Rev.*B **46**, 9128–9131.

Tranquada, J.M., Buttrey, D.J., Rice, D.E. (1993): *Phys.Rev.Lett.* **70**, 445–448.

Tranquada, J.M., Sternlieb, B.J., Axe, J.D., Nakamura, Y., Uchida, S.(1995): *Nature* **375**, 561–563.

Tranquada, J.M., Lorenzo, J.E., Buttrey, D.J., Sachan, V. (1995): *Phys.Rev.*B **52**, 3581–3595.

Tsunetsugu, H., Troyer, M., Rice, T.M. (1995): *Phys.Rev.*B **51**, 16456–16459.

Uemura, Y.J., Le, L.P., Luke, G.M., Sternlieb, B.J., Wu, W.D., Brewer, J.H., Riseman, T.M., Seaman, C.L., Maple, M.B., Ishikawa, M., Hinks, D.G., Jorgensen, J.D., Saito, G., Yamochi, H. (1991): *Phys.Rev.Lett.* **66**, 2665–2668.

Verges, J.A., Louis, E., Lomdahl, P.S., Guinea, F., Bishop, A.R. (1991): *Phys.Rev.*B **43**, 6099–6108.

Wachter, P., Bucher, B., Pittini, R. (1994): *Phys.Rev.*B **49**, 13164–13171.

White, S.R., Noack, R.M., Scalapino, D.J. (1994): *Phys.Rev.Lett.* **73**, 886–889.

Zaanen, J., Gunnarson, O. (1989): *Phys.Rev.*B **40**, 7391–7394.

Superconducting Properties
of the Weakly Interacting Charged Bose Gas

Zygmunt M. Galasiewicz

Institute of Theoretical Physics, University of Wroclaw, Max Born sq. 9, 50-204 Wroclaw, Poland, and
Institute of Low Temperature and Structure Research of the Polish Academy of Sciences, Wroclaw, Poland

Keywords. Other topics in superconductivity, boson degeneracy, high-T_c- superconductors

In 1955 Schafroth [1] considered superconductivity of a charged ideal Bose gas. The $T_c^o - n_c^c$ dependence, where T_c^o denotes the critical temperature for free bosons and n_c^o — the condensate density (superconducting carriers density), has for $T = 0$ the known form $T_c^o \approx (n_c^o)^{2/3}/m^*$, which can be named the Uemura type relation [2]. In some limits it agrees with the experimental data for high–T_c–superconductors [2].

Schafroth's interesting discovery, that an ideal Bose gas exhibits the Meissner–Ochsenfeld effect was manifested by the formula $\mathbf{j} = -(e^2/mc)n_c^o\mathbf{A}$.

A more general model than Schafroth's was considered in refs. [3,4], namely the weakly interacting charged Bose gas. After diagonalization of the Hamiltonian the energy spectrum is described by

$$\varepsilon(p) = \sqrt{(up)^2 + (\frac{p^2}{2m})^2} , \quad m = 2m_e , \tag{1}$$

where

$$u = \frac{2\hbar}{m}\sqrt{\pi n a} , \tag{2}$$

where a denotes the scattering length. In order to use the weak coupling theory $a \ll d = n^{-1/3}$ where d is the mean distance between bosons. The "temperature" t_1 as well as the other dimensionless temperatures are defined as follows

$$T_1 k_B \equiv mu^2 , \quad t_1 = \frac{T_1}{T_c^o} \approx a , \quad t = \frac{T}{T_c^o} , \quad t_c = \frac{T_c}{T_c^o} . \tag{3}$$

One can introduce two different decompositions of the boson density n

$$n = n_c(t_1,t) + n_{int}(t_1) + n_{ex}(t_1,t) = n_s(t_1,t) + n_n(t_1,t) . \tag{4}$$

where n_c describes the density of condensate, n_{int} — the temperature indepenent density describing the contribution to the non–condensed density due to the interaction and n_{ex} is described by the Bose distribution function $n_0(\varepsilon)$ where ε is given by (1). On the other hand the density of the normal component n_n is calculated from the flow properties ($\varepsilon = \varepsilon(p) + \mathbf{pv}$).

$$n_n = -\frac{1}{3V} \sum_{\mathbf{P}} p^2 \frac{\partial n_o}{\partial \varepsilon}\bigg|_{\varepsilon \,=\, \varepsilon(p)} \cdot \tag{5}$$

The density n_{ex} is described by formulae

$$\frac{n_{ex}(t_1, t)}{n} = t^{3/2} \frac{I(t_1/t)}{I(0)} = \frac{n_{ex}^o}{n} \frac{I(t_1/t)}{I(0)} \,,$$

$$n_{ex}(t_1, 0) = 0 \,, \qquad \frac{n_{ex}^o}{n} = t^{2/3} \,,$$

$$I(s) = s^{3/2} \int_0^\infty \frac{x}{e^{sx} - 1} \frac{dx}{\sqrt{1 + \sqrt{1 + x^2}}} \,,$$

$$I(0) = 2.315 \,, \qquad s = t_1/t \,. \tag{6}$$

For n_{int} we have

$$\frac{n_{int}(t_1)}{n} = \frac{\sqrt{2}}{3I(0)} t_1^{3/2} \approx 0.2 t_1^{3/2} \,. \tag{7}$$

From (4), (6) and (7), it follows that

$$\frac{n_c(t_1, t)}{n} = 1 - 0.2 t_1^{3/2} - t^{3/2} \frac{I(t_1/t)}{I(0)} \,. \tag{8}$$

At $t = 0$

$$\frac{n_c(t_1, 0)}{n} = 1 - 0.2 t_1^{3/2} \,. \tag{9}$$

With help of (5) we can express n_n by n_{ex}

$$\frac{n_n(t_1, t)}{n} = \frac{n_{ex}(t_1, t)}{n} - \frac{3}{2} t \sqrt{t_1} \frac{K(t_1/t)}{I(0)} \,, \qquad n_n(t_1, 0) = 0 \,. \tag{10}$$

where

$$K(s) = S \int_0^\infty \frac{dx}{e^{sx} - 1} \frac{\sqrt{\sqrt{1 + x^2} - 1}}{2(\sqrt{1 + x^2})^3} (\sqrt{1 + x^2} + 2) > 0 \,. \tag{11}$$

The integrals $I(s)$ and $K(s)$ are presented in Fig.1.
From (10) we see that for interacting bosons $(t_1 \neq 0)$ $n_{ex} > n_n$.
The expression for $n_s(t_1, t)$ is given by

$$\frac{n_s(t_1, t)}{n} = \frac{n_c(t_1, t)}{n} + 0.2 t_1^{3/2} + \frac{3}{2} t \sqrt{t_1} \frac{K(t_1/t)}{I(0)} \,, \tag{12}$$

For interacting bosons $n_s(t_1, t) > n_c(t_1, t)$.
For non–interacting bosons $(t_1 = 0)$

$$n_n^o = n_{ex}^o \,, \qquad n_c^o = n_s^o \,. \tag{13}$$

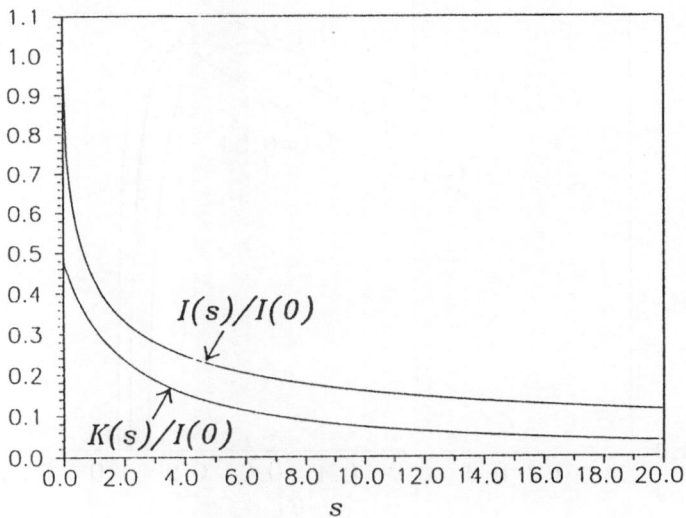

Fig. 1. Plots of the integrals $I(s)$ and $K(s)$.

These relations show some type of degeneracy in the case on non–interacting bosons. The degeneracy is removed by even very weak interaction *i.e.* $t_1 \rightarrow 0$, $t_1 \neq 0$.

We know for helium 4 the values of m, a, n and T_c^o. Hence, with help of (2) and (3) we find that $t_1 = 2.69$ (see [4]). Formula (9) gives

$$\frac{n_c(t_1, t)}{n} = 0.12 = 12\% . \tag{14}$$

Thus, applying the formula derived for a system with weak interactions to a system with strong interactions (helium 4), we get a quite reasonable result, namely 12% of the particles in the Bose–condensate at $T = 0$.

One considers the density of the condensate $n_c(t_1, t)$ as the order parameter (more exactly, $n_c^{1/2}$.

At $t = t_c$

$$n_c(t_1, t_c) = 0 \rightarrow t_1 = t_1(t_c) . \tag{15}$$

Now the superfluid density n_s describes the carrier density. With help of (15) $n_s = n_s(t_c, t)$, which presents the $t_c - n_s$ dependence identified with the Uemura plot. For smaller t_c we have linear behavior followed by a rapid falloff as n_s increases (see Fig.2).

Consideration of the interaction among bosons improves the agreement with Uemura results in comparison to the case of an ideal Bose gas.

According to Ref. [2] the typical effective mass is $m = m^* \approx 5m_e$. According to experimental estimates [6] the lower limit for carrier density (suitable for

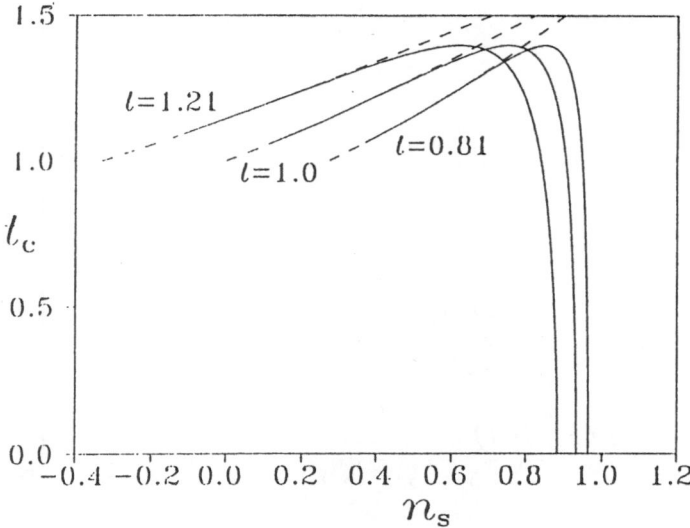

Fig. 2. Uemura plots of the $t_c - n_s$ dependence for three values of the temperature t.

model considered here) is $n_s = 0.0002\text{Å}^{-3}$. Theoretical considerations in [4] give for weak coupling ($t_1 : 0.1 - 0.3$, $t = 1$) $n = 2n_s$, $d = n^{-1/3} \approx 13\text{Å}$. The formula for estimation of the coherence length ξ was taken from [7]. We have

$$\lambda_{\rm L} \approx 1000 \text{ Å}, \quad \xi \approx 1 \text{ Å}, \quad T_c \approx 360 \text{ K}, \quad \xi/d \approx 10^{-1} . \tag{16}$$

For the model considered here, the order of magnitude estimates of the first three patrameters are characteristic for the high–T_c–superconductors. The fourth one describes the bosonic region.

Regarding electrodynamics (Meissner–Ochsenfeld effect) we have for the current density the following expression

$$\mathbf{j} = \frac{e}{V} \sum_{\mathbf{p}} \mathbf{p} n_o(\varepsilon(p) - \frac{e}{mc}\mathbf{pA}) - \frac{e^2}{mc} n\mathbf{A} , \tag{17}$$

where n_o is the Bose distribution function and n is described by (4). For small $\mathbf{A}$ we have (see (5), $\mathbf{v} \to -(e^2/mc)\mathbf{A}$):

$$\frac{e}{V} \sum_{\mathbf{p}} \mathbf{p} n_o(\varepsilon(p) - \frac{e}{mc}\mathbf{pA}) = -\mathbf{A}\frac{1}{3V} \sum_{\mathbf{p}} p^2 \frac{\partial n_o}{\partial \varepsilon}\bigg|_{\varepsilon \, = \, \varepsilon(p)} = \frac{e^2}{mc} n_n \mathbf{A} . \tag{18}$$

In consequence

$$\mathbf{j} = -\frac{e^2}{mc} n_s \mathbf{A} , \qquad n_s = n_c + n_{int} + n_{ex} - n_n . \tag{19}$$

To compare with the Schafroth's result [1], the density of condensate n_c in (19) is replaced in a natural way by the superfluid density n_s. Formula (19) can be derived for boson systems on the basis of microscopic considerations [8] similar to the presentation in [9,10] for fermion systems.

Finally it is worth calling attention to the results of [11] authors who find, as for HeII, 3D XY critical behaviour in $YBa_2Cu_3O_{6.95}$ crystals.

References

1. M.R.Schafroth, *Phys.Rev.* **100**, 463 (1955).

2. Y.J.Uemura *et al.*, *Phys.Rev.Lett.* **62**, 2317 (1989); *ibid.* **66**, 2665 (1991).

3. C.P.Enz and Z.M.Galasiewicz, *Physica* C **214**, 239 (1993).

4. Z.M.Galasiewicz and E.Wolf, *Physica* C **248**, 49 (1995).

5. E.E.Bogoliubov, *Lectures on Quantum Statistics* Vol.2, Part 4 (Macdonald Technical and Scientific, London, 1971).

6. D.R.Harsman and A.P.Mills Jr., *Phys.Rev.*B **45**, 10684 (1992).

7. R.Friedberg, T.D.Lee and H.C.Ren, *Phys.Rev.*B **42**, 4122 (1990).

8. Z.M.Galasiewicz, (*in preparation* 1996).

9. N.N.Bogoliubov, *Usp.Fiz.Nauk* **67**, 549 (1959).

10. Z.(M.)Galasiewicz, *Prog.Theor.Phys.* **23**, 197 (1960).

11. S.Kamal, D.A.Bonn, I.Goldenfeld, P.J.Hirschfeld, R.Liang, and W.A.Hardy, *Phys.Rev.Lett.* **73**, 1845 (1994).

Universal Properties of Multilayer High–Temperature Superconductors: Transition Temperature and a Spatial Modulation of the Gap

Józef Spałek[1,2] and Krzysztof Byczuk[2]

[1] Institute of Physics, Jagiellonian University, ul.Reymonta 4, 30–059 Kraków, Poland
[2] Institute of Theoretical Physics, Warsaw University, ul.Hoża 69, 00–681 Warszawa, Poland*

Abstract. We discuss a three–dimensional model high–temperature superconductor, which encompasses both mono– and multilayer systems as planar non–Fermi liquids. Explicitly, we derive the analytic formulae for the superconducting transition temperature T_c and for the shape of the space profile of the gap for a material composed of a periodic arrangement of CuO_2 planes divided into groups of $p \geq 1$ tightly spaced planes. The results are compared with experiment and provide strong support for an interlayer Cooper–pair hopping between the planes as the source of the large T_c enhancement in multilayer compounds. Our results are universal in the sense that only general properties of the in–plane pairing potential $V_{kk'}$ are required to derive them. The results are applicable to both Landau and nonLandau fermionic liquids such as spin–charge separated (Luttinger) and statistical–spin liquids, each of which is also characterized briefly.

1 Introduction

The normal state of high–temperature superconductors is unusual: basic physical properties such as electrical resistivity, magnetic susceptibility, thermopower or Hall effect (with an exception of the Hall angle) all scale in a wide temperature T range as T^α, where $\alpha = \pm 1$ [1]. This scaling law differs from that for the Landau Fermi liquids for which $\alpha = \pm 2$ with $t \to 0$, where t is the reduced temperature, T/T_c. On the other hand, a well defined two–dimensional Fermi surface has been detected [2], albeit with an anomalous shape of the spectral density for the photoexcited electron. Therefore, the first fundamental question to be addressed is how is it possible to depart from the Landau Fermi–liquid (FL) picture as far as the excitation spectrum is concerned, but preserve the existence (and the volume) of the Fermi surface together with the gapless single–particle excitation across it ?

To describe the anomalous normal properties few alternative routes have been proposed. Probably the simplest of the whole class of models based on the Fermi liquid theory [3], [5] derives the form of the single–particle Green's

---

* E–mail: ufspalek@if.uj.edu.pl, byczuk@fuw.edu.pl

function from the observed experimental properties for one physical quantity and applying it subsequently to calculate all other properties. In such a way the marginal Fermi liquid was invented [3]. An alternative Fermi–liquid approach is the so–called nested Fermi liquid theory [4] for which the nested form of the two–dimensional Fermi surface is explicitly included in the quasiparticle scattering. In the phenomenological approach, one includes the effect of antiferromagnetic paramagnon excitations on both normal properties and on superconducting pairing [5].

An alternative new approach to the strongly correlated electronic systems is based on the concept of the two–dimensional Luttinger liquid (LL) [6]. In this respect the form of the Green's function for the one–dimensional Luttinger liquid has been assumed to remain the same in the two–dimensional liquid (and called the tomographic Luttinger liquid). This approach, supplemented with the interplane Cooper–pair hopping as the source of superconductivity enhancement [7] provides a consistent and tractable model of high–temperature planar superconductors. In this class of models the renormalization–group (RG) approach to correlated fermion systems [7], [8] can be placed. However, the physical implications of this approach to high–T_c systems, as well as inclusion of pairing have not been explored as yet.

In this paper we overview briefly our approach to those strongly correlated fermionic systems [9]–[11]. This work is based on the bosonization scheme generalized to arbitrary spatial dimensions [12], [13]. Within this approach we divide those systems into three universality classes: Fermi, Luttinger and statistical spin liquids, and then derive the properties valid for all three classes, namely the superconducting transition temperature T_c and the spatial variation of the superconducting gap $\Delta \equiv \Delta_{j,\mathbf{k}}$ in the direction perpendicular to the CuO_2 planes ($\mathbf{k} = (k_x, k_y)$ is the two dimensional wave vector and j labels the planes).

2 Fermionic Quantum Liquids as Separated Universality Classes

In the present Section we sketch an effective theory of single–particle states close to the Fermi surface, which is based on the expression of the low–energy excitations as bosons in a d-dimensional interacting gas. This language will be used subsequently to introduce the so–called tomographic Luttinger liquid. Its single–particle Green's function is needed to derive the superconducting characteristics.

To classify the normal states of the quantum fermionic liquids we utilize the language of bosonization, a nonperturbative method of expressing the Hamiltonian of interacting fermionic liquids as a Hamiltonian of free Bose fields. The most important feature of this method is the (assumed) existence of the Fermi surface. To describe those excitations one introduces local fluctuations of the particle density in the form

$$J_\sigma(\mathbf{S}, \mathbf{q}) = \sum_{\mathbf{k}} \Theta(\mathbf{S}, \mathbf{k} + \mathbf{q}) \, \Theta(\mathbf{S}, \mathbf{k}) \, (\psi^+_{\mathbf{k}+\mathbf{q}\sigma} \psi_{\mathbf{k}\sigma} - \delta^{(d)}_{\mathbf{q},0} \bar{n}_{\mathbf{k}\sigma}) \, . \tag{1}$$

In this expression S represents the Fermi surface point, $\mathbf{q}$ is the propagation wave vector of the density fluctuation and the step functions are defined as nonzero only within the squat pill box depicted in Fig.1 of the volume $\Lambda^{d-1}\lambda$ containing point S as its central point, and both the vectors $\mathbf{k}$ and $\mathbf{k} + \mathbf{q}$. The subtraction of the vacuum expectation value $\bar{n}_{\mathbf{k}\sigma} \equiv \langle \Psi_{\mathbf{k}\sigma}^+ \Psi_{\mathbf{k}\sigma} \rangle$ amounts to the normal ordering of the field operators.

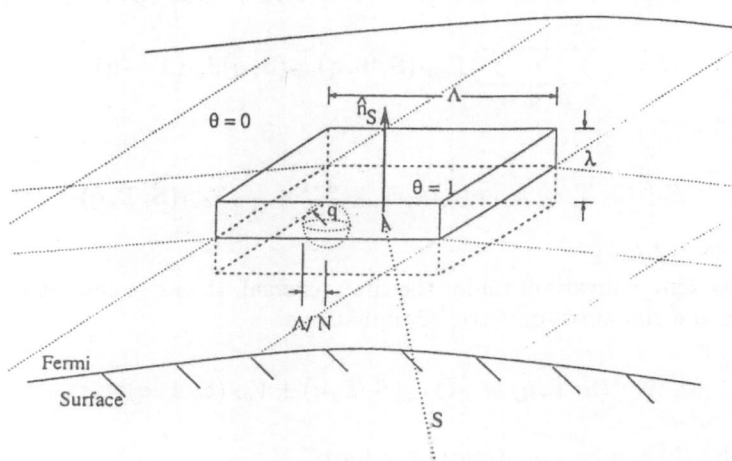

Fig. 1. Schematic representation of the squat pill boxes surrounding each Fermi surface point S. The box has dimension Λ^{d-1} and high $\lambda = \Lambda/\sqrt{N}$ (where $N \to \infty$) and is bisected through the midlpane by the Fermi surface. The quasimomentum $\mathbf{q}$ is small: $|\mathbf{q}| \leq \Lambda/N$. After Ref.12.

The density operators obey the following commutation relations

$$[J_\sigma(\mathbf{S}, \mathbf{q}), J_{\sigma'}(\mathbf{T}, \mathbf{q}')] = \delta_{\sigma\sigma'}\, \delta_{\mathbf{S},\mathbf{T}}^{(d-1)}\, \delta_{\mathbf{q}+\mathbf{q}',0}^{(d)} \Lambda^{d-1} \left(\frac{L}{2\pi}\right)^d \mathbf{q} \cdot \hat{n}_{\mathbf{S}}\,, \qquad (2)$$

where $\hat{n}_{\mathbf{S}}$ is the versor normal to the Fermi surface at point S, and $\delta_{\mathbf{S},\mathbf{T}}^{(d-1)}$ is the $(d-1)$-dimensional Kronecker delta function (Fermi surface is a $(d-1)$-dimensional object in the d-dimensional reciprocal space). The right hand side of (2) is a number called the quantum anomaly, and hence (2) can be brought to the Bose form after rescaling.

To illustrate the usefulness of the operator (1) we start from the effective low–temperature Hamiltonian in the form

$$H = \sum_{\mathbf{S},\sigma} v_F(\mathbf{S}) \int d\mathbf{x}\, \psi_\sigma^+(\mathbf{S}, \mathbf{x}) \left(\frac{\hat{n}_{\mathbf{S}} \cdot \nabla}{i}\right) \psi_\sigma(\mathbf{S}, \mathbf{x}) + \qquad (3)$$

$$\frac{1}{2} \sum_{\mathbf{S},\mathbf{T}} \sum_{\sigma,\sigma'} \int d\mathbf{x}\, d\mathbf{y}\, \psi_\sigma^+(\mathbf{S}, \mathbf{x}) \psi_{\sigma'}^+(\mathbf{T}, \mathbf{y}) V_{\sigma\sigma'}(\mathbf{S}, \mathbf{T}, \mathbf{x} - \mathbf{y}) \psi_{\sigma'}(\mathbf{T}, \mathbf{y}) \psi_\sigma(\mathbf{S}, \mathbf{x})\,,$$

where $v_F(S)$ is the effective (renormalized) Fermi velocity at point S, and $\Psi_\sigma(S, x)$ is the field operator, here defined at a given Fermi surface point. We stress that this Hamiltonian *is not* the usual Hamiltonian of interacting fermions, as the field operators depend additionally on S. The Heisenberg uncertainty principle is not violated since the point S is defined within the volume $\Lambda^{d-1}\lambda$ in k space. For the purpose of this Section the explicit form of the effective potential $VV_{\sigma\sigma'}$ is not necessary.

The main step is taken by expressing (3) in terms of the operators (1), namely

$$H = \frac{1}{2}\sum_{S,T}\sum_{q}\sum_{\sigma\sigma'}\Gamma_{\sigma\sigma'}(S, T, q)J_\sigma(S, q)J_{\sigma'}(T, -q), \qquad (4)$$

where

$$\Gamma_{\sigma\sigma'}(S, T, q) = v_F(S)\frac{1}{\Omega}\delta_{\sigma,\sigma'}\delta_{S,T}^{d-1} + \frac{1}{L^d}V_{\sigma\sigma'}(S, T, q), \qquad (5)$$

and $\Omega = \Lambda^{d-1}(L/2\pi)^d$.

If the system is invariant under the time reversal, then one can introduce the symmetric and the antisymmetric combinations

$$V^{c,s}(S, T, q) \equiv \frac{1}{2}(V_{\sigma\sigma}(S, T, q) \pm V_{\sigma\bar\sigma}(S, T, q)), \qquad (6)$$

and then Eq.(5) can be rewritten in the form

$$H = H^c + H^s \equiv \sum_{\alpha=c,s} H^\alpha, \qquad (7)$$

where

$$H^\alpha = H_0^\alpha + H_{int}^\alpha, \qquad (8)$$

with

$$H_0^\alpha = \frac{1}{2}\sum_{S} v_F(S)\frac{1}{\Omega}\sum_{q} J_\alpha(S, q)J_\alpha(S, -q), \qquad (9)$$

and

$$H_{int}^\alpha = \frac{1}{L^d}\sum_{S,T}\sum_{q} V^\alpha(S, T, q)J_\alpha(S, q)J_\alpha(T, -q), \qquad (10)$$

with

$$J_{c,s}(S, q) \equiv \frac{1}{\sqrt{2}}(J_\uparrow(S, q) \pm J_\downarrow(S, q)). \qquad (11)$$

The operators $J_{c,s}$ represent the total particle density (charge) and longitudinal spin fluctuations, respectively. One can prove that the commutation relation (2) now takes the form

$$[J_\alpha(S, q), J_\beta(T, p)] = \delta_{\alpha\beta}\,\delta_{S,T}^{d-1}\,\delta_{p+q,0}^d\,\Omega\,q\cdot\hat{n}_S, \qquad (12)$$

i.e. preserve essentially the same character, *i.e.* commute to the number. They reflect the independence of the two branches of fluctuations (c and s). In order to introduce the Bose representation explicitly, we rescale the density as follows

$$J_\alpha(\mathbf{S}, \mathbf{q}) = \theta(\hat{n}_\mathbf{S} \cdot \mathbf{q})\sqrt{\Omega \hat{n}_\mathbf{S} \cdot \mathbf{q}}\, a_\alpha(\mathbf{S}, \mathbf{q}) + \theta(-\hat{n}_\mathbf{S} \cdot \mathbf{q})\sqrt{-\Omega \hat{n}_\mathbf{S} \cdot \mathbf{q}}\, a_\alpha^+(\mathbf{S}, -\mathbf{q}), \quad (13)$$

where $\theta(x)$ is the step function and the two terms reflect the particle ($\hat{n}_\mathbf{S} \cdot \mathbf{q} > 0$) and hole ($\hat{n}_\mathbf{S} \cdot \mathbf{q} < 0$) fluctuations. Now

$$\left[a_\alpha(\mathbf{S}, \mathbf{q}), a_\beta^+(\mathbf{T}, \mathbf{p}) \right] = \delta_{\mathbf{S},\mathbf{T}}^{d-1} \delta_{\mathbf{p},\mathbf{q}}^d \delta_{\alpha\beta}. \quad (14)$$

The Hamiltonian (7) takes the following form in the bosonic degrees of freedom:

$$H = \sum_{\alpha=c,s} \sum_{\mathbf{S},\mathbf{T}} \frac{1}{2} \left\{ \sum_\mathbf{q} \theta(\hat{n}_\mathbf{S} \cdot \mathbf{q})\theta(\hat{n}_\mathbf{T} \cdot \mathbf{q})\sqrt{(\mathbf{v}_F(\mathbf{S}) \cdot \mathbf{q})(\mathbf{v}_F(\mathbf{T}) \cdot \mathbf{q})} \times \right.$$

$$\left(\delta_{\mathbf{S},\mathbf{T}}^{d-1} + \Lambda^{d-1} \frac{1}{(2\pi)^d} \frac{V^\alpha(\mathbf{S}, \mathbf{T}, \mathbf{q})}{\sqrt{v_F(\mathbf{S})v_F(\mathbf{T})}} \right) a_\alpha^+(\mathbf{S}, \mathbf{q})\, a_\alpha(\mathbf{T}, \mathbf{q}) +$$

$$\sum_\mathbf{q} \theta(-\hat{n}_\mathbf{S} \cdot \mathbf{q})\theta(-\hat{n}_\mathbf{T} \cdot \mathbf{q})\sqrt{(-\mathbf{v}_F(\mathbf{S}) \cdot \mathbf{q})(-\mathbf{v}_F(\mathbf{T}) \cdot \mathbf{q})} \times$$

$$\left(\delta_{\mathbf{S},\mathbf{T}}^{d-1} + \Lambda^{d-1} \frac{1}{(2\pi)^d} \frac{V^\alpha(\mathbf{S}, \mathbf{T}, \mathbf{q})}{\sqrt{v_F(\mathbf{S})v_F(\mathbf{T})}} \right) a_\alpha^+(\mathbf{S}, -\mathbf{q})\, a_\alpha(\mathbf{T}, -\mathbf{q}) +$$

$$\sum_\mathbf{q} \theta(\hat{n}_\mathbf{S} \cdot \mathbf{q})\theta(-\hat{n}_\mathbf{T} \cdot \mathbf{q})\sqrt{(\mathbf{v}_F(\mathbf{S}) \cdot \mathbf{q})(-\mathbf{v}_F(\mathbf{T}) \cdot \mathbf{q})} \times$$

$$\left(\Lambda^{d-1} \frac{1}{(2\pi)^d} \frac{V^\alpha(\mathbf{S}, \mathbf{T}, \mathbf{q})}{\sqrt{v_F(\mathbf{S})v_F(\mathbf{T})}} \right) a_\alpha(\mathbf{S}, \mathbf{q})\, a_\alpha(\mathbf{T}, -\mathbf{q}) +$$

$$\sum_\mathbf{q} \theta(-\hat{n}_\mathbf{S} \cdot \mathbf{q})\theta(\hat{n}_\mathbf{T} \cdot \mathbf{q})\sqrt{(-\mathbf{v}_F(\mathbf{S}) \cdot \mathbf{q})(\mathbf{v}_F(\mathbf{T}) \cdot \mathbf{q})} \times$$

$$\left. \left(\Lambda^{d-1} \frac{1}{(2\pi)^d} \frac{V^\alpha(\mathbf{S}, \mathbf{T}, \mathbf{q})}{\sqrt{v_F(\mathbf{S})v_F(\mathbf{T})}} \right) a_\alpha^+(\mathbf{S}, -\mathbf{q})\, a_\alpha^+(\mathbf{T}, \mathbf{q}) \right\}. \quad (15)$$

This rather lengthy form of the Hamiltonian has a simple interpretation. Namely, the single–particle ($\sim v_F(\mathbf{S})$) and the interaction ($\sim V^\alpha$) parts appear on the same footing. Secondly, since $\Lambda \sim 1/L \to 0$ as $L \to \infty$, the properties of large systems are determined by an asymptotic behavior of the effective potential in the scaling limit $L \to \infty$. In this respect, only three physically distinct situations are possible as $\mathbf{q} \to 0$ and $L \to \infty$:
   (i) V remains finite (hence $\Lambda^{d-1}V^\alpha \to 0$);
   (ii) V is divergent, but $\Lambda^{d-1}V^\alpha$ remains finite; and
   (iii) V is divergent and so is $\Lambda^{d-1}V^\alpha$.
Depending on this behavior, we will have three universality classes:

(i) The effective Hamiltonian at low energies scales to the free–particle limit with a renormalized Fermi velocity, *i.e.* the system is a *Landau Fermi liquid*;

(ii) the interaction is comparable to the single–particle energy, *i.e.* the system is a *Luttinger liquid*; and

(iii) the interaction diverges essentially, and the bosonization scheme breaks down; *i.e.* we deal with the so–called *statistical spin liquid* [9].

Each of this cases forms a universality class, since the interaction part represents respectively: (i) marginal, (ii) relevant, or (iii) dangerously relevant operator in the renormalization group scheme. The connection with the standard RG approach becomes apparent once one realizes that the interaction term has a form of the same type as the term $\sim |\Psi|^4$ in the Ginzburg–Landau–Wilson functional. Thus the three universality classes (Fermi, Luttinger and the statistical spin liquids) are the only possibilities if the effective interaction is symmetric with respect time reversal and contains no coupling to a singular gauge field. What is even more important, the effective potential takes only the scattering processes in the almost forward direction and must be divergent in the thermodynamic limit in the system of space dimension $d > 1$ (note Λ^{d-1} factor in (15)) in order to make the Fermi liquid solution unphysical.

In connection with the above analysis, one notices that the $2k_F$ scattering processes are absent in (10). They are always irrelevant in the thermodynamic limit unless the Fermi surface has a nesting property, since only then do the $2k_F = Q$ processes represent a finite fraction ($\sim L^{d-1}$) of the Fermi surface. The inclusion of these very important processes will not be discussed here (see, however, some comments at the end of the next section).

3 A Soluble Model of the Luttinger Liquid

To study the superconducting phase of a two–dimensional Luttinger liquid we summarize first the principal features of a soluble model of this nonFermi liquid proposed by us recently [10]. Within this model one determines the single–particle Green function in the normal state, which is needed in subsequent T_c calculations.

In accordance with our analysis in the foregoing section we assume that the effective potential $V_{\sigma\sigma'}$ has the following structure

$$V_{\uparrow\uparrow}(\mathbf{S}, \mathbf{T}, \mathbf{q}) = f_{\uparrow\uparrow}(\mathbf{S}, \mathbf{T}, \mathbf{q}) , \tag{16}$$

$$V_{\uparrow\downarrow} = \begin{cases} \frac{1}{\Lambda^{d-1}} \tilde{g}(\mathbf{S}, \mathbf{q}) & \text{for} \quad \mathbf{S} = \mathbf{T} \\ f_{\uparrow\downarrow}(\mathbf{S}, \mathbf{T}, \mathbf{q}) & \text{for} \quad \mathbf{S} \neq \mathbf{T} , \end{cases} \tag{17}$$

where f, $\tilde{f}$, and $\tilde{g}$ are all nonsingular functions of their arguments. In the present analysis we neglect the regular parts f and $\tilde{f}$ and assume that in the low–energy limit the function $\tilde{g}$ may be expanded only to the first order, *i.e.* we put

$$\tilde{g}(\mathbf{S}, \mathbf{q}) = g_0(\mathbf{S}) + g(\mathbf{S})\hat{n}_{\mathbf{S}} \cdot \mathbf{q} + \cdots . \tag{18}$$

Such a procedure leads to an extremely simple effective Hamiltonian

$$H = \sum_{\alpha=c,s} \sum_{\mathbf{S},\mathbf{q}} (\mathbf{v}_F^\alpha(\mathbf{S}) \cdot \mathbf{q}) \, a_\alpha^+(\mathbf{S},\mathbf{q}) \, a_\alpha(\mathbf{S},\mathbf{q}) \,, \tag{19}$$

where the effective Fermi velocities are $v_F^{c,s}(\mathbf{S}) = v_F(\mathbf{S}) \pm g(\mathbf{S})$ for charge and spin fluctuations are now different. The inequality $v^C \neq v^s$ means that we have separation of the charge and the spin degrees of freedom.

In order to determine the fermion Green's function we have to use the connection between the Fermi and the Bose fields. Within the bosonization scheme we have the relation [12]

$$J_\sigma(\mathbf{S},\mathbf{x}) = \sqrt{4\pi}\, \hat{n}_\mathbf{S} \cdot \nabla \phi_\sigma(\mathbf{S},\mathbf{x}) \,, \tag{20}$$

where

$$J_\sigma(\mathbf{S},\mathbf{x}) = \frac{1}{\sqrt{L^d}} \sum_{\mathbf{q}} e^{i\mathbf{q}\cdot\mathbf{x}} J_\sigma(\mathbf{S},\mathbf{q}), \tag{21}$$

and $\phi_\sigma(\mathbf{S},\mathbf{x})$ is a Bose field. Equivalently, one can introduce the relation between the fermion field operators $\Psi_\sigma(\mathbf{S},\mathbf{x})$ and the Bose fields. It has the form

$$\psi_\sigma(\mathbf{S},\mathbf{x}) = \sqrt{\frac{\Omega}{a}}\, e^{i\mathbf{k}_\mathbf{S}\cdot\mathbf{x}}\, e^{i\frac{\sqrt{4\pi}}{\Omega}\phi_\sigma(\mathbf{S},\mathbf{x})}\hat{O}(\mathbf{S}), \tag{22}$$

where a is a cut–off (of the order of a lattice constant), which is set $a = 0$ at the end of the calculations. In this manner

$$\phi(\mathbf{S},\mathbf{q}) = \sqrt{\frac{\Omega}{4\pi\hat{n}_\mathbf{S} \cdot \mathbf{q}}}\, a_\alpha(\mathbf{S},\mathbf{q}) \,, \tag{23}$$

where $\phi_{c,s} = \frac{1}{\sqrt{2}}(\phi_\uparrow \pm \phi_\downarrow)$. Therefore, the fermion Green's function

$$G_\sigma^>(\mathbf{S},\mathbf{x},t > 0) = \langle \Psi_\sigma(\mathbf{S},\mathbf{x},t)\, \Psi^+(\mathbf{S},0,0)\rangle \tag{24}$$

is expressed as follows

$$G^>(\mathbf{S},\mathbf{x},t) = \frac{\Omega}{a}\, \exp\left[\frac{4\pi}{\Omega^2}\frac{1}{2} G_B^\sigma(\mathbf{S},\mathbf{x},t)\right] \,, \tag{25}$$

where

$$G_B^\sigma(\mathbf{S},\mathbf{x},t) = \langle(\phi_c(\mathbf{S},\mathbf{x},t) + \sigma\phi_s(\mathbf{S},\mathbf{x},t))(\phi(\mathbf{S},0,0) + \sigma\phi(\mathbf{S},0,0))\rangle$$
$$- \langle(\phi(\mathbf{S},0,0) + \sigma\phi(\mathbf{S},0,0))^2\rangle \,. \tag{26}$$

The straightforward but tedious analysis yields

$$G_\sigma^>(\mathbf{S},\mathbf{x},t) = i\Omega\, \frac{e^{i\mathbf{k}_\mathbf{S}\cdot\mathbf{x}}}{\sqrt{\hat{n}_\mathbf{S}\cdot\mathbf{x} - v_F^c(\mathbf{S})t + ia}\,\sqrt{\hat{n}_\mathbf{S}\cdot\mathbf{x} - v_F^s(\mathbf{S})t + ia}} \,. \tag{27}$$

The Green's function for this normal system does not depend on the spin σ. We see that instead of a quasiparticle pole, taking place in the Fermi liquid case, we have a branch cut ranging from v^s to v^c. Note that if the charge and the spin velocities were equal, this function would have a simple pole structure reproducing the standard free–particle result. The singular structure of the function $G_\sigma^>(\mathbf{x}, t)$ survives when we Fourier transform into $G_\sigma^>(\mathbf{k}, \omega)$. This feature can be demonstrated easily by decomposing the argument into components normal and transverse to the Fermi surface at point $\mathbf{S}$, namely $\mathbf{k} \cdot \mathbf{x} = k(\hat{n}_\mathbf{S} \cdot \mathbf{x} + \hat{t}_\mathbf{S} \cdot \mathbf{x})$, where $\hat{t}_\mathbf{S} k$ is the transverse part of $\mathbf{k}$ (tangential to the Fermi surface at point $\mathbf{S}$), and noting that the part $\hat{n}_\mathbf{S} \cdot \mathbf{x}$ can be transformed in the same manner as in the $d = 1$ case [14]. The $\hat{t}_\mathbf{S} \cdot \mathbf{x}$ part has a trivial form, since only the normal components matter in either direct or reciprocal spaces. In other words, as hypothesized by Anderson [6], the singular structure of the Green's function in $d = 1$ and $d = 2$ is the same (the liquid it describes has been termed the *tomographic Luttinger liquid*).

The form (27) of the Green's function will be used to determine the expression for the superconducting transition temperature for this non–Fermi liquid.

A methodological remark is in order at this point. Namely, the Green's function (27) does not contain the anomalous exponent α present in the one–dimensional case when the backscattering processes are included [14]. The general situation will lead to the smooth distribution $\bar{n}_\mathbf{k}$ across the Fermi surface (for $\mathbf{k} = \mathbf{k}_F$). In our simplified picture $\alpha = 0$ and the function $\bar{n}_\mathbf{k}$ has a discontinuity at k_F, the same as in the Fermi liquid case. The general case ($\alpha \neq 0$) should lead to the same type of physics, since the nature of the Green's function singularities (cuts) remains the same.

4   Three–Dimensional Model of High–Temperature Superconductors

We assume that we have a $\mathbf{k}$-dependent pairing potential $V_{\mathbf{k}\mathbf{k}'}$ in the two spatial dimensions (for a single CuO_2 plane) and a coherent (Josephson) electron pair tunneling between the planes. Having in view the application to real high–T_c systems we analyze explicitly the periodic arrangement of groups of planes, each of them containing p tightly spaced identical CuO_2 planes, and separated by a layer distance, but in contact with each other also by, albeit weaker, Cooper pair tunneling. Such a configuration is drawn schematically in Fig.2 (bottom part). The system is thus characterized by the intragroup ($T^{(l)}(\mathbf{k})$) and intergroup ($\tilde{T}(\mathbf{k})$) tunneling amplitudes. Hence, our approach can be applied to both mono- and multilayer high–T_c systems and thus generalizes previous treatments [7], [15]. The space profile calculated below of the superconducting gap is shown schematically in Fig.2 (top part). The goal of this Section is to show that our results for the transition temperature T_c are valid for either Fermi, or Luttinger, or statistical spin liquids. We make the same claim concerning the shape of the space profile of the gap within each group.

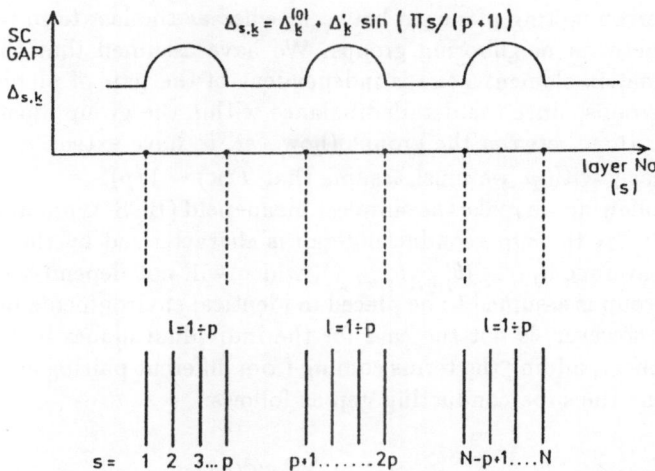

Fig. 2. Schematic plot of the gap magnitude (top, a) for a multilayer situation (bottom, b) involving groups of $p \geq 1$ tightly spaced planes.

We label the single–electron states by the in–plane component of the wave vector $\mathbf{k} = (k_x, k_y)$ and by the spin quantum number $\sigma = \pm 1$, as well as by the two discrete numbers (s, j), where $j = 1, \ldots, p$ characterizes the plane position within the group of tightly spaced planes, and $s = 1, \ldots, N$ labels the groups. The Hamiltonian describing the structure depicted in Fig.2 can be written as follows

$$
\begin{aligned}
H = & \sum_{s=1}^{N} \sum_{\mathbf{k}\sigma} \sum_{j=1}^{p} \varepsilon_{\mathbf{k}} b_{j,\mathbf{k}\sigma}^{+s} b_{j,\mathbf{k}\sigma}^{s} - \frac{1}{N} \sum_{s=1}^{N} \sum_{\mathbf{k}} \sum_{j=1}^{p} V_{\mathbf{k}\mathbf{k}'} b_{j,\mathbf{k}\uparrow}^{+s} b_{j,-\mathbf{k}\downarrow}^{+s} b_{j,-\mathbf{k}'\downarrow}^{s} b_{j,\mathbf{k}'\uparrow}^{s} \\
& - \sum_{s=1}^{N} \sum_{\mathbf{k}} \sum_{j=1}^{p} \sum_{l=1}^{j+l\leq p} T^{(l)}(\mathbf{k}) \left[b_{j,\mathbf{k}\uparrow}^{+s} b_{j,-\mathbf{k}\downarrow}^{+s} b_{j+l,-\mathbf{k}\downarrow}^{s} b_{j+l,\mathbf{k}\uparrow}^{s} + H.C. \right] \\
& - \sum_{s=1}^{N} \sum_{\mathbf{k}} \sum_{j=1}^{p} \sum_{l=1}^{1\leq j-l\leq p} T^{(l)}(\mathbf{k}) \left[b_{j,\mathbf{k}\uparrow}^{+s} b_{j,-\mathbf{k}\downarrow}^{s} b_{j-l,-\mathbf{k}\downarrow}^{s} b_{j-l,\mathbf{k}\uparrow}^{s} + H.C. \right] \\
& - \sum_{s=1}^{N-1} \sum_{\mathbf{k}} \tilde{T}(\mathbf{k}) \left[\left(\sum_{j=1}^{p} b_{j,\mathbf{k}\uparrow}^{+s} b_{j,-\mathbf{k}\downarrow}^{+s} \right) \left(\sum_{j=1}^{p} b_{j,-\mathbf{k}\downarrow}^{s+1} b_{j,-\mathbf{k}\uparrow}^{s+1} \right) + H.C. \right] . \quad (28)
\end{aligned}
$$

The first three terms describe the BCS–type Hamiltonian for Np independent planes (with an additional Coulomb interaction U for the cases of the Luttinger or the statistical spin liquids), each representing a two–dimensional metallic system; the fourth and the fifth terms express the pair tunneling (with the matrix element $T^{(l)}(\mathbf{k})$) between the planes in the same group, separated by

$(l-1)$ superconducting planes in between, whereas the last term represents the tunneling between neighboring groups. We have assumed that the intergroup tunneling matrix element $\widetilde{T}(\mathbf{k})$ is independent of the pair of planes selected in those two groups, since the detailed balance within the group must be achieved much faster than between the groups (however, to have extensive properties as scaling linearly with p, we must assume that $\widetilde{T}(\mathbf{k}) \sim 1/p$).

In the following we make the simplest mean–field (BCS–type) approximation, *i.e.* assume that the superconducting gap is characterized by the nonvanishing anomalous average $d_{j,\mathbf{k}} \equiv \langle b^s_{j,\mathbf{k},\uparrow} b^s_{j,-\mathbf{k},\downarrow} \rangle$, which will not depend on group index s, as each group is assumed to be placed in identical environments in the $N \to \infty$ limit (this, however, is not the case for the individual planes in the individual group). In effect, adding the terms coming from different pairing channels in (28) we can define the superconducting gap as follows:

$$
\Delta_{j,\mathbf{k}} = \frac{1}{N} \sum_{\mathbf{k}'} v_{\mathbf{k}\mathbf{k}'} d_{j,\mathbf{k}'} + \sum_{j=1}^{p} 2\widetilde{T}(\mathbf{k}) d_{j,\mathbf{k}}
$$
$$
+ \sum_{l=1}^{j+l \leq p} T^{(l)}(\mathbf{k}) d_{j+l,\mathbf{k}} + \sum_{l=1}^{1 \leq j-l \leq p} T^{(l)}(\mathbf{k}) d_{j-l,\mathbf{k}} . \tag{29}
$$

The self–consistency condition requires that

$$
d_{j,\mathbf{k}} = \Delta_{j,\mathbf{k}} \chi_{j,\mathbf{k}} . \tag{30}
$$

The pair–electron susceptibility $\chi_{j,\mathbf{k}}$ in this universal relation depends on the nature of the normal state of the system, *i.e.*, to which universality class discussed above the system belongs. In the case of the Fermi liquid it takes the usual form

$$
\chi_{j,\mathbf{k}} = \frac{1}{2E_{j,\mathbf{k}}} \tanh\left(\frac{\beta E_{j,\mathbf{k}}}{2}\right) , \tag{31}
$$

where the quasiparticle energy is $E_{j\mathbf{k}} = \sqrt{(\varepsilon_{\mathbf{k}} - \mu)^2 + |\Delta_{j,\mathbf{k}}|^2}$, and $\beta = 1/k_B T$ is the inverse temperature in energy units. In the case of the spin–charge separated Luttinger liquid discussed in the foregoing section, the function $\chi_{j,\mathbf{k}}$ is [16]

$$
\chi_{j,\mathbf{k}} = \frac{|\Delta_{j,\mathbf{k}}|^2}{\sqrt{\nu_-^2 + 4|\Delta_{j,\mathbf{k}}|^4}} \frac{\tanh(\beta E_{j,\mathbf{k}}/2)}{E_{j,\mathbf{k}}}
$$
$$
+ \frac{1}{\pi} \int_{-\pi}^{\pi} dx \tanh\left(\frac{\beta x}{2}\right) \frac{h_s h_c}{h_s^2 h_c^2 + |\Delta_{j,\mathbf{k}}|^4} , \tag{32}
$$

where $E_{j,\mathbf{k}} = \frac{1}{2}\left[\nu_+ + \sqrt{\nu_-^2 + 4|\Delta_{j,\mathbf{k}}|^4}\right]$, $h_{c,s} \equiv \sqrt{|x^2 - (v_{c,s}k)^2|}$, and $\nu_\pm \equiv (v_c^2 \pm v_s^2)k^2$.

Finally, for the statistical spin liquid the mean–field approximation leads also to Eq.(29), in which the right–hand side is multiplied only by the factor

$A_\mathbf{k} \equiv (1 - \bar{n}_\mathbf{k}/2)$, which near the Fermi energy has the value 2/3 (by taking this value, the further analysis is the same as for the Fermi liquid case) [17].

In what follows we consider the situation with coupling between nearest neighbours, i.e. assume that $T^{(l)}(\mathbf{k}) = T(\mathbf{k})\delta_{l,l\pm 1}$. In that situation the gap modulation within an individual group of p planes is determined by the system of equations

$$\Delta_{1\mathbf{k}} - \frac{1}{N} \sum_{\mathbf{k}'} V_{\mathbf{kk}'} \chi_{1,\mathbf{k}'} \Delta_{1,\mathbf{k}'} = T(\mathbf{k})\chi_{2,\mathbf{k}}\Delta_{2,\mathbf{k}} + 2\widetilde{T}(\mathbf{k}) \sum_{j'=1}^{p} \chi_{j',\mathbf{k}}\Delta_{j',\mathbf{k}} \,,$$

$$\Delta_{j,\mathbf{k}} - \frac{1}{N} \sum_{\mathbf{k}'} V_{\mathbf{kk}'} \chi_{j,\mathbf{k}'} \Delta_{j,\mathbf{k}'} = T(\mathbf{k})(\chi_{j+1\mathbf{k}}\Delta_{j+1,\mathbf{k}} \quad (33)$$

$$+ \chi_{j-1,\mathbf{k}}\Delta_{j-1,\mathbf{k}}) + 2\widetilde{T}(\mathbf{k}) \sum_{j'=1}^{p} \chi_{j',\mathbf{k}}\Delta_{j',\mathbf{k}} \,,$$

$$\Delta_{p,\mathbf{k}} - \frac{1}{N} \sum_{\mathbf{k}'} V_{\mathbf{kk}'} \chi_{p,\mathbf{k}'} \Delta_{p,\mathbf{k}'} = T(\mathbf{k})\chi_{p-1,\mathbf{k}}\Delta_{p-1,\mathbf{k}} + 2\widetilde{T}(\mathbf{k}) \sum_{j'=1}^{p} \chi_{j',\mathbf{k}}\Delta_{j',\mathbf{k}} \,.$$

The shape of the space profile of $\Delta_{j,\mathbf{k}}$ (i.e., its dependence on j) should not depend strongly on the system temperature, but only on the form of the boundary conditions. Therefore, we solve this system of equations for $T \to T_c$, where the generalized susceptibility can be linearized in the following way:

$$\chi_{j,\mathbf{k}} \simeq \frac{\tanh\left(\frac{\beta}{2}\varepsilon_\mathbf{k}\right)}{2\varepsilon_\mathbf{k}} + O(\Delta_{j,k}^2) \,, \quad (34)$$

for FL, and

$$\chi_{j\mathbf{k}} \simeq \frac{1}{\pi} \int_{v_s k}^{v_c k} dx \, \tanh\left(\frac{\beta x}{2}\right) \frac{1}{h_s h_c} + O(\Delta_{j,k}^4) \,, \quad (35)$$

for LL. Thus to zero order in the gap parameter the susceptibility does not depend on the layer index j, and the system reduces to the linear equations. In order to solve them we introduce two additional fictitious layers labelled $j = 0$ and $j = p + 1$. The boundary conditions are expressed in terms of the vanishing gap on the limiting layers, i.e. we set

$$\Delta_{0,\mathbf{k}} = \Delta_{p+1,\mathbf{k}} = 0 \,. \quad (36)$$

These boundary conditions express the physical equivalence of all p layers within the group, i.e., we do not introduce the surface layers, which differ from those inside [18]. In such a situation, the system (33) can be rewritten in a closed form

$$\Delta_{j,\mathbf{k}} - \frac{1}{N} \sum_{\mathbf{k}'} V_{\mathbf{kk}'} \chi_{\mathbf{k}'} \Delta_{j,\mathbf{k}'} = T(\mathbf{k})\chi_\mathbf{k}(\Delta_{j+1,\mathbf{k}} + \Delta_{j-1,\mathbf{k}}) + Z_\mathbf{k} \,, \quad (37)$$

where $Z_{\mathbf{k}} \equiv 2\tilde{T}(\mathbf{k})\chi_{\mathbf{k}} \sum_{j'=1}^{p} \Delta_{j',\mathbf{k}}$, and $j = 1, \ldots, p$. The last term in (37) does not depend on j, and therefore we study first the homogeneous part, $i.e.$ we put $Z_{\mathbf{k}} = 0$. Then, the general solution is taken in the form

$$\Delta_{j,\mathbf{k}} = \Delta_{\mathbf{k}}^{(+)} e^{i\alpha j} + \Delta_{\mathbf{k}}^{(-)} e^{-i\alpha j}, \tag{38}$$

where α is a constant determined from the boundary conditions (36), which can be written in the form

$$\begin{pmatrix} 1 & 1 \\ e^{i\alpha(p+1)} & e^{-i\alpha(p+1)} \end{pmatrix} \begin{pmatrix} \Delta_{\mathbf{k}}^{(+)} \\ \Delta_{\mathbf{k}}^{(-)} \end{pmatrix} = 0. \tag{39}$$

The vanishing determinant of the above matrix provides a nontrivial solution only when $\alpha = n\pi/(p+1)$, where n is an integer. Substituting the solution back to (37) with $Z_{\mathbf{k}} = 0$, we obtain $\Delta_{\mathbf{k}}^{(+)} = -\Delta_{\mathbf{k}}^{(-)} \equiv \Delta_{\mathbf{k}}$, and hence for $T(\mathbf{k}) > 0$ we have

$$\Delta_{j,\mathbf{k}} = 2i\Delta_{\mathbf{k}} \sin\left(\frac{n\pi j}{p+1}\right). \tag{40}$$

Physically, the admissible solution is that with $n = 1$, as this solution has no nodes inside p layers. One should notice also that this solution can be modified and adopted to the monolayer systems, for which one takes $p = N \to \infty$, and $\tilde{T}(\mathbf{k}) = 0$. In that situation solution (38) yields the equation for the gaps $\Delta_{\mathbf{k}}^{(\pm)}$

$$\Delta_{\mathbf{k}}^{(\pm)}(1 - 2T(\mathbf{k})\chi_{\mathbf{k}} \cos \alpha) = \frac{1}{N} \sum_{\mathbf{k}'} V_{\mathbf{k}\mathbf{k}'} \chi_{\mathbf{k}'} \Delta_{\mathbf{k}'}^{(\pm)}. \tag{41}$$

Also, periodic boundary conditions replacing (39) yield the solutions $\alpha = n2\pi/N$. The largest gap is obtained for $\alpha = 0$ for $T(\mathbf{k}) > 0$ or $\alpha = \pi$ for $T(\mathbf{k}) < 0$. In the former case we have a homogeneous gap, whereas in the latter we have an alternating sign from layer to layer for given $\mathbf{k}$.

Returning to our solution (40) for the multilayer case we obtain from (41) the self–consistent equation for $\Delta_{\mathbf{k}}$

$$\Delta_{\mathbf{k}} = \frac{1}{1 - 2T(\mathbf{k})\chi_{\mathbf{k}} \cos\left(\frac{\pi}{p+1}\right)} \frac{1}{N} \sum_{\mathbf{k}} V_{\mathbf{k}\mathbf{k}'} \chi_{\mathbf{k}'} \Delta_{\mathbf{k}'}. \tag{42}$$

The in–plane pairing potential is thus enhanced by the denominator. The solution of the full eq. (37) is found by superposing the general solution (40) of the homogeneous part of (37) with a particular solution of the full equation, $i.e.$ by taking

$$\Delta_{j,\mathbf{k}} = \Delta_{\mathbf{k}}' + \Delta_{\mathbf{k}} \sin\left(\frac{\pi}{p+1}j\right), \tag{43}$$

where now $\Delta_{\mathbf{k}}$ is still to be determined from the full equation and $\Delta_{\mathbf{k}}'$ is treated as a small perturbation. The part $\Delta_{\mathbf{k}}'$ does not depend on j because the periodic boundary conditions can be taken for $N \to \infty$ groups of p layers. Substituting the solution (43) into (37) and omitting higher–order terms (*i.e.* those $\sim \Delta'_{\mathbf{k}} \cdot \tilde{T}(\mathbf{k})$), we have the equation for $\Delta_{\mathbf{k}}$

$$\Delta_{\mathbf{k}} = \frac{1}{1 - 2 \left[\tilde{T}(\mathbf{k}) f(p) + T(\mathbf{k}) \cos \left(\frac{\pi}{p+1} \right) \right] \chi_{\mathbf{k}}} \frac{1}{N} \sum_{\mathbf{k}'} V_{\mathbf{k}\mathbf{k}'} \chi_{\mathbf{k}'} \Delta_{\mathbf{k}'} , \qquad (44)$$

where

$$f(p) \equiv \frac{\sin \left(\dfrac{\pi}{2} \dfrac{p}{p+1} \right)}{\sin \left(\dfrac{\pi}{2} \dfrac{1}{p+1} \right)} . \qquad (45)$$

The intergroup hopping $\tilde{T}$ additionally enhances the pairing potential $V_{\mathbf{k}\mathbf{k}'}$. Finally, the equation for $\Delta'_{\mathbf{k}}$ is

$$\Delta'_{\mathbf{k}} = \frac{1}{1 - 2T(\mathbf{k}) \chi_{\mathbf{k}}} \frac{1}{N} \sum_{\mathbf{k}'} V_{\mathbf{k}\mathbf{k}'} \chi_{\mathbf{k}'} \Delta'_{\mathbf{k}'} . \qquad (46)$$

This equation yields a transition temperature lower than the value obtained from (44). Therefore, we put $\Delta'_{\mathbf{k}} = 0$, since the system can have only one transition temperature for the onset of long–range ordering.

Equation (44) describes both the monolayer and multilayer ($p > 1$) systems. The proper limit $p \to \infty$ is obtained if we assume $\tilde{T}(\mathbf{k}) \sim 1/p$; this means that effectively only border layers between the two neighboring groups interact effectively.

5 Transition Temperature for a Quasiplanar System

To discuss the implications of Eq.(36) we consider first the case with $p = 1$, *i.e.*, a material with the interlayer hopping $\tilde{T}(\mathbf{k}) = (t')^2/t$, for which t is the magnitude of the bare single–particle hopping in the plane. In that case a simple estimate of the denominator of (44) for $T = T_c$ in the Fermi liquid case is:

$$1 - \tilde{T}(\mathbf{k}) \chi_{\mathbf{k}} \sim 1 - r \frac{t'}{t} \cdot \frac{t'}{k_B T_c} , \qquad (47)$$

where r is a numerical factor of the order of unity. Hence, the pair tunneling process strongly enhances the pairing potential and $k_B T_c \sim r(t')^2/t$ [7].

To calculate explicitly the value of T_c we consider explicitly the $p \geq 1$ case in a model potential $V_{\mathbf{k}\mathbf{k}'} = V$ in the energy regime below $\omega_D \ll \mu$. Introducing the density of states $N(0)$ at the Fermi level we find from (44) the equation for T_c:

$$1 = VN(0) \int_{-\omega_D}^{\omega_D} d\varepsilon \frac{\chi(\varepsilon)}{1 - 2 \left[\tilde{\alpha} f(p) + \cos(\frac{\pi}{p+1}) \right] \chi(\varepsilon)(t')^2/t} , \qquad (48)$$

where $\tilde{\alpha}$ is the ratio of intergroup to intragroup tunneling amplitudes. In the extreme situation, when the interplanar tunneling provides a dominant contribution to the effective pairing, we can make the assumption $\omega_D/k_BT_c \ll 1$ and thus approximate $\tanh(x) \approx x$. In the case of a Luttinger liquid, (48) reduces to

$$1 = VN(0) \int_0^{\beta\omega_D/2} dx \, \frac{I(x)}{1 - \beta T(p)I(x)}, \tag{49}$$

where $T(p) \equiv [\tilde{\alpha}f(p) + \cos\left(\frac{\pi}{p+1}\right)](t')^2/t$, and

$$I(x) = \frac{1}{\pi x} \int_{v_s/v_c}^1 du \, \frac{\tanh(xu)}{\sqrt{u^2 - (v_s/v_c)^2}\sqrt{1 - u^2}} \, .$$

In the limit $\beta\omega_D \ll 1$ the last integral is of the elliptic form, which has the property $\int_a^b dz/(\sqrt{z-a}\sqrt{b-z}) = \pi$; hence $I(x) = 1$. In effect, we obtain a remarkably simple expression for T_c

$$T_c = VN(0)\frac{\omega_D}{2} + \frac{1}{2}\frac{(t')^2}{t}\left[\cos\left(\frac{\pi}{p+1}\right) + \tilde{\alpha}f(p)\right]$$

$$\equiv T_c^o(p) + T_c'\cos\left(\frac{\pi}{p+1}\right). \tag{50}$$

The identical formula can be derived easily in the Fermi liquid case by noting that under the approximation scheme $\chi_\mathbf{k} \simeq 1/(2k_BT_c)$. Hence, the expression (50) is universal and contains combined effects of the intrinsic in-plane pairing and the two interplanar Josephson tunneling processes. In the $p = 1$ case one has explicitly that $T_c = VN(0)\omega_D/2 + (t')^2/(2t) \sim (t')^2/t$.

To test the relative roles of the two tunneling contributions we have fitted the tabulated [1] T_c values to the expression (50). Table 1 displays a detailed comparison, together with the T_c^0 and T_c' values. In that table we have also included some of the systems not studied so far, as well as the aymptotic value for $p \to \infty$. The quality of the formula fitting has been displayed in Fig.3 for the known superconducting families.

Table 1. Critical temperatures for various single and multilayer superconductors

Material	p	T_c^{exp}	T_c^o	T_c'	Theory: T_c
$Bi_2Sr_2CuO_6$	1	0–20	10	148	10
$Bi_2Sr_2CaCuO_8$	2	85	10	148	84
$Bi_2Sr_2Ca_2Cu_3O_{10}$	3	110	10	148	112
—	∞	–	10	148	158
$Tl_2Ba_2CuO_6$	1	0–80	41	126	41
$Tl_2Ba_2CaCuO_8$	2	108	41	126	103
$Tl_2Ba_2Ca_2Cu_3O_{10}$	3	125	41	126	128
—	∞	–	41	126	167
$Tl_2Ba_2CuO_5$	1	0–50	24	121	24
$Tl_2Ba_2CaCu_2O_7$	2	80	24	121	84
$Tl_2Ba_2Ca_2Cu_3O_9$	3	110	24	121	107
$Tl_2Ba_2Ca_3Cu_4O_{11}$	4	122	24	121	122
—	∞	–	24	121	145
$HgBa_2CuO_4$	1	94	94	59	94
$HgBa_2Ca_2Cu_3O_8$	3	135	94	59	135
—	∞	–	94	59	153

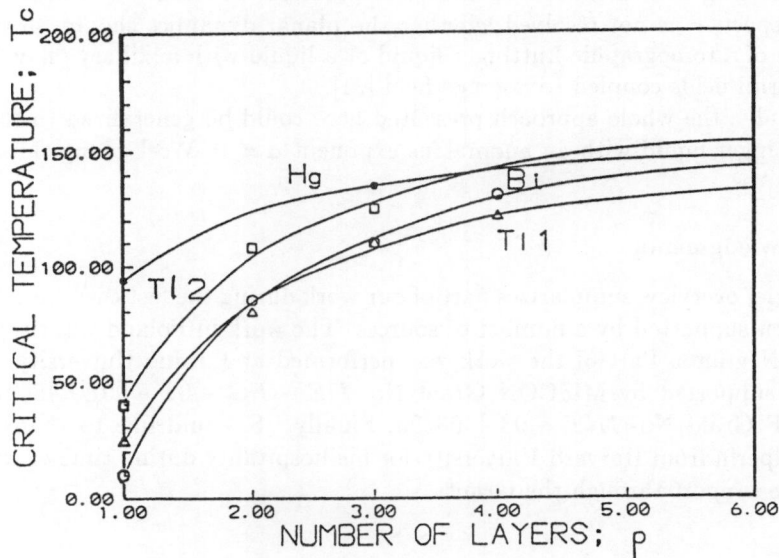

Fig. 3. Transition temperature as a function of p, number of tightly spaced planes. The points represent experimental values taken from Tab.1, whereas the solid lines provide the fitting according to Eq.(50).

The value of T_c^o is practically constant for $p > 1$. This means that the coherence length achieved by the intragroup tunneling $T(\mathbf{k})$ exceeds the intergroup distance even for $p = 2$ and thus the intergroup pairing is not so important for all multilayer systems with $p > 1$. It would be interesting to inquire whether the systems with $p = 1$ differ remarkably from those with $p > 1$. Such a question requires a detailed study of the microscopically derived Ginzburg–Landau functional with inclusion of the tunneling process considered above. With the present approach, one can account for the tunneling processes the two–dimensional Ginzburg–Landau functional [19] for a nonFermi liquid.

6 Concluding Remarks

The results obtained here are universal in two respects. First, only the knowledge of general properties of the in–plane part of the pairing potential are required. For the sake of simplicity, the numerical results for T_c have been obtained for the $\mathbf{k}$-independent potential. Also, a gapless form of $\Delta_\mathbf{k}$ has been assumed since at T_c the magnitude of Δ is zero, and hence its angular dependence should be irrelevant.

The Equation (36) determines the gap anisotropy in the CuO_2 a plane. The simplest way of achieving a d– or s-wave solution is to assume that the hopping matrix elements T and $\tilde{T}$ are $\mathbf{k}$-independent and that the in–plane pairing potential $V_{\mathbf{kk'}}$ is $\mathbf{k}$-dependent. Such anisotropy is provided for example by real space pairing with inclusion of three–site processes in the $t - J$ model [20]. In this respect, it is not resolved whether the planar dynamics should be treated as that of a tomographic Luttinger liquid or a liquid with auxiliary (slave) Bose and Fermi fields coupled to a gauge field [21].

Finally, the whole approach presented here could be generalized to the case of Luttinger liquid with an anomalous exponent $\alpha \neq 0$. Work along this line is in progress.

Acknowledgments

This brief overview summarizes part of our work during the last two years, which has been supported by a number of sources. The work in Poland was supported by KBN grants. Part of the work was performed at Purdue University, where it was supported by MISCON Grant No. $DE - FG - 02 - 90ER45427$, and by NSF Grant No. $INT - 93 - 08323$. Finally, JS would like to thank Prof. B.I.Halperin from Harvard University for his hospitality during time of writing, and the support through the Grant.

References

1. G.Burns, "High–Temperature Superconductivity — an Introduction", (Academic Press, New York 1992); see also: *Physica* C **235-240** (1994).

2. ZX.Shen and D.S.Dessau, *Phys.Repts.* **235**, 1–162 (1995); K.J.Gofron, Ph.D. Thesis, Univ.of Illinois at Chicago, 1993 (unpublished).

3. C.M.Varma, P.B.Littlewood, S.Schmitt–Rink, E.Abrahams, and A.E.Ruckenstien, *Phys.Rev.Lett.* **63**, 1996 (1989); for discussion within the Fermi liquid theory context see: N.Mitani and S.Kurihara, *Physica* C **192**, 230 (1992).

4. A.Virtoszek and J.Ruvalds, *Phys.Rev.*B **42**, 4064 (1990).

5. D.Pines, these Proceedings.

6. P.W.Anderson and Y.Ren, in: "High Temperature Superconductivity", ed.by K.Bedell *et al.* (Addison–Wesley, New York 1990) pp.3–33; P.W.Anderson, *Science* **256**, 1526 (1992).

7. P.W.Anderson, *Physica* B **199&200**, 8 (1994); S.Chakravaty, A.Sudbo, P.W.Anderson, and S.P.Strong, *Science* **261**, 337 (1993).

8. C.Nayak and F.Wilczek, *Nucl.Phys.*B **417**, 359 (1994);

9. K.Byczuk and J.Spałek, *Phys.Rev.*B **51**, 7934 (1995);

10. K.Byczuk and J.Spałek, submitted for publication (1995).

11. K.Byczuk and J.Spałek, *Phys.Rev.*B **53**, in press (1996).

12. A.Houghton and J.B.Marston, *Phys.Rev.*B **48**, 7990 (1993); A.Houghton, H.J.Kwon, and J.B.Marston, *Phys.Rev.*B **50**, 1351 (1994).

13. F.D.M.Haldane, Proc.Int.School of Physics "Enrico Fermi", Course 121, Varenna 1992, ed.by R.Schrieffer *et al.* (North–Holland, Amsterdam 1994), p.5.

14. Y.Ren and P.W.Anderson, *Phys.Rev.*B **48**, 16662 (1993); M.Fabrizio and A.Parola, *Phys.Rev.Lett.* **70**, 226 (1993).

15. A.Sudbo, *J.Low.Temp.Phys.* **97**, 403 (1994); see also: S.P.Kruchinin, *Mod.Phys.Lett.* **9**, 379 (1995); T.Schneider, Z.Gedlik, and Z.Ciraci, *Europhys.Lett.* **14**, 261 (1991).

16. A.Sudbo, *Phys.Rev.Lett.* **74**, 2575 (1995) (we consider here the situation with an anomalous exponent $\alpha = 0$ only, *i.e.* the spin–charge separated liquid with a Fermi liquid scaling).

17. J.Spałek and W.Wojcik, *Phys.Rev.*B **37**, 1532 (1988); J.Spałek, *Physica* B **163**, 621 (1990); For a review see: K.Byczuk, J.Karbowski, J.Spałek, and W.Wójcik, "Superconductivity and Strongly Correlated Electron Systems", ed.by C.Noce *et al.* (World Sci., Singapore 1994).

18. These boundary conditions do not reflect the situation when the surface layer is perturbed, *i.e.* oxidized to a different degree. Then we may have in the extreme cases either a normal (dead) surface layer or surface superconductivity. The ratio $2\Delta/k_B T_c$ depends always on the layer number.

19. V.N.Muthukumar, Deabanand Sa, and M.Sadar, *Phys.Rev.*B **52**, 9647 (1995).

20. J.Spałek, *Phys.Rev.*B **37**, 533 (1987).

21. P.A.Lee and N.Nagaosa, *Phys.Rev.*B **46**, 5261 (1992); M.U.Ubens and P.Lee, *Phys.Rev.*B **50**, 438 (1994); *ibid.*, **49**, 6852 (1994).

Driven Motion of Vortices in Superconductors

G. W. Crabtree,[1] G. K. Leaf,[2] H. G. Kaper,[2] V. M. Vinokur,[1]
A. E. Koshelev,[1] D. W. Braun,[2] and D. M. Levine[2]

[1] Materials Science Division
[2] Mathematics and Computer Science Division
Argonne National Laboratory, Argonne, IL 60439 USA

Abstract. The driven motion of vortices in the solid vortex state is analyzed
with the time-dependent Ginzburg-Landau equations. In large-scale numerical
simulations, carried out on the IBM Scalable POWERparallel (SP) system at Argonne National Laboratory, many hundreds of vortices are followed as they move
under the influence of a Lorentz force induced by a transport current in the presence of a planar defect (similar to a twin boundary in $YBa_2Cu_3O_7$). Correlations
in the positions and velocities of the vortices in plastic and elastic motion are
identified and compared. Two types of plastic motion are observed. Organized
plastic motion displaying long-range orientational correlation and shorter-range
velocity correlation occurs when the driving forces are small compared to the
pinning forces in the twin boundary. Disorganized plastic motion displaying no
significant correlation in either the velocities or orientation of the vortex system
occurs when the driving and pinning forces are of the same order.

1   Introduction

The driven motion of vortices in superconductors is now attracting substantial
interest for its scientific and technological value [1]–[17]. Scientifically, vortices
provide a well defined system of interacting strings or pancakes, which can form
a liquid, lattice, or disordered glass in equilibrium. Each vortex can be subjected
to a controllable external force—the Lorentz force, $F_L = J \times \Phi/c$—through the
application of a transport current J. (Φ is a vector, whose magnitude is equal
to the flux quantum, oriented parallel to the local magnetic induction, B.) This
external Lorentz force is resisted by the pinning forces, which arise from material
defects and act on the vortices. The defects may be point defects, line defects,
or planar defects, and they may be naturally present or artificially induced. The
interplay of pinning, interaction, and driving forces, operating within the various
phases, leads to richly detailed dynamics and many new and interesting dynamic
phenomena. One of the experimental attractions of vortices as a dynamical
system is the extensive control over the relevant parameters. For example, the
density of vortices can be changed by several orders of magnitude by a simple
change in the applied field, the driving force can be controlled through the
transport current, and the pinning forces can be varied by controlled irradiation
with electrons, protons, neutrons, or heavy ions.

On the technological side, vortex dynamics plays a central role in the success of superconducting applications. Elementary electrodynamics requires that vortex motion be associated with energy dissipation, a detrimental feature for practical use of superconductors. If superconductors are to find extensive applications, the dynamics of vortices must be understood as a prerequisite to prevention and control of dissipation.

The nature of vortex motion differs in each of the equilibrium states. In the liquid state, the shear modulus is zero, allowing neighboring vortices to slide past each other with finite relative velocity. The velocity change between neighboring vortices is controlled by the shear viscosity and is described by a suitable form of hydrodynamics [18]. In the solid state, the motion of vortices is dramatically different. The finite shear modulus prevents the relative shear motion of neighboring vortices by imposing an elastic energy penalty on shear distortions. Consequently, the average velocity of neighboring vortices is identical as long as the elastic limit of the shear modulus is not exceeded. Local elastic distortions may occur, but the shear modulus prevents these distortions from growing because the elastic energy cost becomes too great. The motion may be described by an average velocity of the vortex system, with local elastic fluctuations relative to this average velocity. This *elastic motion* is a continuous process and can be described by a set of partial differential equations involving the elastic displacements of the moving lattice.

When the shear yield stress of the vortex lattice is exceeded, the shear forces are too strong to be accommodated elastically, and another type of solid-state motion occurs which is qualitatively different from elastic motion. The elastic bonds between vortices are broken, and neighboring vortices can have different velocities in the steady state. Remarkably, the velocity differences are not spread over many vortex spacings by viscous processes as in the liquid. Rather, discontinuities in the velocity profile occur, which cannot be described by the usual partial differential equations of hydrodynamics or elasticity theory. We refer to this kind of motion, where a given vortex may see different neighbors during the course of the motion, as *plastic motion*. Plastic motion has recently been recognized experimentally [3]–[5], [8], [10]–[12], [15] and theoretically [1, 2, 6, 7, 9, 16, 17] as a fundamentally important feature of driven vortex dynamics.

In this paper we explore the nature of driven motion of the vortex solid. We present results of numerical simulations of vortex motion near a planar defect, like a twin boundary in $YBa_2Cu_3O_7$, showing important fundamental characteristics of both plastic and elastic motion. We show how the the driven motion of a vortex solid evolves from organized plastic motion at low driving forces, where the planar defect presents a barrier to vortex motion, to disorganized plastic motion at intermediate driving forces, and finally to elastic motion at high driving forces. The defining characteristics of each type of motion revealed by the simulations are compared and discussed.

2   Numerical Simulations

The numerical simulations presented here are based on the time-dependent Ginzburg-Landau (TDGL) equations of superconductivity. These equations, first written down by Schmid [19], were critically analyzed in the context of the microscopic Bardeen-Cooper-Schrieffer theory by Gor'kov and Eliashberg [20]. In the zero-electric potential gauge, they may be summarized as follows:

$$\frac{\hbar^2}{2m_s D}\frac{\partial\psi}{\partial t} = -\frac{\delta\mathcal{L}}{\delta\psi^*}, \quad \frac{\sigma}{c^2}\frac{\partial\mathbf{A}}{\partial t} = -\frac{\delta\mathcal{L}}{\delta\mathbf{A}} - \frac{1}{4\pi}\nabla\times\nabla\times\mathbf{A}, \qquad (1)$$

where $\mathcal{L}$ is the density of the Helmholtz free-energy functional,

$$\mathcal{L} = a|\psi|^2 + \frac{b}{2}|\psi|^4 + \frac{1}{2m_s}\left|\left(\frac{\hbar}{i}\nabla - \frac{e_s}{c}\mathbf{A}\right)\psi\right|^2. \qquad (2)$$

Here, ψ is the complex order parameter, and $\mathbf{A}$ the vector potential; the other symbols have their usual meaning. Vortices are identified with zeros of the order parameter. We use link variables to preserve gauge invariance in the discretization of the field equations; details of the approximation procedure and the computational method can be found in [21].

The simulations refer to a superconducting slab, infinite and homogeneous in the z direction, periodic in the y direction, and finite in the x direction. The unit of length in all calculations is the penetration depth, λ. A magnetic field is applied in the positive z direction, so the problem is essentially two-dimensional. The superconductor occupies a rectangular region measuring 32×48 (units of λ) in the (x,y) plane. Periodicity is imposed in the y direction. A transport current is applied in the positive y direction, so the Lorentz force is in the positive x direction. At the free surfaces, the boundary condition is $\mathbf{J}_s\cdot\mathbf{n} = 0$, where $\mathbf{J}_s$ is the supercurrent density and $\mathbf{n}$ the normal unit vector.

The transport current is induced by applying a magnetic field $H_l = H + \Delta H$ at the left, $H_r = H - \Delta H$ at the right free surface, where $\Delta H > 0$. Ampère's law requires a current (per unit length in z) of magnitude $J = 2(c/4\pi)\Delta H$ in the y direction.

We take $\kappa = 4$ and adopt a grid of 256×384 mesh points. (κ is the Ginzburg-Landau parameter, $\kappa = \lambda/\xi$, where λ is the penetration depth, ξ the coherence length.) Thus, two mesh widths correspond to one coherence length. No thermal fluctuations are included, so the simulations reflect the motion of the vortex solid.

The twin boundary is simulated as a planar slab, two coherence lengths thick, parallel to the z-axis and making an angle of 45° with the sample boundaries. This geometry, often encountered in single crystals of $YBa_2Cu_3O_7$, was inspired by the interesting barrier effects observed in magneto-optical images of twinned crystals [17], [22]–[27]. The twin boundary is modeled by locally reducing the condensation energy, with random Gaussian fluctuations to provide the experimentally observed pinning opposing vortex motion within the plane [28]. The average condensation energy in the twin boundary is 56% of the bulk value, and the standard deviation of the fluctuations is 25% of the bulk value.

The computational procedure during each simulation was as follows. First, a small field was applied, to establish the Meissner state. After 200 time steps, the field was increased suddenly to $1.13H_c$, to bring the system into the vortex state. Simultaneously, a transport current was imposed by adjusting the applied magnetic field on either side of the slab. Simulations were run with three transport currents—referred to hereafter as *weak, intermediate*, and *strong*—corresponding to approximately 2%, 4%, and 8% of the depairing current. The time-dependent Ginzburg-Landau equations were iterated for 1.832×10^6 time steps, to establish the steady state. At that point, all transient effects had been eliminated, and data recording was initiated. The equations were iterated for an additional 0.580×10^6 time steps, and the values of the order parameter and vector potential at each grid point were recorded at regular time intervals. The number of vortices throughout the recording period was 455, 561, and 651 for the weak, intermediate, and strong current, respectively, with a variation of less than 1% in each case.

The simulations were carried out on the IBM Scalable POWERparallel (SP) system at Argonne National Laboratory (128 processors, 128 Mbytes per processor, theoretical peak performance 16 Gflops). On 16 processors, a simulation of 2.412×10^6 time steps required approximately 100 hours.

The output of a simulation is a sequence of snapshots showing the time evolution of the order parameter, vector potential, and other calculated quantities at each grid point. The spatial variation of the order parameter is analyzed at each time step, to determine the positions of the vortices. The aggregate of these positions over time yields the vortex trajectories during the period of observation. On such a trajectory plot, a moving vortex appears as a line whose length and direction indicate its average velocity.

3 Results

3.1 Weak Current

Figure 1 shows the positions of the vortices for the weak current case for one of the time steps, after steady state has been reached. Delaunay triangulation is shown to highlight the structure of the vortex pattern. The location of the twin boundary is marked by the diagonal dotted line ending near the upper right corner. The vortex system is highly ordered spatially, as expected for a system in the solid state. The lattice structure accommodates the twin boundary

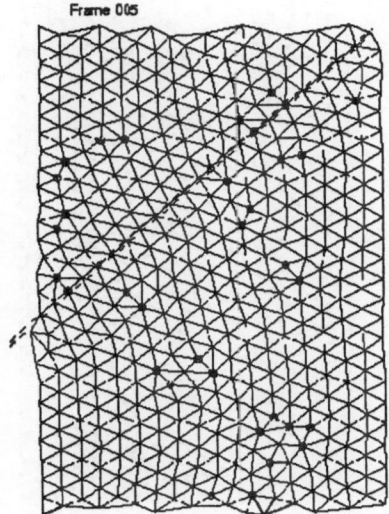

Figure 1: Delaunay triangulation of the vortex positions for one time step in the observation period at weak current. Open and solid circles mark the positions of vortices with five and seven neighbors, respectively.

by orienting one of its close-packed directions with the boundary plane. This is natural, since the lower condensation energy on the twin boundary attracts vortices, making it energetically favorable to maximize the local density. The incommensurablity between the density of vortices on the twin boundary and in the bulk is resolved by dislocations adjacent to the twin boundary.

The accommodation of the structure to the twin boundary conflicts with another accommodation to the right and left edges of the sample [29]. The competition can be seen clearly in Figure 1, where the close-packed directions shift from parallel to the twin boundary near the twin boundary to parallel to the edges near the edges. The shift in orientation of 15° is accommodated by defects in the vortex lattice structure, indicated in Figure 1 by solid or open circles at vortices with seven or five neighbors, respectively. The orientation may shift abruptly, as near the left edge just above the twin boundary, or gradually, as in the center of the sample.

The trajectories of the vortices in weak current are shown in Figure 2. The vortices in the twin boundary are stationary, being pinned against motion parallel or perpendicular to the boundary plane by the random potential. Thus, at weak currents the twin boundary is an impenetrable barrier to vortex motion. The motion occurs external and parallel to the boundary, illustrating a form of

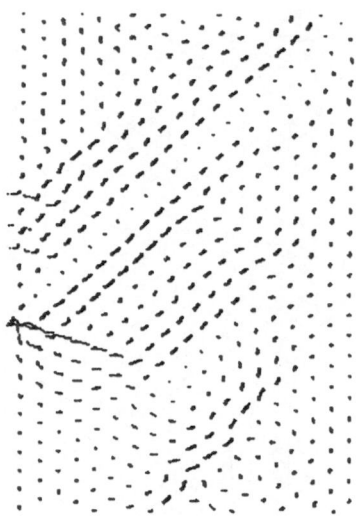

Figure 2: Vortex trajectories at weak current.

twin boundary guidance explained in detail in [17]. At weak current, the twin boundary dominates both the local structure and the motion of vortices.

Close examination of Figure 2 reveals that vortex motion occurs primarily along the close-packed directions of the lattice. This feature is a key element in understanding plastic vortex motion and we will use it often in the following discussion. In particular, it plays an essential role in creating the guided motion that occurs in Figure 2. The guided motion occurs because (i) the twin boundary defines the close-packed directions in the driven vortex system, (ii) this orientational order persists over long range, up to the dimension of the simulated sample, and (iii) vortex motion is restricted to the close-packed directions. It is the combination of all these elements which allows the twin boundary to determine the velocity direction of vortices at distant points. If these elements are absent, the guided motion is severely inhibited or missing, as will be demonstrated explicitly by the results at intermediate current.

A second characteristic feature of Figure 2 is the occurrence of velocity discontinuities. These discontinuities are most obvious at the twin boundary, where the velocity suddenly jumps from zero to approximately its highest value in one lattice spacing. This is quite different from the hydrodynamic motion of liquids, where the velocity profile grows monotonically from zero at the boundary, reaching its highest value deep in the liquid. Additional discontinuities occur in Figure 2 far from any local structural feature. Four rows above the twin boundary, the velocity abruptly jumps from a high value to nearly zero, and there are discontinuous velocity changes two rows and seven rows below the boundary. Farther below the twin boundary, after a region of little or no motion, two adjacent rows of vortices suddenly flow at substantial velocity parallel to the twin boundary. The discontinuities associated with these two rows have no apparent communication with the twin boundary or with the guided motion adjacent to the boundary. They illustrate the collective nature of the plastic response of the vortices to the particular driving and pinning forces in the simulation.

The plastic motion in Figure 2 displays velocity discontinuities of *direction* as well as *magnitude*. Near the left edge of the sample, just below the twin boundary, there are several rows of vortices moving to the lower right with substantial speed. These vortices border on another group moving to the upper right with approximately equal speed. The discontinuity in direction is dramatic: the velocity change occurs in one vortex spacing with no transition region.

This velocity direction discontinuity may be understood in terms of the principle of motion restricted to close-packed directions. The lattice accommodates the twin boundary by orienting one of its close-packed directions along the boundary, as described above. Since the twin boundary is a barrier to vortex flow, the other two close-packed directions are effectively blocked as paths for motion. If any motion is to occur, it must be along the close-packed direction parallel to the twin boundary. However, just below the left end of the twin boundary, the barrier effect is absent, and all close-packed directions are available for vortex motion. The vortices choose to move to the lower right, because it is the close-packed direction oriented nearest to the direction of the Lorentz force.

Despite the velocity discontinuities, there is a great deal of correlation in the vortex motion in Figure 2. The four rows of vortices above the twin boundary move with approximately equal average velocity, as do the two rows just below the boundary and the fifth to seventh rows below. These correlations of neighboring velocities are easy to understand qualitatively as an effect of the shear modulus. Elastic energy is minimized if neighboring vortices move at the same velocity, so that the shear bonds are not stretched. In spite of this mechanism, the velocity correlations are relatively short range, extending less far than the orientational correlation of the lattice.

Summarizing the observations at weak current, the observed plastic motion is highly organized in several ways. It respects the local structure of the lattice by restricting motion to the close-packed directions. This restricted motion maintains the long-range orientational order of the lattice, while breaking the long-range translational periodicity. The motion is further organized by a high correlation of velocities extending over relatively short ranges compared to the range of orientational correlation. The regions of highly correlated velocity locally minimize the elastic shear energy and terminate suddenly in velocity discontinuities of both magnitude and direction. Finally, the motion is collective, each vortex coordinating its movements with others to minimize the configurational energy through maintaining long-range orientational order and, to a lesser extent, local translational periodicity.

3.2 Intermediate Current

A qualitatively different kind of plastic motion occurs at intermediate current, shown in Figure 3. Here the twin boundary pinning forces are comparable with

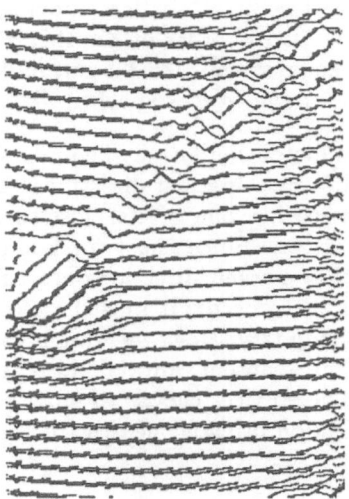

Figure 3: Vortex trajectories at intermediate current.

the driving forces, and the vortices in the boundary are no longer stationary. A new kind of guidance occurs, where vortices move parallel to the boundary but internal to it [17]. This internal guidance is most easily seen at the lower left of the twin boundary, but it also occurs elsewhere along the boundary over shorter distances. Internal guidance occurs in regions of the boundary where the random pinning wells are relatively deep compared to the bulk, but comparable in depth to neighboring wells. The driving force is sufficient to overcome the relatively low local barriers between wells, but insufficient to overcome the larger barriers blocking access to the bulk.

The high correlation among vortex trajectories near the twin boundary, which was apparent at weak current, is missing in Figure 3. There are crossing trajectories in the twin boundary, which indicate that different vortices do not necessarily follow the same path when encountering the same pinning configuration at different times. Their motion depends not only on the pinning configuration, but also on the local vortex configuration at the time of the encounter.

In Figure 3 the boundary has lost its structure, no longer appearing as an extended object to the vortices. Rather, it is a line of random pinning wells, some of which are strong enough to trap vortices and some of which are too weak to do so. Without local structure, there are no well-defined close-packed directions and no structural features to guide the motion of vortices. The randomness associated with the relative sizes of the pinning and Lorentz forces at intermediate current destroys the coherence of the boundary and is ultimately responsible for the disorder which characterizes the plastic motion in Figure 3. If there is no random element, as in the bulk of the sample where there is no pinning, the motion is highly ordered.

Far from the twin boundary, where pinning is absent, a new order appears in the vortex motion. Figure 3 shows a remarkable uniformity in the vortex trajectories. The vortices all move in nearly the same direction with the same speed. Further, the direction of motion is nearly the Lorentz force direction, not the twin boundary direction. This is quite different from the situation at weak current, where the direction of motion is determined by the close-packed directions and the velocities showed many discontinuities in magnitude. The motion of Figure 3 is the beginning of elastic motion, where all vortices move with the same average velocity. The effect of the twin boundary on the vortex velocities is greatly reduced from that at weak current. There is only local influence in the vicinity of the twin boundary, and it upsets the elastic order imposed by the Lorentz force, rather than defining the orientational order which controls the Lorentz force. Intermediate current represents competition between the Lorentz force and the pinning forces. Neither is dominant, and the unstructured velocities of the vortices near the twin boundary reflect the incoherent nature of their response.

3.3   Strong Current

The vortex trajectories (during the first one-fifth of the observation period) at strong current are shown in Figure 4. The Lorentz force dominates the pinning forces. The direction of motion of the vortices is primarily parallel to the Lorentz force, even in the vicinity of the twin boundary. The internal motion in the twin boundary is gone, except for a small section near the left edge of the frame. Elsewhere, vortex velocities deviate only slightly as they routinely break through the boundary. The plastic motion that occurred at the twin boundary at intermediate current is nearly completely replaced by elastic motion. The boundary no longer interrupts the orderly pattern of the trajectories that occurs on either side, as it does at intermediate current, where the uniform velocity pattern does not continue across the boundary. Here, the trajectories can be traced across the boundary, making the elastic motion across the boundary coherent.

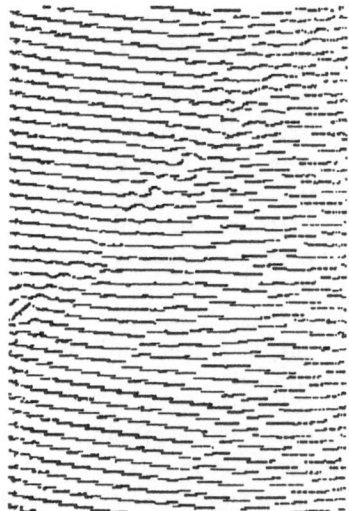

Figure 4: Vortex trajectories during the first one-fifth of the observation period at strong current. Partial trajectories are shown for clarity, to avoid including so many lines that the paths of individual vortices cannot be followed.

4   Discussion and Conclusions

The simulations at weak, intermediate, and strong current reveal several important features of plastic and elastic motion in the presence of extended defects. Two types of plastic motion have been identified: organized motion, where the twin boundary defines the orientational and dynamic structure of the vortex lattice over long distances, and disorganized motion, where there are no long-range correlations in the positions or velocities of the vortices. The key feature determining the kind of plastic motion which occurs is the relative strength of the driving Lorentz force and the opposing pinning forces. When the pinning forces dominate, the twin boundary appears as an extended defect which establishes the orientation of the vortex lattice by defining the close-packed directions. The orientation so defined extends over a long range. The simulations reveal that motion within the lattice structure is restricted to the close-packed directions. This key characteristic explains many of the features of organized plastic flow. Near the twin boundary, two of the close-packed directions are blocked by the barrier effect of the boundary, leaving only one available direction for motion. This produces the guided motion of vortices by the boundary. Because the vortex solid maintains the close-packed directions defined by the twin boundary over long distances, the predominant direction of motion is parallel to the boundary over the whole frame.

The simulations demonstrate discontinuities in the magnitude and direction of the velocities of vortices in plastic motion. The directional discontinuities are controlled by the principle of motion along close-packed directions. They occur at the end of the twin boundary, where the barrier effect suddenly disappears, and are equivalent to an exchange of one close-packed direction for another which is more nearly aligned with the Lorentz force direction. Discontinuities in the magnitude of the velocity separate regions of correlated motion, containing vortices which move with approximately the same average velocity. The size of these regions is a measure of the velocity correlation length, which is intermediate between the intervortex spacing and the size of the simulated sample. By contrast, the orientational correlation length, which is revealed in the Delaunay triangulation, is substantially longer than the velocity correlation length and is comparable to the size of the simulated sample.

In disorganized plastic motion, both the orientational and velocity correlation lengths are comparable to or shorter than the intervortex spacing. As a result, there is little or no effective correlation. Discontinuities in magnitude and direction of the vortex velocities occur on neighboring vortices, and there is no apparent structure discernible in the vortex trajectories. Vortex velocities include many directions, rather than only the close-packed directions as in organized plastic flow. These qualitative differences from organized plastic flow have their origin in the breakdown of the local structure imposed by the twin boundary. At intermediate current, the driving Lorentz force is strong enough to overcome some of the pinning forces, destroying the extended nature of the twin boundary. The boundary is not capable of defining a local structure for the vortices, so there are no close-packed directions to define the orientation of the vortex system or the allowed directions of vortex motion. We propose that disorganized plastic motion is a result of a random element in the pinning configuration. Here, the random element is the pinning strength which, at intermediate current, competes with the Lorentz force for dominance over the vortex motion. Randomness in the position of pinning sites also produces disordered plastic motion [30]. At lower or higher current, the random element is missing, because the Lorentz force is either decisively smaller or larger than the pinning forces, and a more ordered driven state appears.

The most ordered of all the driven solid vortex states is elastic motion, which occurs at strong current. Here, both the position and velocity correlation are of longer range than the simulated sample size. Translational periodicity as well as orientational order are preserved in the moving lattice. Elastic motion occurs when the Lorentz force dominates the pinning forces. The velocity direction is the Lorentz force direction, a feature that does not occur in either type of plastic motion. The principle of motion along close-packed directions, which defined many of the features of organized plastic motion, can be seen in elastic motion as well. The lattice has re-oriented to make the Lorentz force direction a close-packed direction.

The large-scale time-dependent Ginzburg-Landau simulations presented here demonstrate a powerful new tool for exploring the nature of driven motion of vortices. Parallel processing provides the capability to track the simultaneous motion of hundreds of vortices, sufficient to observe the discontinuities and correlations which characterize the plastic and elastic flow patterns in the vortex solid. The simulations provide complete microscopic information on position and velocity, which cannot be obtained experimentally. This type of detailed information allows thorough statistical analyses to explore new concepts like dynamic correlation lengths and their relation to the defining parameters of the system, such as the pinning configuration and the driving Lorentz force. Simulations building on the results presented here can be expected to provide important new insights into the nature of dynamic vortex states.

We thank W. K. Kwok and U. Welp for many productive and stimulating discussions. This work was supported by the U.S. Department of Energy under contract #W-31-109-ENG-38, through the Office of Basic Energy Sciences–Materials Science (GWC, VMV) and the Office of Computational and Technology Research–Mathematical, Information, and Computational Sciences Division subprogram (GKL, HGK, DWB, DML) and by the U.S. National Science Foundation Science and Technology Center for Superconductivity under contract #DMR 91-20000 (AEK).

References

[1] A. Brass, H. J. Jensen, and A. J. Berlinsky, Phys. Rev. B **39** 102 (1989).

[2] S. N. Coppersmith and A. J. Millis, Phys. Rev. B **44** 7799 (1991).

[3] S. Bhattacharya and M. J. Higgins, Phys. Rev. B **49** 10005 (1994).

[4] W. K. Kwok, J. A. Fendrich, C. J. van der Beek, and G. W. Crabtree, Phys. Rev. Lett. **73** 2614 (1994).

[5] G. D'Anna, M. V. Indenbom, M.-O. Andre, and W. Benoit, Europhys. Lett. **25** 225 (1994).

[6] R. Kato, Y. Enomoto, and S. Maekawa, Physica C **227** 387 (1994).

[7] A. E. Koshelev and V. M. Vinokur, Phys. Rev. Lett. **73** 3580 (1994).

[8] A. C. Marley, M. J. Higgins, and S. Bhattacharya, Phys. Rev. Lett. **74** 3029 (1995).

[9] A. I. Larkin, M. C. Marchetti, and V. M. Vinokur, Phys. Rev. Lett. **75** 2992 (1995).

[10] P. Thorel, R. Kahn, Y. Simon, and D. Cribier, Journal de Physique **34** 447 (1973).

[11] U. Yaron, P. L. Gammel, D. A. Huse, R. N. Kleiman, C. S. Oglesby, E. Bucher, B. Batlogg, D. J. Bishop, K. Mortensen, and K. N. Clausen, Nature (London) **376** 753 (1995).

315

[12] J. M. Harris, N. P. Ong, R. Gagnon, and L. Taillefer, Phys. Rev. Lett. **74** 3684 (1995).

[13] G. Blatter, M. V. Feigel'man, V. B. Geshkenbein, A. I. Larkin, and V. M. Vinokur, Rev. Mod. Phys. **66** 1125 (1995).

[14] L. Balents and M. P. A. Fisher, preprint.

[15] A. Duarte, E. F. Righi, C. A. Bolle, F. d. l. Cruz, P. L. Gammel, C. S. Ogelsby, B. Batlogg, and D. J. Bishop, preprint.

[16] D. W. Braun, G. W. Crabtree, H. G. Kaper, A. E. Koshelev, G. K. Leaf, D. M. Levine, and V. M. Vinokur, Phys. Rev. Lett. (1996) (in press).

[17] G. W. Crabtree, G. K. Leaf, H. G. Kaper, V. M. Vinokur, A. E. Koshelev, D. W. Braun, D. M. Levine, W. K. Kwok, and J. A. Fendrich, Physica C (1996) (in press).

[18] M. C. Marchetti and D. R. Nelson, Phys. Rev. B **42** 9938 (1990).

[19] A. Schmid, Phys. kondens. Materie **5** 302 (1966).

[20] L. P. Gorkov and G. M. Eliashberg, Sov. Phys. – JETP **27** 328 (1968).

[21] W. D. Gropp, H. G. Kaper, G. K. Leaf, D. M. Levine, M. Palumbo, and V. M. Vinokur, J. Comp. Phys. **123** (1996) (in press).

[22] C. A. Duran, P. L. Gammel, R. Wolfe, V. J. Fratello, D. J. Bishop, J. P. Rice, and D. M. Ginsberg, Nature (London) **357** 474 (1992).

[23] M. Turchinskaya, D. L. Kaiser, F. W. Gayle, A. J. Shapiro, A. Roitburd, V. Vlasko-Vlasov, A. Polyanskii, and V. Nikitenko, Physica C **216** 205 (1993).

[24] V. K. Vlasko-Vlasov, L. A. Dorosinskii, A. A. Polyanskii, V. I. Nikitenko, U. Welp, B. W. Veal, and G. W. Crabtree, Phys. Rev. Lett. **72** 3246 (1994).

[25] U. Welp, T. Gardiner, D. Gunter, J. A. Fendrich, G. W. Crabtree, V. K. Vlasko-Vlasov, and V. I. Nikitenko, Physica C **235-240** 241 (1994).

[26] C. A. Duran, P. L. Gammel, D. J. Bishop, J. P. Rice, and D. M. Ginsberg, Phys. Rev. Lett. **74** 3712 (1995).

[27] U. Welp, T. Gardiner, D. O. Gunter, B. W. Veal, G. W. Crabtree, V. K. Vlasko-Vlasov, and V. I. Nikitenko, Phys. Rev. Lett. **74** 3713 (1995).

[28] W. K. Kwok, U. Welp, G. W. Crabtree, K. G. Vandervoort, R. Hulscher, and J. Z. Liu, Phys. Rev. Lett. **64** 966 (1990).

[29] F. Ternovskii and L. N. Shekhata, Sov. Phys. – JETP **35** 1202 (1972).

[30] H. J. Jensen, Y. Brechet, and A. Brass, J. Low Temp. Phys. **74** 293 (1989).

Superconducting Fraunhofer Microscopy
The One Vortex Problem

D. K. Finnemore

Ames Laboratory and Department of Physics and Astronomy, Iowa State University, Ames, Iowa 50011, U.S.A.

Abstract. A new kind of microscopy based on the Fraunhofer diffraction effect has been developed to determine the location of a single Abrikosov vortex that is trapped in a superconducting thin film, which in turn is one leg of a Josephson junction. With proper manipulation of currents, the vortex can be pushed to any desired location in the thin film. This microscopy has been used to measure the elementary pinning force on a single vortex in several different materials. Using these methods, it has been possible to follow a vortex as it begins to thermally depin and hops to different sites. Thermal depinning always seems to occur when the bulk superfluid density is about 4% of the zero temperature value.

1 Introduction

The fundamental idea of this work is to develop a new kind of microscopy that permits one to specify the location of an isolated Abrikosov vortex using simple current and voltage measurements. Other methods of imaging that use lasers or electron microscopes are rather complicated and our desire here is to have a very simple device so that it could be used in a flux shuttle or logic device. The concept is similar to a sophomore physics experiment in optics in which photons are directed through a slit and a Fraunhofer pattern is observed. In this experiment, electrons are directed through a Josephson junction and a diffraction pattern is observed in the critical current (I_c) vs magnetic field (B_y) applied parallel to the junction. If there are no vortices trapped in the junction, a Fraunhofer pattern is observed. If there is a vortex trapped in the junction, the Fraunhofer pattern is distorted and there is a unique connection between the location of the vortex and the shape of the diffraction pattern (Miller et al. 1985). Because this effect allows one to determine the location of the vortex, we have called the effect superconducting Fraunhofer microscopy. Since the initial discovery (Miller et al. 1985), the method has been used to measure the temperature dependence of the elementary pinning force (f_p) on an isolated vortex in Pb-Bi alloys (Hyun et al. 1989). Subsequently, methods were developed to study f_p in pure Pb (Li et al. 1991), Pb with Pb_3Au precipitates (Sanders et al. 1993), and in pure Nb (Sok et al. 1994). The purpose of the work reported here is to review the progress in this field with special emphasis on the study of thermal depinning (Sanders et al. 1993, Sok et al. 1994).

2 Concept

To operate as a microscope so that one can determine the position of a vortex, it is essential that the vortex have a net component of flux parallel to the plane of the junction to alter the phase difference across the junction. Hence we always put a thick normal metal layer between the two superconducting layers so the vortex will either leak out the edge of the junction or have a jog between the two S-layers. In a typical superconductor-insulator-superconductor (SIS) junction, distorted Fraunhofer patterns are not observed because the vortex in the two S-layers are very tightly coupled and the vortex goes straight through the junction perpendicular to the plane of the junction. In the early work in this field, Pb-Bi/Ag/Pb SNS junctions were used, typically having crossed S-layers 50 μm wide and 0.5 μm thick with a 0.4 μm thick N-layer. With the N-layer this thick, the pinning energies are comparable to the coupling energy between the vortex in each of the two S-layers so that the vortex will have a jog or leak out of the film. We have always found that it is much easier to move a vortex around with a Lorentz force if the vortex leaks out the edge of the junction rather than going through both films. If the vortex leaks out the edge of the film, one can calculate the I_c vs B_y curves by adding the applied field to the field of the vortex (Miller et al. 1985). In doing this calculation, the effect of the edges of the film are taken into account using image vortices. The relative phase of the electrons on either side of the junction is calculated from the total field, the local current density, J_c, is calculated from the Josephson relation, and the total I_c is calculated by integrating over the entire area of the junction.

3 Accuracy of the Position Determination

All of the studies reported so far have been for cross-strip Josephson junctions that are approximately 50 μm square. For these junctions, one can easily detect a motion of 0.5 μm anywhere near the center of the junction. As the vortex moves toward the edge of the junction, the image effects tend to cancel the effects of the vortex. At about five penetratation depths (or about 1 μm) from the edge, the accuracy falls to about half the accuracy in the center. The way we normally proceed is to calculate the predicted diffraction pattern for about 100 different locations in the junction and make a first estimate of the location by comparing the observed diffraction pattern with these calculated ones. Starting with this location as a first guess, the computer then does a least square fit to specify the exact location. If there are multiple vortices in the junction, there is no "one vortex" diffraction pattern that matches observed pattern. In this case, the junction is warmed above the transition temperature, T_c, and the experiment is started again.

4 Nucleation of a Vortex

Two different methods have been used to nucleate a vortex, by field cooling and by a transport current in one leg of the junction. In all of the early work, vortices were introduced into the junction by cooling through the transition temperature in a magnetic field applied perpendicular to the plane of the junction, B_z. In a typical run with 50 μm wide junctions, a vortex will be trapped cooling in a field of about 2 μT. In the initial discovery of this effect, the entire measurement was made with a junction in which the first vortex nucleated near the center of the junction, jogged slightly and exited through the other S-layer. This vortex gave very little distortion of the diffraction pattern. This vortex was so tightly pinned that the transport current needed to depin the vortex was large enough to nucleate a second vortex at the edge of the film. By cooling in a field a couple of μT higher, two vortices were trapped, one near the center as above and a second the leaked out the edge of the junction. All of the measurements were then made by pushing this second vortex around. Unfortunately, this field cooling method often gives vortices that jog and go through both S-layers. A better method to nucleate the vortex is to use the magnetic field of a transport current in one leg of the junction. This generally gives a vortex that threads through only one of the S-layers with the magnetic field leaking out the edge of the junction. To push the vortex around the junction after nucleation, a transport current in the y-axis leg of the junction will push the vortex in the x-direction via a Lorentz force. A current in x-axis leg of the junction will induce a magnetic field in the barrier region that in turn induces a current across the other strip and thus pushes the vortex in the y-direction. By a combination of currents, the vortex can be pushed in any desired direction.

5 SNIS Junctions

With a SNS junction, the impedance is so low that a superconducting quantum interference device is needed to measure the voltage-current $(V-I)$ curves. To raise the junction impedance, a family of junctions were developed in which Al was used as the N-layer and this layer was oxidized to give $Pb/Al/Al_2O_3/Pb$ (Li et al. 1991) or $Nb/Al/Al_2O_3/Nb$ SNIS junctions (Sanders et al. 1993) with a resistance of a few milliohms. The voltages of the $V-I$ curves rise from zero with a linear slope rather than the square root singularity of the resistively-shunted-junction behavior of SNS junctions, but otherwise, all the physics of the diffraction patterns is the same. This means that very conventional microvolt electronics can be in these experiments.

6 Elementary Pinning Force

For vortices trapped in a Pb-Bi thin film, pinning forces were on the order of 10^{-4} dyne/cm and had a temperature dependence of $f_p \sim (1-T/T_c)^{3/2}$ (Hyun et al. 1989). For pure Pb, the elementary pinning forces are an order of magnitude smaller (Li et al. 1991). For pure Nb, f_p is comparable to pure Pb (Sok et al. 1994).

7 Thermal Depinning

If a vortex is trapped in some specific location and the junction is slowly warmed toward T_c, the order parameter and the superfluid density (that goes as the square of the order parameter) gradually diminish in the superconductor. Because it is the gradient in the superfluid density that causes the pinning, eventually the vortex will thermally depin. For both Pb and Nb junctions and for all pinning sites studied, the vortex thermally depins when the superfluid density is approximately 4% of the $T = 0$ value. When it depins, it moves through a very definite sequence of pinning sites, typically 4–6, and exits the film when the superfluid density is below 2% of the $T = 0$ value of superfluid density. The regularity of this result has led us to believe that any extended region in which the superfluid density drops below 4% will be a weak link and flux will flow freely.

Acknowledgments

Ames Laboratory is operated for the U.S. Department of Energy by Iowa State University under Contract No. W-7405-Eng-82. The work at Ames was supported by the Director for Energy Research, Office of Basic Energy Sciences.

References

Hyun, O.B., Clem, J.R., Finnemore, D.K. (1989): Phys. Rev. B **40**, 175
Li, Q., Clem, J.R., Finnemore, D.K. (1991): Phys. Rev. B **27**, 2913
Miller, S.L., Biagi, K.R., Clem, J.R., Finnemore, D.K. (1985): Phys. Rev. B **31**, 2684
Sanders, S.C., Sok, J., Finnemore, D.K., Li, Q. (1993): Phys. Rev. B **47**, 8996
Sok, J., Finnemore, D.K. (1994): Phys. Rev. B **46**, 3179

The Impact of Tailored Defects on Length Scales and Current Conduction in High-T_c Superconductors

J. R. Thompson,[1,2] D. K. Christen,[1] M. Paranthaman,[1] L. Krusin-Elbaum,[3] A. D. Marwick,[3] L. Civale,[4] R. Wheeler,[5] J. G. Ossandon,[6] P. Lisowski[7] and J. Ullmann[7]

[1] Oak Ridge National Laboratory, Oak Ridge TN 37831-6061 USA
[2] Department of Physics, Univ. of Tennessee, Knoxville TN 37996-1200 USA
[3] IBM Watson Research Center, Yorktown Heights, NY 10598-0218 USA
[4] Centro Atomico Bariloche, 8400 Bariloche, Argentina
[5] Materials Science and Engineering, Ohio State University, Columbus, OH 43210
[6] University of Talca, Talca, Chile
[7] Los Alamos National Laboratory, Los Alamos NM 87545

Abstract This article surveys the formation and effects of "tailored" defects, having controlled numbers and several different morphologies, in high temperature superconductors. Defects can affect the equilibrium properties, such as the superconducting length scales ξ (the coherence length) and λ (the London penetration depth). Very importantly, defects provide vortex pinning that supports the conduction of a macroscopic current density. The article introduces these topics and illustrates them with specific examples.

Keywords: defects, vortex pinning, persistent current density, magnetization, irradiation effects, flux creep

1 Introduction

Given the complexity in structure and other features of high-T_c superconductors, it is natural to wonder, "Why complicate the materials further by deliberately putting defects in them?" There are several motivations for doing so. First, *real* materials inevitably contain microscopic imperfections, so it is useful to modify their concentration controllably, as a method for inferring the properties of an *ideal* superconductor. Second and more importantly, the disorder from defects is essential for the conduction of a large, loss-free critical current density. The situation is somewhat analogous to the doping of semiconductors, in which the choice and concentration of dopants dominates their conductive properties and enables the fabrication of useful devices.

So what are "tailored" defects? For our purposes here, they are defects with controlled densities and specific morphologies -- shapes, sizes, and orientations. The most direct method of forming defects after-the-fact in already synthesized materials is by irradiation, using particles of different mass, charge, and energy. This approach has the advantage that one can compare properties before and after materials modification, thereby making it easier to isolate to effects of the added defects.

Defects modify a superconductor in two major ways: (1) they produce scattering of the conduction electrons and shorten the electronic mean free path (mfp) ℓ, and (2) they change locally the superconducting order parameter Δ, typically depressing Δ, T_c, and the condensation energy density $F_c = H_c^2/8\pi$, where H_c is the thermodynamic critical field.

In this survey, we first present an example demonstrating significant changes, due to mfp effects, in the superconducting length scales: the London penetration depth λ and the coherence length ξ. Then we consider several examples of tailored defects and their interaction with vortices, the quantized lines of magnetic flux that permeate a superconductor in the mixed state. In the common situation that a superconductor conducts a macroscopic current in the presence of a magnetic field, there exists a Lorentz force on each vortex line. Since movement of vortices in a superconductor is viscous and dissipates energy, the flux lines must be locally anchored in the material to have loss-free current conduction. This anchoring (i.e., vortex pinning) depends strongly on the nature of the defects as well as the properties of the superconductor itself. In a classical isotropic superconductor, one can think of a vortex as a rather rigid rod composed of supercurrents circulating about a core and extending in space to a distance $\sim\lambda$; the core itself is a region with depressed order parameter of diameter $\sim 2\xi$. In markedly anisotropic materials[1] like the layered high-T_c superconductors (HTS), one must consider the orientation of magnetic field relative to the CuO planes. With applied magnetic field H $\parallel$ c- crystallographic axis (i.e., perpendicular to the CuO planes), a vortex tends to segment into "pancake vortices," especially in more anisotropic cases like the Bi-Sr-Ca-Cu-O and Tl-Ca-Ba-Cu-O oxides.[2] In this survey, we consider mostly the situation with field H $\parallel$ c-axis, so that induced, macroscopic currents flow within the strongly conductive CuO planes. From a perspective of potential technological applications, this orientation constitutes the limiting case, as pancakes (or their extension to vortex lines) can slide relatively easily parallel to the CuO planes. (With H $\perp$ c, the layered structure itself provides very effective "intrinsic pinning"[3,4] that can support large transport currents in high fields at elevated temperatures.)

There has been an interesting evolution of defect structures in high temperature superconductors. (1) The first and all subsequent materials contain naturally occurring defects; (2) then point-like defects and "blob" defects, e.g., from neutron irradiation, were generated; (3) then columnar defects were formed via irradiation with heavy ions; (4) and most recently, highly splayed columnar defects have been formed, using deeply penetrating GeV energy protons. Below we give an illustration of each of these. In parallel with these developments have been many efforts and significant progress in creating defects chemically, e.g., by addition of Y_2BaCu-oxide "211" particles to $Y_1Ba_2Cu_3O_{7-\delta}$ materials.[5] The remainder of this survey, however, is devoted to the topic of irradiation-induced defects.

An enormous literature has evolved in the area of vortices in high-T_c superconductors -- it is quite hopeless to discuss more than a tiny fraction of this work. Other reviews have considered this general topic from various perspectives. Recent reviews by Blatter et al.[6] and by Brandt[7] provide a comprehensive overview of the theoretical background for this work. Experimental surveys of various topics

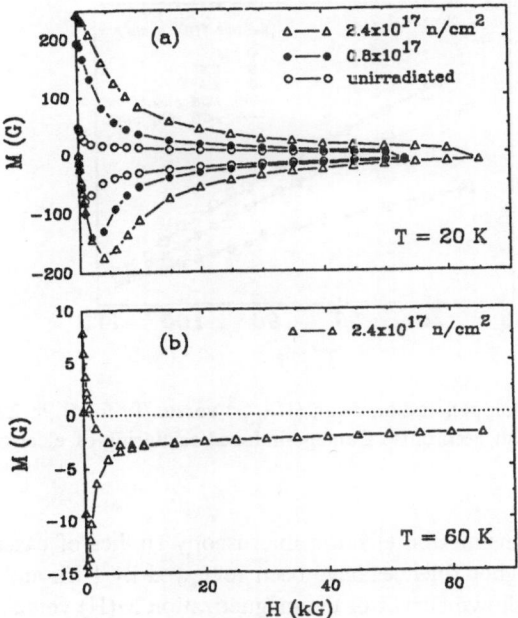

Fig. 1. The magnetization M versus applied magnetic field H, for $Bi_{1.8}Pb_{0.3}Sr_2Ca_2Cu_3O_{10}$ materials at temperatures T of (a) 20 K and (b) 60 K. Figure (a) includes results for the materials as-formed and neutron-irradiated with the fluences shown. In (b), the material is magnetically reversible for fields above ~10 kG.

include those of Civale,[8] Weber and Crabtree,[9] Malozemoff,[10] the series edited by Ginsberg[11] and by Narlikar.[12] Finally, the expanded book by Tinkham[13] provides an overall, very useful, first principles treatment of superconductivity.

2 Influence of defects on the superconducting length scales λ and ξ

For 3-D superconductors, there is well developed theory[14] that predicts specific changes in the superconducting length scales when the mean free path ℓ becomes comparable with the coherence length ξ. Qualitatively, scattering of conduction electrons limits the spatial extent of the Cooper pair, i.e., reduces ξ. Consequently, the upper critical field $H_{c2} = \phi_0/2\pi\xi^2$ and its slope dH_{c2}/dT increase as ℓ decreases; concurrently, the London penetration depth increases. This theory has been supported by experiments[14] on three dimensional superconductors, such as the A15 materials V_3Si and Nb_3Sn. We are not aware of corresponding theory for highly anisotropic superconductors.

Since HTS's have very short coherence lengths, they are generally regarded as clean limit materials for which $\ell >> \xi$. Recently, however, a study by Ossandon et al.[15] gave clear evidence for significant increases in H_{c2} and λ in $Bi_{1.8}Pb_{0.3}Sr_2Ca_2Cu_3O_{10}$ (BiPb-2223) following damage by neutron irradiation. For moderate neutron fluences, damage occurs in the form of collision cascades -- localized "blobs" of highly disordered material -- and accompanying smaller scale,

324

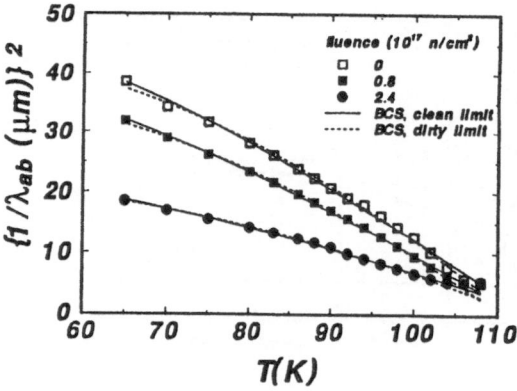

Fig. 2. Plot of the London penetration depth $(1/\lambda_{ab})^2$ versus T, for BiPb-2223 materials, as formed and irradiated with neutrons. Damage-induced scattering of electrons produces the increase in λ.

point-like defects. Transmission electron microscopy studies of cascades and more complex disorder at higher fluences have been reviewed by Kirk and Weber.[16]

In Fig. 1 are shown curves of the magnetization M(H) versus magnetic field H. At low temperature, M is hysteretic, as seen in Fig. 1a and discussed below. At higher temperatures as in Fig. 1b, the mixed state M is reversible for a wide range of magnetic field; this is the equilibrium magnetization. From studies of the equilibrium magnetization M of randomly oriented, polycrystalline material, Ossandon et al.[15] determined the characteristic lengths for a BiPb-2223 specimen, as-prepared and neutron-damaged at two fluences. Several analyses were employed, with each of the underlying theoretical formulations based on different expressions for the free energy in the mixed state. Application of thermodynamics, which is valid for the equilibrium (but not irreversible) magnetization, then leads to expressions containing the characteristic lengths. For example, standard London-limit theory,[17] assuming only static interactions between vortices, predicts a logarithmic field dependence M $\propto \ln(H)$. More recently, Bulaevskii et al.[18] and Kogan et al.[19] have shown that vortex fluctuations contribute to the free energy, due to entropic effects. This modifies the simple London relation by introducing a correction term. With fluctuations included and assuming the magnetic field is applied perpendicular to the CuO layers, one has

$$\frac{\partial M_o}{\partial \ln(H)} = \frac{\phi_o}{32\pi^2\lambda_{ab}^2(T)}[1-g(T)], \tag{1}$$

where

$$g(T) = 32\pi^2 k_B T \lambda_{ab}^2(T)/(\phi_o^2 s) \tag{2}$$

Here λ_{ab} is the magnetic penetration depth corresponding to screening by supercurrents in the ab planes and s is the interlayer spacing. For BiPb-2223 containing trilayers of three adjacent CuO sheets, s is the separation of trilayers sets, 1.8 nm. Note that at a characteristic temperature $T^* < T_c$, one has $g(T^*) = 1$; consequently M is independent of field and has the value $M^* = k_B T^*/\phi_o s$. The field-independence of the mixed state magnetization at temperature T^* is an important feature of vortex fluctuation theory as well as nonpertubative scaling theory.[20,21,22] The formalism of Hao-Clem[23] complements the fluctuation analysis by incorporating a more detailed structure of the vortex. The H-C treatment provides values for the Ginzburg-Landau parameter $\kappa(T) \equiv \lambda/\xi$ and the thermodynamic critical field $H_c(T)$. The product of these two gives $H_{c2}(T) = \kappa(T)H_c(T)\sqrt{2}$. Also $\lambda(T)$, $\xi(T)$ and $H_{c1}(T)$ can be obtained from the usual G-L expressions.

Some results of this analysis are shown in Fig. 2, a plot of $1/\lambda^2$ versus temperature, for neutron fluences of 0, 0.8, and 2.4×10^{17} n/cm^2. The systematic shift is quite clear. At the same time, there was no marked change in T_c nor did the neutron-generated defects affect the most fundamental property, the superconductive condensation energy F_c. The 3D theory provides the following relation[14] for the slope of the upper critical magnetic field and for λ^2, scaled to their values in the clean limit with $\ell >> \xi$:

$$(dH_{c2}/dT)/(dH_{c2}/dT)_{CL} = (\lambda/\lambda_{CL})^2 = [1 + 0.88(\xi_0/\ell) \propto [1 + \xi_0(a\phi)]. \quad (3)$$

Here the subscript CL denotes the clean limit, $\lambda = \lambda_{ab}$, and ξ_0 is the BCS coherence length. The last proportionality follows from the fact that the mfp ℓ is inversely proportional to neutron fluence ϕ with (unknown) factor "a". This scaling is verified for the Bi-2223 in Fig. 3, which shows the predicted linear dependence on defect density and the *same* scaling for both the slope of H_{c2} and for λ^2. Overall, these results show rather directly a significant impact of defects on the fundamental superconducting length scales in high-T_c superconductors.

3 Impact of defects on HTS's irreversible properties

The flow of a macroscopic current in a superconductor can be loss-free, provided that there are no energy-dissipating mechanisms. Aside from weakly linked grain boundaries and problems of the sort, a major source of dissipation is movement of vortices. Hence effective vortex pinning is essential -- yet, two features conspire to make this difficult. First, the coherence length ξ is short and the London penetration depth is large, meaning that the scale of pinning energy U_0 is small. Second, the transition temperature T_c is high, meaning that the potential operating temperature is high. Consequently the probability $\sim \exp[-U/k_B T]$ of thermal excitation is large enough that vortices are frequently depinned. This thermally activated "giant flux creep"[24] means that the circulating currents in a magnetization experiment decay in time.

As realized by Anderson and Kim,[25] magnetic relaxation arises from thermal activation (flux creep) of flux-lines or flux bundles over an average energy

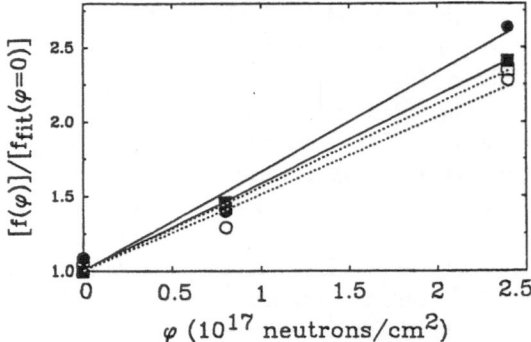

Fig. 3. The ratios of $\lambda^2(T\to 0)$ ($\square$, $\circ$) and dH_{c2}/dT ($\blacksquare$, $\bullet$), relative to their clean limit values prior to irradiation, plotted versus neutron damage fluence ϕ. The similarity of behavior follows from Eq. 3. Symbols: circles are results from a fluctuation analysis; squares from a Hao-Clem analysis.

barrier U. This same effect leads to a reduction of the *apparent* "critical current density" to its observable, i.e., persistent value J_p. In high temperature superconductors, the relaxation rate is so large that the temperature dependence of J_p is dominated by flux-creep effects.

In the simplest model that was first introduced by Anderson, the net barrier is linearly reduced by the presence of a bulk current density, which, by Maxwell's equations, is necessarily related to a gradient in flux-line density. Thus

$$U(J) = U_0[1 - (J/J_c)], \tag{4}$$

where U_0 and J_c are the (temperature-dependent) barrier height and true critical current density that would be observed in the absence of flux creep. Combining this expression with an Arrhenius probability leads to the flux-creep relation

$$J_p(T) = J_c[1 - (k_BT/U_0)\ln(t/t_{eff})], \tag{5}$$

where t_{eff} is an effective attempt time. The linear dependence on J is a reasonable approximation for the case that J is near J_c. However, current decay is so pronounced in HTS materials that one often has $J \ll J_c$; furthermore, observed properties of the superconductor seem to require an effective barrier that grows rapidly as J decreases and perhaps diverges as $J \to 0$.

At first sight, it is difficult to reconcile a diverging U(J) with the Anderson picture of a single particle in a well. However, this feature arises naturally when long elastic vortices are considered.[26] Both vortex glass[27,28] and collective pinning-collective flux creep[29,26] theories provide an inverse power law dependence:

$$U(J) = (U_0/\mu)[(J_c/J)^\mu - 1], \tag{6}$$

with characteristic exponent μ. Note that U depends implicitly on temperature through the quantities U_0 and J_c. Combining this with the relation[6] that U $\approx$ $k_B T ln(t/t_0)$ then leads to the "interpolation formula:"

$$J_p(T) = J_c/[1 + (\mu k_B T/U_0)ln(t/t_{eff})]^{1/\mu} \qquad (7)$$

In the limit of small arguments, this relation has the same form as the Anderson-Kim expression and thus interpolates between it and the case with $J << J_c$.

As noted above, the presence of a bulk current density implies a gradient in flux density. If the currents circulate as in a typical magnetization experiment, they produce a magnetic moment whose sign depends on the sense of current flow; by Lenz's Law, the moment has one orientation when the applied field increases and has the opposite sign when the field decreases. (Here we assume that the field change is large enough that the flux profile penetrates to the center of the sample.) Thus the magnetic moment and associated magnetization are hysteretic in field, as seen in Fig. 1a. This is the basis of the Bean critical state model.[30,31] This model provides that the circulating current density J_p is proportional to the magnetic hysteresis ΔM = $[M(H\downarrow)-M(H\uparrow)]$ via a geometrical factor. For a cylinder of radius R conducting azimuthal bulk currents, one has $J_p = 15\Delta M/R$, with M expressed in units of Gauss, R in cm, and current density in A/cm^2. Similar relations hold for other geometries, where R is replaced by some characteristic transverse length. This topic and several other aspects of experimental magnetization studies can be found in a recent monograph.[32]

4 Point-like defects in $Y_1Ba_2Cu_3O_{7-\delta}$

Point-like defects refer to imperfections on an atomic scale, such as vacancies, interstitials, and lattice site exchanges. Entropy considerations insure that they are presence at some concentration in materials grown at high temperatures. It is useful, however, that additional point defects can be produced in controlled numbers by particle irradiation, under conditions that the energy transfer to the crystal is the order of the energy for vacancy formation. In practice, this is often accomplished with MeV energy electrons[33] or light ions such as protons. Early work[34] showed that irradiation with MeV protons produced large increases in the persistent current density J_p of crystals of YBCO. However, the Irreversibility Line (IL), the line in the H-T plane at which the critical current density goes to zero, was not changed by these point-like defects.

As an example, let us consider the persistent current density in a single crystal of YBCO. Experimental results for $J_p(T)$ versus T, measured in a field H $\parallel$ c = 1 tesla, are shown in Fig. 4. The three data sets were obtained for one crystal, irradiated with 3 MeV protons to fluences of 0, 0.3, and 0.6 x 10^{16} protons/cm^2. From the figure, we see several interesting features. (1) The addition of point-like defects enhances J_p by approximately an order of magnitude to levels approaching 10^7 A/cm^2 at 5 K. (2) The persistent current density falls off smoothly and quasi-exponentially with T in the range $\leq$ 60 K. (3) At higher temperatures, $J_p(T)$ decreases more and more rapidly as one approaches the (same) irreversibility

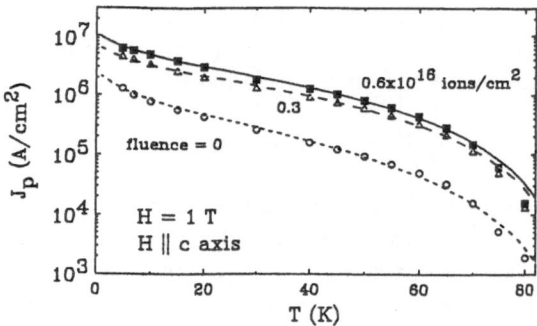

Fig. 4. The persistent current density J_p for single crystal $Y_1Ba_2Cu_3O_{7-\delta}$ versus temperature. Irradiation with 3 MeV protons to the fluences shown created point-like defects and enhanced J_p significantly. The lines are model calculations based on collective pinning theory; see the text.

temperature for the three fluences. Similar results were obtained for other applied fields.

Figure 4 shows substantial increases in J_p due to static disorder from point-like defects. To interpret the widely observed quasi-exponential temperature dependence,[35] we model[36] $J_p(T)$ using the interpolation formula discussed above, incorporating simple BCS-like temperature dependencies for $J_c(T)$ and $U_0(T)$. The model results are the lines in Fig. 4, which describe the experimental results reasonably well. The values of the parameters, e.g., $U_0(0) \approx 150$ K $\approx$ condensation energy in a coherence volume $4\pi\xi_{ab}^2\xi_c/3$, are very similar for all fluences, except of course for $J_c(0)$ that depends on defect density. Furthermore, the same restricted set of parameters describes well the current decay rate $S(T) = dlnJ/dlnt$ with time t. Finally, the model values of the characteristic exponent μ are consistent with detailed determinations of μ from long-term creep studies.[37] For proton-irradiated crystals at intermediate temperatures, we have $\mu \approx 3/2$. This is just the value predicted by collective creep theory[38] for hopping of small vortex bundles. Other features such as the detailed values of μ for other experimental conditions, are described reasonably well by this theory. Hence a reasonably consistent picture emerges for this archetypical case, the high-T_c superconductor YBCO containing point-like defects.

In closing this section, we note that the current density J_p often can be increased in other HTS materials by point defects and neutron-generated defects. This is illustrated in Fig. 1a, where the magnetic hysteresis ΔM and underlying current density J_p of BiPb-2223 progressively increases with neutron fluence; see the reviews of Civale[8] and Kirk and Weber[16] and references therein for a more extensive treatment. Unfortunately, as an HTS becomes more anisotropic, the vortex lines progressively segment into pancake vortices. In the limit, each pancake must be pinned *individually*. This requires a high density of (randomly distributed) point defects in every layer of superconductor. This feature, along with a statistical summation of pinning forces, strongly limits the effectiveness of point defects, especially at high fields and temperatures. Next we discuss tailored defects of another geometry that overcome many of these problems -- columnar defects.

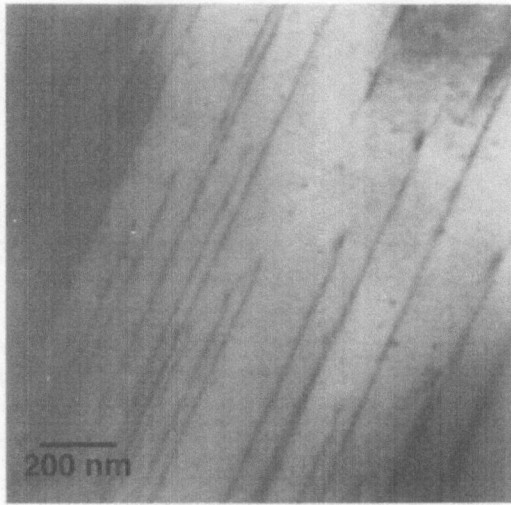

Fig. 5. A transmission electron (TEM) micrograph of columnar defects in a $Y_1Ba_2Cu_3O_{7-\delta}$ crystal, formed by irradiation with 1 GeV Au-ions. The columns are oriented nearly perpendicular to the CuO planes of the superconductor.

5 Columnar defects

These tailored defects were first produced in HTS materials by irradiation with *heavy* ions of GeV energy, such as Xe,[39] Sn,[40] Pb,[41,42] Cu,[43] Ag,[44] etc. Here the dominant channel for energy loss by the incident ion is inelastic excitation of electrons in the crystal; by contrast, point-like defects are produced by Rutherford scattering interactions with individual nuclei in the crystal. Marwick et al.[45] have described many salient aspects of columnar defect formation. In passing through the material, a heavy ion deposits several keV/Å of energy into the electronic system. Provided the energy loss exceeds a threshold level of ~2 keV/Å, this amorphizes the crystal along the ion path, although some details of the process remain unclear. What is clear is that each ion produces a long, highly disordered track with a diameter of 50-100 Å. An example is shown in Fig. 5, a TEM micrograph of columnar defects in YBCO produced by Au^{+30} ions of 1 GeV energy. Since the superconducting order parameter is highly suppressed in the column, it forms a potential well for the normal core of a vortex line. The resulting flux pinning is highly effective, in part for topological reasons: the geometry of a column and a vortex are similar, meaning that a vortex can be pinned over a longer length -- hence more strongly. Since each ion produces one columnar defect that can pin one vortex, it is natural to express the ion fluence and defect density in units of flux density. Thus we multiply the ion fluence, e.g., $\phi = 2.3\times10^{11}$ cm^{-2} by the flux quantum $\phi_0 = 2.07\times10^{-11}$ tesla-cm^2 to obtain the "matching field" $B_\phi = 4.7$ tesla.

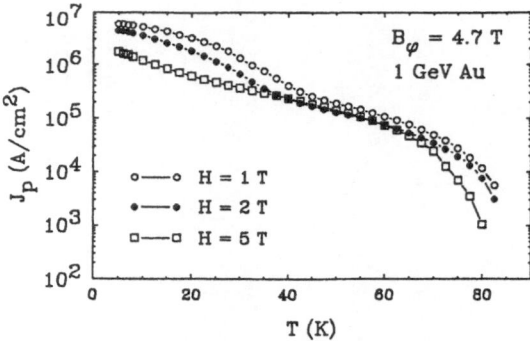

Fig. 6. The persistent current density J_p of a $Y_1Ba_2Cu_3O_{7-\delta}$ crystal containing parallel, Au-ion-generated columnar defects with a density $B_\phi = 4.7$ tesla. For fields $H < B_\phi$, the current density drops significantly as T increases toward the depinning temperature of ~40 K. The structure in J_p disappears for fields $H > B_\phi$ where there are more vortices than columns.

In Fig. 6 are shown illustrative results for $J_p(T)$ for a YBCO crystal. The crystal was irradiated with 1 GeV Au ions at an angle of 2° from the c-axis (to prevent channeling), to a dose $B_\phi = 4.7$ T. The magnitude of current density is high. One of the most notable features of the data is the marked drop in J_p at T = 30-40 K. This drop is associated with the columnar defects: as seen in Fig. 4, there is no corresponding feature in the unirradiated or proton irradiated crystals. Furthermore, as the applied field H increases, the drop occurs at progressively lower temperatures and disappears altogether for $H > B_\phi$. We associate this behavior with the "accommodation field" $B^*(T)$ predicted by Nelson and Vinokur.[46] Below this field, each vortex can be accommodated by a column and is pinned individually; above it, however, vortex-vortex interactions dominate and pinning occurs collectively (or plastically, according to the theory). Thus the current density decreases, as observed in the figure, when the system crosses $B^*(T)$. At low temperature, we have $B^* \approx B_\phi$. At a sufficiently high temperature $T_{dp} \approx 40$ K, entropic effects decrease the free energy of binding, even in the absence of vortex-vortex interactions; then pinning is again collective. In this regime above ~40 K, the values for J_p are very similar for fields spanning B_ϕ, i.e, H = 1-6 T. This similarity extends upward in temperature until the irreversibility line is approached. Further details will be presented elsewhere.[47]

As noted above, columnar defects are particularly effective in improving vortex pinning in highly layered HTS materials. For example, heavy ion irradiation of $Bi_2Sr_2Ca_1Cu_2O_8$ and $Bi_2Sr_2Ca_2Cu_3O_{10}$ produced large increases in J_p at intermediate temperatures.[48,49,50] In addition, the IL was elevated in temperature by more than 20 K.

How can a vortex move in the presence of columnar defects? Due to fluctuations or some weak point along the column, the vortex can bulge outward and form a half loop. If the loop is so large that the vortex touches an adjacent column, then it forms a "double kink" with part of the line pinned by the first column and the remainder pinned by the second. If the columns are parallel, however, there is nothing to prevent the double kink from spreading, so that the line transfers entirely

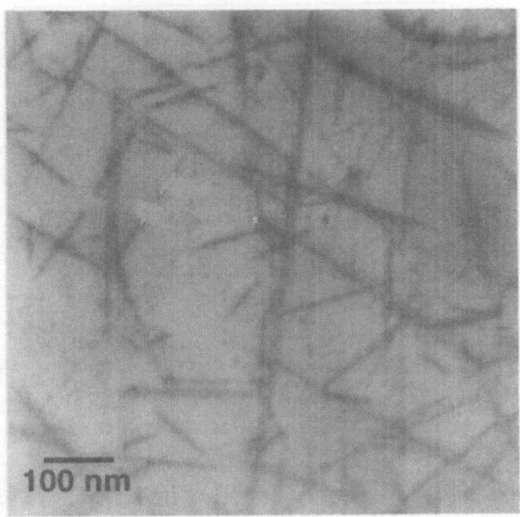

Fig. 7. A TEM micrograph of angularly dispersed columnar defects in $Bi_2Sr_2Ca_1Cu_2O_8$, produced by irradiation with 0.8 GeV protons. The protons induce fission of heavy nuclei such as Bi; then the resulting fission fragments create linear tracks.

to the second column. Hence the vortex moves in the direction of the Lorentz force and dissipates energy. Hwa et al.[51] suggested that one can counteract this process by forcing an entanglement of many vortices, using an angular spread or *splay* in the orientation of the columns. Experimentally, Civale et al.[52] compared the current densities in YBCO crystals irradiated with two ion species: Au and Sn. The tracks formed by the Sn ions had a greater angular spread, due to larger angle of Rutherford scattering for these lower mass particles; this effect is accentuated near the end of the ion path. Relative to the parallel columns from Au ions, the resultant natural splay gave significantly higher values for J_p and smaller flux creep rates. In the next section, we consider the formation of highly splayed columnar defects through yet another mechanism of track formation.

6. Splayed columnar defects via deeply penetrating ions

While heavy ions are very useful for creating columnar defects, they have a major disadvantage: their limited range in matter. Typically, heavy ions penetrate to a depth of ~ 30 μm, which is less than the typical thickness of Ag cladding in composite wires and tapes. Recently, we have shown that irradiation with nearly GeV energy protons increases the persistent current density in superconducting $Bi_2Sr_2Ca_1Cu_2O_8$ (Bi-2212)/Ag composites.[53] The mechanism for current enhancement is this: the 0.8 GeV protons, when travelling through a high-T_c superconductor, collide with and induce fission of its constituent nuclei that are sufficiently massive. The resulting fission fragments recoil with high energy, which often is large enough to form columnar defects. As discussed above, such columnar defects are particularly effective in pinning vortices, thus enhancing the persistent

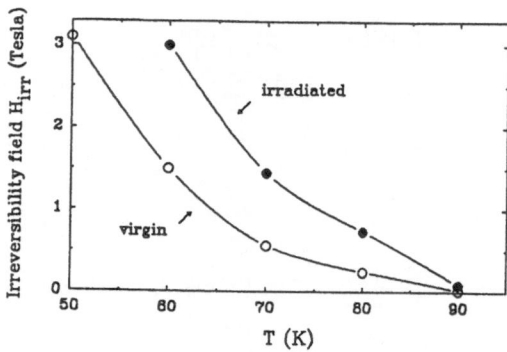

Fig. 8. The irreversibility line of polycrystalline $HgBa_2CuO_{4+\delta}$, prior to (virgin) and following irradiation with 0.8 GeV protons (3×10^{16} ions/cm²). The IL is displaced to higher temperatures and fields; higher fluences should lead to further enhancements.

current density J_p and elevating the irreversibility line $H_{irr}(T)$ to higher temperatures and magnetic fields.

This method is especially attractive in that the ~GeV protons penetrate deeply into a composite (projected range ≈ 1/2 m) and can generate damage tracks far below the surface. The TEM image in Fig. 7, a view along the incident ion direction, shows that the tracks are randomly oriented, due to an approximately isotropic distribution of fission fragments; the track length is 5-10 μm. Fortunately, previous studies on Bi-2212 crystals showed that this system is relatively insensitive to angular misalignment between the columns and the field direction, particularly at low and intermediate temperatures where the current density is large.[54] The splay in orientation may assist in pinning in some regimes, due to forced entanglement of vortices.

Irradiation of $Bi_2Sr_2Ca_1Cu_2O_8$/Ag tapes with GeV energy protons produced large enhancements in the persistent current density. Furthermore, this led to lower rates of current decay[55] (smaller flux creep) than in the as-formed material. This current stabilization is obtained, even though the current density itself is much higher in the irradiated tape. When evaluated at the same filling factor H/B_ϕ, the creep rates $S(T)$ for the proton-irradiated tape ($B_\phi \approx 1.3$ T) are comparable with those observed for a single crystal of Bi-2212 irradiated to a fluence $B_\phi = 5$ Tesla of Sn-ion-generated columnar defects; this Sn-fluence produces a maximum current density J_p. In addition, the irreversibility line is elevated considerably compared with the as-formed tape, but lies below the IL obtained for the Sn-irradiated crystal. It is further significant that these irradiations enhance the in-field *transport* current density, which must cross strongly linked grain boundaries, in monocore and multi-filamentary BSCCO/Ag tapes.[56]

The simple structure and high T_c values of the Hg-based HTS materials make them attractive candidates for potential applications. To date, however, there have been few reports[57] of success in achieving effective vortex pinning in this family of compounds. We have shown, however, that irradiation with 0.8 GeV protons, which causes Hg nuclei to fission, significantly increases pinning in polycrystalline

$HgBa_2CuO_{4+\delta}$ materials. A fluence of 3×10^{16} ions/cm^2 (corresponding roughly to $B_\phi \sim 1/2$ T) produced an order of magnitude increase in magnetic irreversibility ΔM. Furthermore, the irreversibility line displaced upward in temperature. This is shown in Fig. 8, a plot of H_{irr} versus temperature. The increase is significant, given that the density in tracks is relatively low. This result and corresponding results for the Bi-2212/Ag tapes suggests that still higher proton fluences may further expand the potential working region in field-temperature space.

Conclusions

Defects strongly affect the properties of high temperature superconducting materials. Tailored defects of controlled densities and some specific morphologies can be created using particle irradiation methods. We have shown an example in which disorder from neutron damage shifts the superconducting length scales ξ and λ in BiPb-2223. Regarding the conduction of macroscopic currents, defects provide essential pinning sites for vortices. At low temperatures, both point-like and columnar defects can support a high current density J_p; however, at elevated temperatures, the linear tracks are more effective, particularly in highly anisotropic materials like Bi-Sr-Ca-Cu-O. Randomly oriented columnar defects have been formed in this and Hg-based cuprates by irradiation with protons of GeV energy, via a fission process. Unlike the heavy ions, GeV protons can penetrate many cm thicknesses of material, offering the possibility of processing clad or thick superconducting composites.

A portion of the work of JRT was supported by the Science Alliance at the University of Tennessee. The research and technology development were cosponsored by the U. S. Department of Energy, Division of Materials Sciences and Division of Advanced Utility Concepts, both under contract DE-AC05-84OR21400 with Lockheed Martin Energy Systems, Inc.

REFERENCES

1. M. Tinkham, Physica C 235-240, 3 (1994).
2. John R. Clem, Phys. Rev. B 43, 7837 (1991).
3. M. Tachiki and S. Takahashi, Sol. State Comm. 72, 1083 (1989); Physica C 162-164, 241 (1989).
4. D. K. Christen, C. E. Klabunde, R. Feenstra, D. H. Lowndes, D. P. Norton, J. D. Budai, H. R. Kerchner, J. R. Thompson, and S. Zhu, in Superconductivity and Its Applications, edited by Yi-Han Kao, P. Coppens, and Hoi-Sing Kwok (American Institute of Physics, New York, 1991), AIP Conf. Proc. 219, 336-342..
5. M. Murakami, Mod. Phys. Lett. B 4, 163 (1990).
6. G. Blatter, M.V. Feigel'man, V.B. Geshkenbein, A.I. Larkin, and V.M. Vinokur in "Vortices in High Temperature Superconductors," Rev. Mod. Phys., 66, 1125 (1994) and references therein.
7. E. H. Brandt, Physica C 195, 1 (1992); Reports on Prog. in Phys. (in press, 1995).
8. L. Civale in Processing and Properties of High-T_c Superconductors, vol I - Bulk Materials, edited by S. Jin (World Scientific, Singapore, 1992), p. 299.

9.   H. W. Weber and G. W. Crabtree in *Studies of High Temperature Superconductors, vol. 9*, edited by A. V. Narlikar (Nova Scientific, New York, 1992), pp. 37-79.

10.   A. P. Malozemoff, Physica C **185-189**, 264 (1991).

11.   *Physical Properties of High Temperature Superconductors* edited by D. M. Ginsberg (World Scientific, Singapore).

12.   *Studies of High Temperature Superconductors*, edited by A. V. Narlikar (Nova, Commack, NY).

13.   M. Tinkham, *Introduction to Superconductivity, 2-nd Ed.* (McGraw-Hill, New York, 1995).

14.   T. P. Orlando, E. J. McNiff, Jr., S. Foner, and M. R. Beasley, *Phys. Rev. B 19*, 4545 (1979) and references therein.

15.   J. G. Ossandon, J. R. Thompson, Y. C. Kim, Yang Ren Sun, D. K. Christen and B. C. Chakoumakos, Phys. Rev. B 51, 8551 (1995).

16.   M. A. Kirk and H. W. Weber in *Studies of High Temperature Superconductors, vol. 10*, edited by A. V. Narlikar (Nova, New York, 1992), p. 253.

17.   V. G. Kogan, M. M. Fang, and Sreeparna Mitra, *Phys. Rev. B 38*, 11958 (1988).

18.   L. N. Bulaevskii, M. Ledvij, and V. G. Kogan, *Phys. Rev. Lett.* **68**, 3773, (1992).

19.   V. G. Kogan, M. Ledvij, A. Yu. Simonov, J. H. Cho, and D. C. Johnston, *Phys. Rev. Lett.* **70**, 1870 (1993).

20.   S. Ullah and A. T. Dorsey, *Phys. Rev. B* **44**, 262 (1991).

21.   Z. Tesanovic, L. Xing, L. Bulaevskii, Qiang Li and M. Suenaga, *Phys. Rev. Lett.* **69** 3563 (1992). Also: Z. Tesanovic and A. V. Andreev, *Phys. Rev. B* **49**, 4064 (1994).

22.   Qiang Li, K. Shibutani, M. Suenaga, I. Shigaki, and R. Ogawa, *Phys. Rev. B* **48**, 9877 (1993); Qiang Li, M. Suenaga, L. N. Bulaevskii, T. Hikata, and K. Sato, *Phys. Rev. B* **48**, 13865 (1993).

23.   Z. Hao, J. R. Clem, M. McElfresh, L. Civale, A. P. Malozemoff, F. Holtzberg, *Phys. Rev. B* **43**, 2844 (1991).

24.   Y. Yeshurun and A. P. Malozemoff, Phys. Rev. Lett. **60**, 2202 (1987).

25.   P. W. Anderson and Y. B. Kim, Rev. Mod. Phys. **36**, 39 (1964).

26.   M. V. Feigel'man, V. B. Geshkenbein, A. I. Larkin, and V. M. Vinokur, Phys. Rev. Lett. **63**, 2301 (1989).

27.   M. P. A. Fisher, Phys. Rev. Lett. **62**, 1415 (1989).

28.   D. S. Fisher, M. P. A. Fisher and D. A. Huse, Phys. Rev. B **43**, 130 (1991).

29.   M. V. Feigel'man and V. M. Vinokur, Phys. Rev. B **41**, 8986 (1990).

30.   C. P. Bean, Rev. Mod. Phys. **8**, 31 (1964).

31.   W. A. Fietz and W. W. Webb, Phys. Rev. **178**, 657 (1969).

32.   J. R. Thompson, D. K. Christen, H. R. Kerchner, L. A. Boatner, B. C. Sales, B. C. Chakoumakos, H. Hsu, J. Brynestad, D. M. Kroeger, J. W. Williams, Yang Ren Sun, Y. C. Kim, J. G. Ossandon, A. P. Malozemoff, L. Civale, A. D. Marwick, T. K. Worthington, L. Krusin-Elbaum, and F. Holtzberg, in *Magnetic Susceptibility of Superconductors and Other Spin Systems*, edited by R. A. Hein, T. Francavilla, and D. Liebenburg (Plenum, New York, 1992), pp. 157-176.

33.   J. Giapintzakis, W. C. Lee, J. P. Rice, D. M. Ginsberg, I. M. Robertson, R. Wheeler, M. A. Kirk, M.-O. Ruault, Phys. Rev. B **45**, 10677 (1992).

34.   L. Civale, A. D. Marwick, M. W. McElfresh, T. K. Worthington, A. P. Malozemoff, F. Holtzberg, J. R. Thompson, and M. A. Kirk, Phys. Rev. Lett. **65**, 1164 (1990).

35. S. Senoussi, M. Oussena, G. Collin, and I. A. Campbell, Phys. Rev. B **37**, 9792 (1988).

36. J. R. Thompson, Yang Ren Sun, L. Civale, A. P. Malozemoff, M. W. McElfresh, A. D. Marwick, and F. Holtzberg, Phys. Rev. B **47**, 14440 (1993).

37. J. R. Thompson, Yangren Sun, and F. H. Holtzberg, Phys. Rev. B **44**, 469 (1991).

38. M. V. Feigel'man, V. B. Geshkenbein, A. I. Larkin, and V. M. Vinokur, Phys. Rev. Lett. **63**, 2301 (1989).

39. D. Bourgault, M. Hervieu, S. Bouffard, D. Groult, and B. Raveau, Nucl. Instrum. Meth. in Phys. Res. **B42**, 61 (1989).

40. L. Civale, A. D. Marwick, T. K. Worthington, M. A. Kirk, J. R. Thompson, L. Krusin-Elbaum, Y. Sun, J. R. Clem and F. Holtzberg, Phys. Rev. Letters, **67**, 648 (1991).

41. V. Hardy, J. Provost, D. Groult, M. Hervieu, B. Raveau, S. Durcok, E. Pollert, J. C. Frison, J. P. Chaminade, and M. Pouchard, Physica C **191**, 85 (1992).

42. M. Konczykowski, F. Rullier-Alebenque, E.R. Jacoby, A. Shaulov, Y. Yeshurun, and P. Lejay, Phys. Rev. B **44**, 7167 (1991).

43. H. Kumakura, H. Kitaguchi, K. Togano, H. Maeda, J. Shimoyama, S. Okayasu, and Y. Kazumata, J. Appl. Phys. **74**, 451 (1993); Jpn. J. Appl. Phys. **31**, L1408 (1992).

44. R. C. Budhani, M. Suenaga, and S. H. Liou, Phys. Rev. Lett. **69**, 3816 (1992).

45. A. D. Marwick, L. Civale, L. Krusin-Elbaum, R. Wheeler, J. R. Thompson, T. K. Worthington, M. A. Kirk, Y. R. Sun, H. R. Kerchner, and F. Holtzberg, Nucl. Instrum. Meth. B **80/81**, 1143-9 (1993).

46. D. R. Nelson and V. M. Vinokur, Phys. Rev. Lett. **68**, 2398 (1992); Phys. Rev. B **48**, 13060 (1993).

47. L. Krusin-Elbaum, L. Civale, J. R. Thompson, and C. Feild (unpublished).

48. W. Gerhaeuser, G. Ries, H.-W. Neumueller, W. Schmidt, O. Eibl, G. Seaman-Ischenko, and S. Klaumuenzer, Phys. Rev. Lett. **68**, 879 (1992).

49. J.R. Thompson, Y.R. Sun, H.R. Kerchner, D.K. Christen, B.C. Sales, B.C. Chakoumakos, A.D. Marwick, L. Civale, and J.O. Thomson, Appl. Phys. Lett. **60**, 2306 (1992).

50. L. Civale, A.D. Marwick, R. Wheeler IV, M.A. Kirk, W.J. Carter, G.N. Riley Jr., and A.P. Malozemoff, Physica C **208**, 137 (1993).

51. T. Hwa, P. LeDoussal, D.R. Nelson, and V.M. Vinokur, Phys. Rev. Lett. **71**, 3545 (1993).

52. L. Civale, L. Krusin-Elbaum, J. R. Thompson, R. Wheeler, A. D. Marwick, M. A. Kirk, Y. Sun, F. Holtzberg, and C. Feild, Phys. Rev. B **50**, 4102 (1994).

53. L. Krusin-Elbaum, J. R. Thompson, R. Wheeler, A. D. Marwick, C. Li, S. Patel, D. T. Shaw, P. Lisowski, and J. Ullmann, Appl. Phys. Letters, **64**, 3331 (1994).

54. J. R. Thompson, Y. R. Sun, H. R. Kerchner, D. K. Christen, B. C. Sales, B. C. Chakoumakos, A. D. Marwick, L. Civale, and J. O. Thomson, Appl. Phys. Lett. **60**, 2306 (1992).

55. J. R. Thompson, L. Krusin-Elbaum, Y. C. Kim, D. K. Christen, A. D. Marwick, R. Wheeler, C. Li, S. Patel, and D. T. Shaw, P. Lisowski and J. Ullmann, IEEE Trans. Appl. Superconductivity **5**, 1876 (1995).

56. Safar, J. H. Cho, S. Fleshler, M. P. Maley, J. O. Willis, J. Y. Coulter, J. L. Ullmann, P. W. Lisowski, G. N. Riley, M. W. Rupich, J. R. Thompson, and L. Krusin-Elbaum, Appl. Phys. Lett. **67**, 130 (1995).

57. J. Schwartz, S. Nakamae, G. W. Raben, Jr., J. K. Heuer, S. Wu, J. L. Wagner, and D. G. Hinks, Phys. Rev. B **48**, 9932 (1993).

Coexistence of Superconductivity and Magnetism in HTSC Materials? μSR and Magnetooptical Studies

A. Golnik[1], C. Bernhard[2], J. Budnick[3], M. Kutrowski[1*], Ch. Niedermayer[2], and T. Szumiata[1**]

[1] Institute of Experimental Physics, Warsaw University, Hoża 69, PL–00-681 Warsaw, Poland
[2] Fakultät für Physik, Universität Konstanz, D-78434 Konstanz, Germany
[3] Physics Department, University of Connecticut, Storrs, 06268 CT, USA

Abstract. The evidences for coexistence of superconductivity and magnetic moments of Cu ions in HTSC materials are discussed in frames of different theoretical models. Main attention is paid to the evidences obtained by Muon Spin Rotation (μSR) spectroscopy. The same models are used to analyze the negative result of series of magnetooptical studies of 123 and 214 films, which were directed to determine the magnitude of spin–carrier exchange interaction. The detection sensitivity of 10^{-5} was achieved in the measurements of magnetic circular dichroism, but in the films of good quality no measurable MCD signal was observed.

1 Introduction

The mechanisms responsible for high temperature superconductivity (HTSC) are still far from being well understood. Many people believe that one of the crucial points on the way to establish the commonly accepted model of HTSC is the exploration of the phase diagram of the HTSC materials versus their carrier concentration.

It is common, that for all classes of HTSC materials, their "parent", antiferromagnetic compounds were found (La_2CuO_4 for the "214" group, $[Re]Ba_2Cu_3O_6$

---

* Present address: Institute of Physics, Polish Academy of Sciences, Al.Lotników 32/46, 02-668 Warsaw, Poland
** Present address: Department of Physics, Technical University of Radom, Malczewskiego 29, 26-600 Radom, Poland

for "123" group, yttrium substituted compounds for Bi and Tl based HTSC). The transition from this "parent" insulating and antiferromagnetic phase, to the metallic and superconducting state is controlled by the changes of the charge carrier concentration. This concentration could be continuously varied, when changing the stoichiometry of the material. For "123" compounds and La_2CuO_{4-y} this is the oxygen content, for the other groups it is the concentration of dopants. In series of our Muon Spin Rotation (μSR) experiments we have studied the magnetic phase diagram of the HTSC materials (for reviews, see e.g. [1]). For carrier concentrations, which were close to metal-insulator transition, we observed the μSR evidences for the coexistence of both superconductivity and magnetism in the same samples. The experimental results and various microscopic pictures of such coexistence will be discussed in the first part of this paper.

Some of the models laying behind this microscopic pictures link the superconductivity in HTSC materials to the interactions of magnetic origin between charge carriers. The essential parameter in such models is the exchange integral of the carrier - Cu spin interaction (p-d exchange). Many theoretical estimates of this parameter give the values exceeding $0.5\,eV$. Our magnetooptical experiments were aimed on the experimental determination of this quantity. The negative result of our search for magnetic circular dichroism will be discussed in the second part of the paper in the framework of the previously introduced microscopic pictures.

2 Muon Spin Rotation experiments

The Muon Spin Rotation spectroscopy was found to be very good method for investigation different kinds of magnetic ordering in the HTSC materials and determination of their magnetic phase diagram. Further discussion will be based mainly on our μSR experiments performed at the Paul-Scherrer Institute (PSI) in Villigen, Switzerland.

2.1 Muon Spin Rotation spectroscopy

Positive, spin polarized muons are implanted into the studied materials and stop there within 10^{-11} s, practically without loosing their initial polarization. In HTSC materials muons have been found to stop close to oxygen atoms forming there a kind of "muoxyl" bonding [2]. The spin of the muon starts to precess in the local internal magnetic field at the muon site, but the positive muon decays to positron and two neutrinos with the characteristic time of $2.197\,\mu s$. The violation of parity conservation in weak interactions causes the asymmetry of the muon decay, so the decay positron is emitted preferentially in the direction of muon spin at the time of decay. This decay anisotropy is detected by the scintillator counters, which are placed around the sample and register the positrons emitted in certain directions with respect to the initial muon spin polarization.

The output of the experiment are the histograms of counts as a function of time elapsed between the start pulse (given by incoming muon in the counter placed before the sample) and the detection of the decay positron in the counter at certain direction. The asymmetry part of this histogram (usually plotted as a μSR spectrum) contains the information on the evolution of the muon spin polarization versus time. Namely the frequency of the oscillations determines the value of the local magnetic field at the muon site and the depolarization function gives the information on the shape field distribution or on the dynamics of the field fluctuations.

Here we will discuss only the zero-field (ZF-)μSR experiments i.e. those done without applying the external magnetic field. In this case the clear μSR oscillations might be observed only, when the internal fields at different stopping sites have the same magnitude, but not necessarily the same orientation (which is the common case for ferromagnetic or antiferromagnetic materials). The reason is, that the precession frequency is given by the magnitude of the internal field, whereas the direction of the field determines only the amplitude of a precession cone. The amplitude of observed oscillation is then given by the angular average over all directions. In case of their random distribution (e.g. powder average) the counters placed forward and backward to the initial muon spin polarization will detect the 2/3 of the muon asymmetry (transverse components) to exhibit oscillations and the 1/3 of the signal (longitudinal component) to remain constant (in the static case).

In the real case, however, both components of the ZF-μSR signal would be depolarized. The depolarization of the 2/3 oscillating signal might be considered as an analogue of the transverse depolarization time T_2 in NMR experiments and should be related to the interplay between the broadening of static field distribution (in homogeneities) and the dynamic processes (e.g. spin fluctuations). The 1/3 "constant" component would be depolarized by the dynamic processes only, those responsible for energy relaxation (analogously to the T_1 time).

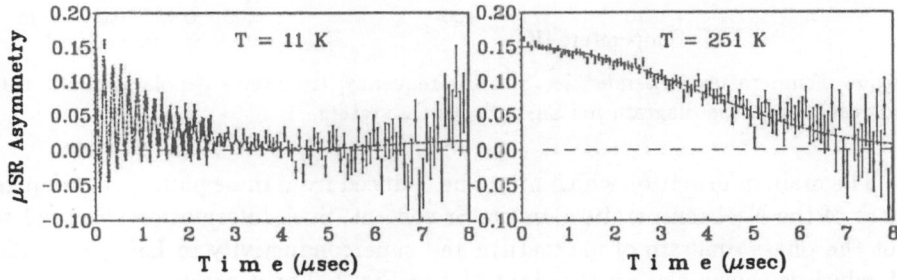

Fig. 1. Example of ZF-μSR spectra for the La_2CuO_4 sample at 11 K and 256 K.

An example of spontaneous ZF-μSR oscillations might be seen in the left plot of Fig. 1. The observed frequency of 5.64 MHz corresponds to the internal magnetic field at the muon site equal to 41.6 mT.

Another very important case of ZF-μSR is the Kubo-Toyabe signal, which is observed when the internal fields have a zero average, but the distribution of a

Gaussian shape. The Kubo-Toyabe dependence resembles the single oscillation of 2/3 component with the recovery of 1/3 component at long times. The example of slow Kubo-Toyabe μSR signal characteristic for HTSC materials is shown in the right plot of Fig. 1. The origin of such signal are the local fields due to the nuclear spins of Cu atoms, which are much weaker than those from the spin moments of Cu d-electrons.

2.2 The μSR studies of magnetism in HTSC materials

As presented already in Fig. 1, the antiferromagnetism of La_2CuO_4 shows up in μSR experiments as a clear spontaneous oscillations. Similar oscillations were observed also for $La_{2-x}Sr_xCuO_4$ with $x \leq 0.07$. Temperature dependencies of frequencies and depolarization's rates of the μSR signals are shown in Fig. 2. Please note that it is the recompilation of our previous results from [3,4].

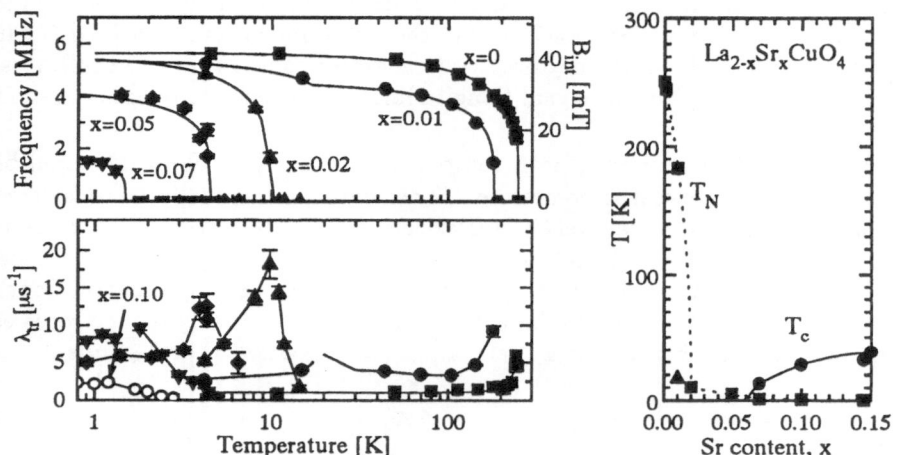

Fig. 2. Temperature dependencies of μSR frequency, transverse depolarization's rate and resulting phase diagram for $La_{2-x}Sr_xCuO_4$ system.

The main information which might be deduced from these plots, is the dependence of the Néel temperature on the Sr content. Such information was used to plot the phase diagram of magnetism and superconductivity in $La_{2-x}Sr_xCuO_4$ [4], which is shown also on the right plot in Fig. 2. The dependency of μSR frequency (and so of the local field) versus temperature can not be described by the mean field approach (Brillouin function). This problem was carefully studied by Le et al. [5] and interpreted in terms of spin-wave theory for anisotropic magnetism.

Secondly, the only difference between long range antiferromagnetism (for $x < 0.02$) and spin-glass phase (for $0.02 < x < 0.06$) found in our μSR experiments is the change of the character of T_N versus x dependence at $x = 0.02$ (and somewhat higher depolarization rates for spin-glass samples at low temperatures). The zero temperature limit of the internal field diminishes only slightly, i.e. from about 40 mT for $0 < x < 0.02$ to about 30 mT at $x = 0.05$. The oscillating character of μSR spectra does not change. That means that locally (in the vicinity of the muon stopping site) the spin-glass and long-range antiferromagnetic phases must be very similar.

Even for the superconducting sample ($x = 0.07$, $T_c = 14$ K) we observed at low temperatures a kind of strongly damped oscillations (see Fig. 3). For superconducting samples with higher x the traces of magnetism show up at very low temperatures ($T \lesssim 1$ K) as an exponentially damped component superimposed on the much slower Kubo-Toyabe signal (Fig. 3) (the second, as before, results from the Cu nuclear spins). The high damping rate of the exponential component must be caused by the magnetic moments of electron origin (nuclear moments are too small). The exponential damping means that either the fields are not static or their distribution is Lorenzian.

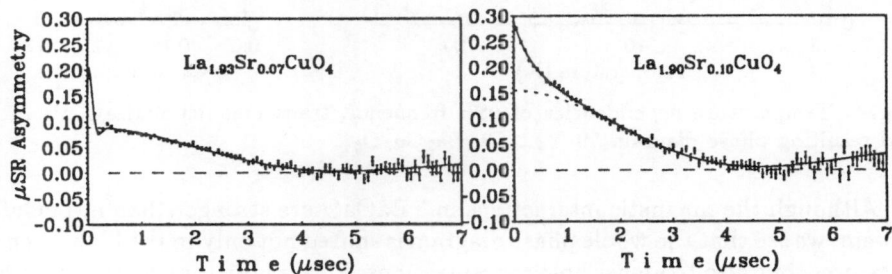

Fig. 3. μSR spectra for superconducting La$_{2-x}$Sr$_x$CuO$_4$ samples at $T = 35$ mK.

The another feature overlooked in our previous papers but nowadays carefully studied by Borsa et al. [6,7] is the anomaly, trace of which might be seen in Fig. 2 around 18 K (for the sample with $x = 0.01$) in both frequency and depolarization rate dependencies. Such anomaly observed before in the series of NQR experiments for $0 < x \leq 0.02$ was interpreted in frames of reentrant spin-glass behavior.

Very similar features were found also for YBa$_2$Cu$_3$O$_{6+x}$ system [8]. The phase diagram must be then drawn versus oxygen content, which, however, does not correspond univocally to the carrier concentration. The actual carrier concentration in the sample (and so the phase diagram) depends on the conditions of sample preparation (quenched versus slowly cooled samples). This seems to be caused by the tendency of oxygen to form some kind of superstructure, in case when its diffusion is allowed in the "slow-annealing" process. (This effect of ordering of oxygen vacancies in "chains" of YBa$_2$Cu$_3$O$_{6+x}$ causes also the appearance of two characteristic plateaus at 60 K and 90 K in the T_c versus x dependence.)

The hole doping of YBa$_2$Cu$_3$O$_6$ might be achieved not only by oxygen- but also by Ca- doping, The hole concentration corresponds roughly to $x/2$ [9] (similarly as for $La_{2-x}Sr_xCuO_4$ system). Figure 4 presents the preliminary results of the recent μSR experiments for $Y_{1-x}Ca_xBa_2Cu_3O_6$ system [10].

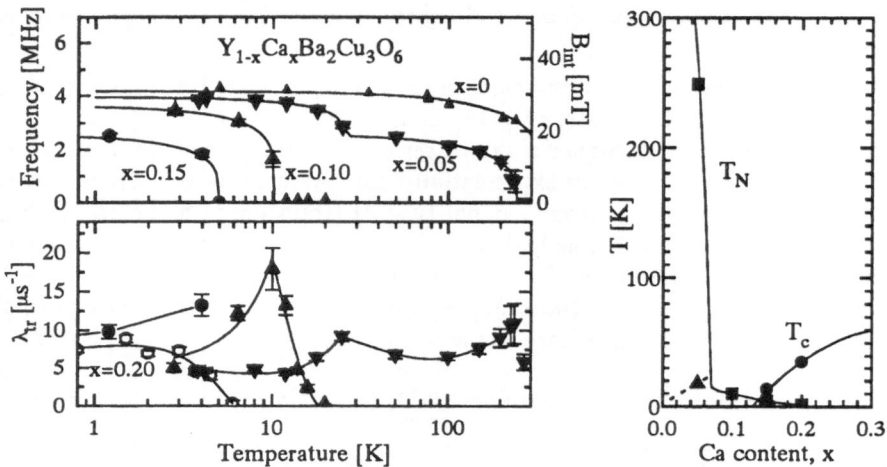

Fig. 4. Temperature dependencies of μSR frequency, transverse depolarization's rate and resulting phase diagram for $Y_{1-x}Ca_xBa_2Cu_3O_6$.

Although the magnetic interactions in YBaCuO are stronger than in LaCuO system, we see that the whole phase diagram is shifted not only to the higher temperatures, but also to higher hole concentrations. Please note that reentrant spin-glass anomaly is now visible around $T = 26$ K for $x = 0.05$ in both: frequency (local field) and depolarization's rate dependencies. Again for superconducting samples we observe either strongly damped oscillation (for $x = 0.15$, $T_c = 13$ K sample) or strongly, exponentially damped signal (for x=0.2, $T_c = 36$ K sample).

2.3 Question of coexistence

The main experimental facts, which could be interpreted as signature of coexistence of superconductivity and magnetic correlation in HTSC are summarized below:

a. In superconducting samples of LaSrCuO, YBaCuO and YCaBaCuO the exponentially damped component of the zero-field μSR depolarization function was found at sufficiently low temperatures. It must be caused by the magnetic moments of electron origin, which slow down their random fluctuations by some kind of magnetic correlation's.

b. For the $La_{1.93}Sr_{0.07}CuO_4$ (with $T_c = 14$ K) and $Y_{1-x}Ca_xBa_2Cu_3O_6$ ($T_c = 13$ K) at the lowest temperatures the exponentially damped signal converts to strongly damped oscillations.

c. In the samples with highest T_c ($T_c > 32$ K for $La_{2-x}Sr_xCuO_4$ and $T_c > 50$ K for $YBa_2Cu_3O_{6+x}$) the exponentially damped signal was not observed. The whole μSR asymmetry observed in zero field experiments could be described by slow Kubo-Toyabe dependence due to Cu nuclear moments.

d. The fractions of muons influenced by strong fields decreases with increasing x and temperature.

The first question was, whether the observed features were not caused by some impurity phases or by crystallographic phase separation, like that observed in superoxydated La_2CuO_4 [11]. The careful X-ray analysis of our samples did not show any traces of impurity phases or chemical phase-separation. The independent μSR experiments of Brewer et al. [12] confirmed, that "there is true microscopic coexistence of AFM and SC" in $YBa_2Cu_3O_{6+x}$ with x between 6.38 and 6.48. (The inhomogeneities of oxygen concentration in their samples was estimated to be smaller than 0.01.)

3 Microscopic pictures

The main question of the coexistence problem is, whether the superconductivity coexists with some kind of magnetic correlation's in the same microscopic regions of the sample or both phases are spatially separated. The absence of chemical separation (discussed above), does not rule out the possibility of spontaneous phase separation, where the only difference between two phases is the concentration of charge carriers. The "hole rich" phase would be superconducting and "hole pure" - magnetic. Similar phase separation was also predicted for the Mott-Anderson (metal-insulator) transition in semiconductors [13]. For HTSC systems, Emery et al. [14] found theoretically, that the diluted holes should be unstable against such phase separation.

There is a strong experimental evidence of such spontaneous phase separation found by Yu et al. [15] in the experiments on transient photo-induced superconductivity. The authors derived from the experiment that photoexcited carriers condense in hole droplets and observed the existence of minimum metallic conductivity as predicted by Mott-Anderson theory. Please remember that the photo-induced superconductivity might be also persistent or metastable [16,17].

Although, it might be expected, that such spontaneous "condensation" of charge carriers should be stabilized by the fluctuations of dopant concentration, it is also possible, that the phase separation has a dynamic character. More complicated models of phase separations were reported recently.

In order to explain the reentrant spin-glass behavior of $La_{2-x}Sr_xCuO_4$ system with $x < 0.02$, it was proposed [18,7], that the mobile holes form walls separating the undoped uncoupled antiferromagnetic domains. The anomaly of the internal fields and relaxation rates (like that seen on Fig. 2 and 4) would be then caused by the localization of the doped holes and freezing of their spins. (Please note that the results of Bernhard et al. [10] for $Y_{1-x}Ca_xBa_2Cu_3O_6$ seem to be the first report of such behavior for the system other than $La_{2-x}Sr_xCuO_4$.) For higher temperatures the holes would move and so the Cu spins in clusters should fluctuate. (It seems to be consistent with the behavior of μSR depolarization time seen in Fig. 4.) By increasing the hole doping (region of $0.02 < x < 0.08$ for $La_{2-x}Sr_xCuO_4$ system) the cluster size is believed to decrease and finally the hole rich walls become superconducting. Recently the theoretical models and experimental arguments were proposed for wall-like phase separation even in superconducting samples (see e.g. [19]). The μSR data from the superconducting region (exponential depolarization) strongly suggest that the phase separation should be dynamic in the time scale of the muon life-time. It may look, however, like static in the neutron scattering experiments, which deal with much shorter time scale.

The another question is, whether the Cu spins disappear in the superconducting phase (as predicted e.g. by Schrieffer's spin-bag theory [20]) or are still present but strongly fluctuating due to the presence of hole spins (like in Aharony model of magnetic attraction [21]). Our μSR results can not give definite answer, since the lack of magnetic signals in the perfectly superconducting samples (see point c in 2.3) means only that the Cu spins (if present) do not freeze (or slow down their fluctuations to the time scale of muon life-time). This question, however, was the motivation for our magnetooptical studies.

4 Magnetooptical studies

The characteristic feature for all HTSC materials, which is also responsible for magnetic behavior of their insulating phases, is the relatively strong exchange-interaction between magnetic moments localized on CuO planes. (It is mainly d-d superexchange, involving the $3d^9$ orbitals of Cu ions.) Such a strong d-d superexchange is expected to be accompanied by analogously strong p-d exchange between localized spins and charge carriers (holes, mainly from O-2p orbitals). Although significant fraction of theoretical works ascribe the mechanism responsible for high-temperature superconductivity to such p-d exchange (e.g. [21]), the experimental studies of this interaction in HTCS materials are not very extensive. Therefore, we decided to start experimental investigation of this p-d exchange interaction using magnetooptical methods. We expected to found some magnetic circular dichroism in the region of the $3d^92p^1 \rightarrow 3d^{10}$ charge transfer transition.

The early investigations of circular dichroism for HTCS thin films were done on a search for spontaneous effects due to anyon superconductivity [22-24] and therefore were performed on the superconducting films. Since these experiments were conflicting, later studies on spectral dependence of magnetic circular dichroism were done for both superconducting and insulating films [25,26]. These results seem to be still inconclusive and very sample dependent.

We expected that the best conditions for the studies of p-d interactions by magnetooptical methods will be available for the sample stoichiometry close to the metal-insulator transition. Antiferromagnetism would be there weakened sufficiently, so that the magnetic moments of Cu ions can be affected by external field, but the field itself would not be screened by the sample superconductivity.

Our first trial were the investigations of the 1-2-3 thin films as a function of the oxygen stoichiometry. We used $TmBa_2Cu_3O_{6+x}$ films of the thickness between 0.2 and 0.5 μm deposited on MgO substrate by magnetron sputtering. The oxygen content was changed by stepwise vacuum annealing and monitored by the a-c susceptibility, resistance and optical absorption studies. The stoichiometry before and after annealing process was additionally checked in RBS experiments. Typical results of characterization obtained for the sample of the thickness 230 nm and starting composition of $x \approx 0.9$ ($T_c = 82$ K) are presented in Fig. 5 and 6. Numbering of curves denotes the sequence of annealing steps.

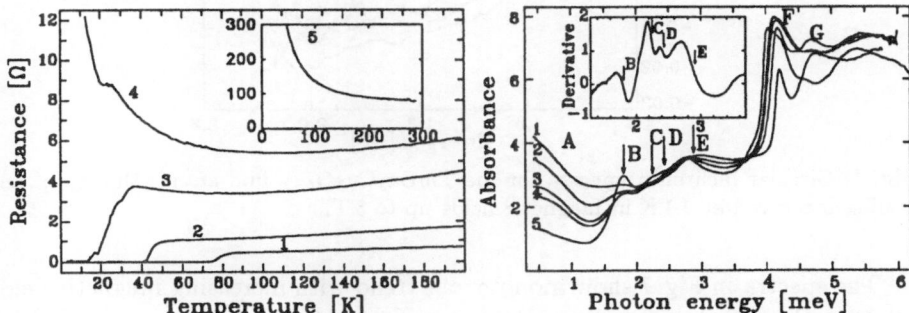

Fig. 5. Temperature dependence of the $TmBa_2Cu_3O_{6+x}$ film resistance after sequential steps of annealing. Note different scale for the curve 5 presented in the insert.

Fig. 6. Absorption spectra for the same film after same steps of annealing. The insert shows the derivative of curve 1. The absorption features are labeled by capital letters (see text).

We were able to identify in our spectra several absorption structures known from the reflectivity measurements of 123 materials [27] i.e.:

A The Drude-like free-carrier absorption, intensity of which scales with carrier concentration.

B,F The two fundamental absorption edges connected with CuO_2 planes (B) and CuO chains (F) with intensity strongly dependent on x.

C Structure with weak dependence of position and intensity, reported already in literature but without clear assignment.

E Usually interpreted as a transition from $O-2p_\pi$ $Cu-3d_{xy}$ states (laying deeply under the Fermi level) to the upper Hubbard band ($Cu-3d^{10}$ configuration). Its position varies with change of oxygen composition but its intensity does not.

G Transition involving O and Ba states.

The absorption structure denoted by D was not found previously by analysis of reflectivity spectra. It does not vary drastically with oxygen concentration. The separation of the structures C and D can be seen more clearly on the insert of Fig. 6, which shows the numerical derivative of the spectrum 1.

Preliminary measurements of magnetic circular dichroism shown in Fig. 7 looked very promising. (The measurements were made with the polarization modulator based on Fresnel rhomb and rotating polarizer.)

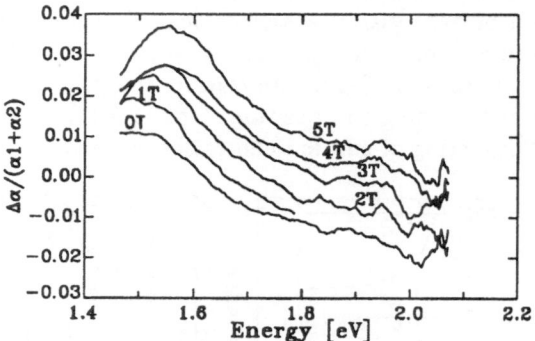

Fig. 7. Circular dichroism spectra for the $TmBa_2Cu_3O_{6+x}$ film after 4-th step of annealing measured at 2.1 K in magnetic fields up to 5 T.

The spectra in Fig. 7 show monotonous trend with increasing magnetic field, however, the signal does not vanish for zero field. Although, the nonzero dichroism for zero field was also reported by Izumi et all. [25], it was hard to explain it theoretically. Therefore, we started to increase the sensitivity of our system and finally with the use of elastooptical modulator we achieved the sensitivity of about 10^{-5}. (By the total signal $(I_{\sigma+} + I_{\sigma-})$ being of the order of 1 V the sensitivity of detection for the modulation signal $(I_{\sigma+} - I_{\sigma-})$ was better than few nV.)

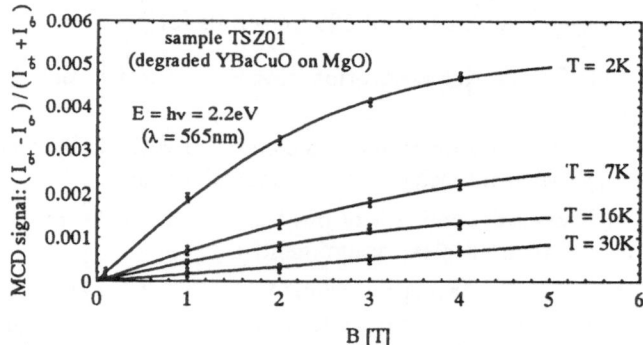

Fig. 8. Magnetic circular dichroism at its maximum versus magnetic field for degraded TSZ01 sample (YBaCuO film on MgO substrate). Solid lines - the Brillouin function for the spin 1/2.

We were able to measure the MCD spectra due to chromium impurities in $SrLaAlO_4$ substrates, the MCD of the order of 10^{-3} on some degraded (never superconducting) YBaCuO film (see Fig. 8), but for good quality samples we never succeeded to measure the MCD spectrum, which could be assigned to the $3d^9 2p^1 \rightarrow 3d^{10}$ charge transfer transition.

In the simplest case, when the charge transfer absorption spectrum would be split in both circular polarization's in magnetic field, we expect the MCD spectrum proportional to the derivative of the absorption spectrum. The example of such spectra for the $La_{1.955}Sr_{0.045}CuO_4$ film deposited on $SrTiO_3$ substrate is shown in Fig. 9.

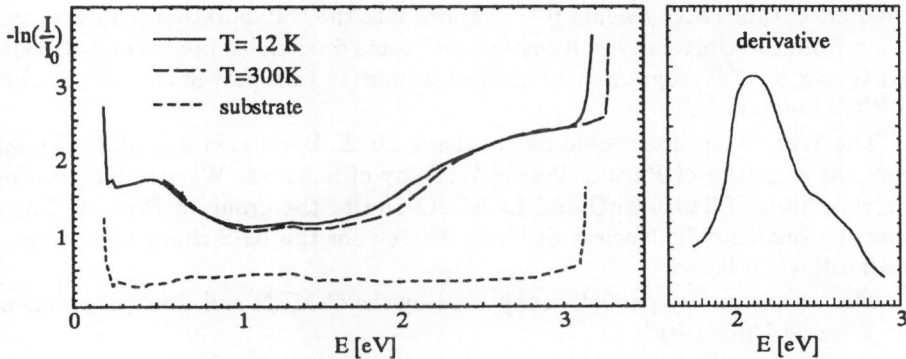

Fig. 9. Absorption spectra for $La_{1.955}Sr_{0.045}CuO_4$ layer deposited on $SrTiO_3$ substrate and substrate itself - left plot, derivative of absorption spectrum of the layer - right plot.

Again the absorption spectrum is very similar to those measured for the $La_{2-x}Sr_xCuO_4$ system by M. Suzuki [28], and shows the charge-transfer absorption edge around 2.1 eV. The maximum for the derivative spectrum lies by 2.13 eV. Taking the theoretical value of the p-d exchange integral of about 0.5 eV one could estimate the edge splitting by $T = 12\,K$ and $B = 4\,T$ to be about 50 meV and the MCD value at 2.13 eV to be 10^{-1}. In the experiments we found, however, that it is smaller than 10^{-4}.

Of course this fact might be interpreted as a sign of the phase separation. The "hole rich" cluster walls would be spatially separated from the clusters, where magnetic moments exist, so the p-d exchange would be reduced and not observed. The Cu spins in the walls must be then depressed like in the Schrieffer's model [20].

There might be however another explanation. Even when the two spin states of the holes will be split by the p-d exchange interaction with ordered Cu moments (or by the magnetic field itself), in order to observe the MCD one needs to have different selection rules for the optical transitions from this two states for both circular polarizations. This requires however the non zero spin-orbit coupling on the oxygen p orbitals to couple the spin degrees of freedom with the circular polarized light in this electric-dipole allowed $3d^9 2p^1 \rightarrow 3d^{10}$ transition. The spin-orbit coupling on the O2p orbitals of CuO layers is however believed to be negligible.

In summary we can state that the magnetic circular dichroism of the good quality HTSC films is smaller than 10^{-4}, but the MCD itself seems not to be a good tool for studies of p-d exchange interaction in this materials.

Acknowledgments

The first author (A.G.) would like to appreciate the collaboration with the μSR group from the University of Konstanz and some financial support from this site (DPG and BMFT programs), which enabled him to take part in the μSR studies of HTSC materials.

The Warsaw authors would like to thank Dr. L. Baczewski and M.Z. Cieplak from the Institute of Physics Polish Academy of Sciences, Warsaw for growing the thin films of TmBaCuO and LaSrCuO and to the group of Prof. A. Turos from the Institute for Nuclear Studies, Warsaw for the RBS characterization of the TmBaCuO layers.

The work was also partially supported by the G-MEN and BST programs at the Warsaw University.

References

1.   A. Golnik: Acta Physica Polonica, A84 16 (1993)
2.   W.K. Dawson, C.H. Halim, S.P. Weathersby, J,A. Flint, J.C. Lam, T.J. Hoffman, C. Boekema, K.C.B. Chan, R.L. Lichti, D.W. Cooke, M.S. Jahan, J.E. Crow: Hyp. Int. 63 219 (1990)
3   J.I. Budnick, A. Golnik, Ch. Niedermayer, E. Recknagel, M. Roßmanith, A. Weidinger, B. Chamberland, M. Filipkowski, D.P. Yang: Phys. Lett. A124 103 (1987); J.I. Budnick, B. Chamberland, D.P. Yang, Ch. Niedermayer, A. Golnik, E. Recknagel, M. Roßmanith, A. Weidinger: Europhys. Lett. 5 651 (1988)
4   A. Weidinger, Ch. Niedermayer, A. Golnik, R. Simon, E. Recknagel, J.I. Budnick, B. Chamberland, C. Baines: Phys. Rev. Lett. 62 102 (1989)
5   L.P. Le, G.M. Luke, B.J. Sternlieb, Y.J. Uemura, J.H. Brewer, T.M. Riseman, D.C. Johnston, L.L. Miller, Y. Hidaka, H. Murakami: Hyp. Int. 63 279 (1990)
6   F. Borsa, P. Carretta, J.H. Cho, F.C. Chou, Q. Hu, D.C. Johnston, A. Lascialfari, D.R. Torgeson, R.J. Gooding, N.M. Salem, K.J.E. Vos: Phys. Rev. B (to be published)
7   D.J. Johnston et al.: this conference
8   see e.g. A. Weidinger, Ch. Niedermayer, H. Glückler, A. Golnik, G. Nowitzke, E. Recknagel, H. Eickenbusch, W. Paulus, R. Schöllhorn, J.I. Budnick: Hyp. Int. 63 147 (1990)
9   H. Casalta, H. Alloul, J.-F. Maruco: Physica C204 331 (1993)
10   C. Bernhard et al.: in preparation
11   J.D. Jorgensen, B. Dabrowski, S. Pei, D.G. Hinks, L. Sonderholm: Phys. Rev. B38 11377 (1988)

12   J.H. Brewer, J.F. Carolan, W.N. Hardy, B.X. Yang, P. Schleger, R. Kadono, J. Kempton, R.F. Kiefl, S.R. Kreitzman, G.M. Luke, T.M. Riseman, D.Ll. Williams, K. Chow, P. Dosanjh, B. Gowe, R. Krahn, M. Norman: Physica C162-164 33 (1989)

13   see e.g., N.F. Mott, M. Kaveh: Adv. Phys. 31 329 (1985); N.F. Mott: Adv. Phys. 39 55 (1990)

14   V.J. Emery, S.A. Kivelson, H.Q. Lin, Phys. Rev. Lett. 64 475 (1990)

15   G. Yu, C.H. Lee, A.J. Heeger, N. Herron, E.M. McCarron, Lin Cong, G.C. Spalding, C.A. Nordman, A.M. Goldman: Phys. Rev. B45 4964 (1992); G. Yu, C.H. Lee, D. Mihailovic, A.J. Heeger, C. Fincher, N. Herron, E.M. McCarron: Phys. Rev. B48 7545 (1993)

16   V.I. Kudinov, A.I. Kirilyuk, N.M. Kreines, R. Laiho, E. Lähderlanta: Phys. Lett. A151 356 (1990)

17   H. Szymczak, R. Szymczak: this conference

18   F.C. Chou, F. Borsa, J.H. Cho, D.C. Johnston, A.Lascialfari, D.R. Torgenson, J.Ziolo: Phys.Rev.Lett. 71 2323 (1993)

19   V.J. Emery, this conference

20   J.R. Schrieffer, X.-G. Weng, S.-C. Zhang: Phys. Rev. Lett. 60 944 (1988)

21   A. Aharony, R.J. Birgenau, A. Conigilio, M.A. Kastner, H.E. Stanley: Phys. Rev. Lett. 60 1330 (1988)

22   K.B. Lyons, J. Kwo, J.F. Dillon Jr., G.P. Espinosa, M. McGlashan-Powell, A.P. Ramirez and L.F. Scheemeyer, Phys. Rev. Lett. 64 2949 (1990)

23   S. Spielman, K. Fesler, C.B. Eom, T.H. Geballe, M.M. Fejer and A. Kapitulnik, Phys. Rev. Lett. 64 123 (1990); S. Spielman, J.S. Dodge, L.W. Lombardo, C.E. Eom, M.M. Fejer, T.H. Geballe and A. Kapitulnik, Phys. Rev. Lett. 68 3472 (1992)

24   H.J. Weber, D. Weitbrecht, D. Brach, H. Keiter, A.L. Shelankov, W. Weber, Th. Wolf, J. Geerk, G. Linker, G. Roth, P.C. Splittgerber-Hünnekers, G. Güntherodt: Solid State Commun. 76 511 (1990)

25   H. Izumi, K. Ando, N. Koshizuka and K. Ohata: Physica C199 171 (1992)

26   A.A. Milner, N.F. Kharhenko, O.V. Miloslavskaya, A.T. Pugachev, A.T. Stetsenko and A.N. Chirkin: Physica C209 (1993) 225; R.V. Pisarev et al.: Fiz. Nisk. Temp. 18 Suppl., no. S1 (1992) 33; N.F. Kharchenko et al.: Fiz. Nisk. Temp. 18 Suppl., no. S1 (1992) 25

27   M.K. Kelly, P. Barboux, J.-M. Tarascon and D.E. Aspnes: Phys. Rev. B. 40 (1989) 6797

28   M. Suzuki: Phys. Rev. B39 2312 (1989)

List of Unpublished Lectures

List of Lectures read but not included to the present Proceedings

1. **New High Temperature Superconductors and Physics.**
   C.W.Chu (*Texas Cent.for Supercond. and Dept.Phys., Univ. Houston, TX, USA*)

2. **What is the Meaning of the Critical Current Density in Polycrystalline Forms of High Temperature Superconductors?**
   David Larbalestier (*Appl.Supercond.Cent., Dept.Mat.Sci.& Engng and Dept.Phys., Univ.of Wisc., Madison, WI, USA*)

3. **Power Frequency AC Losses in High T_c Superconductors and Their Composites.**
   M.Suenaga (*Brookhaven Natl.Lab., Upton, NY, USA*)

4. **Thermal Conductivity of High Temperature Superconductors and a New Interpretation of the Normal–State Thermal Transport.**
   A.Jeżowski (*Inst.Low Temp.& Struct.Res., Polish Ac.Sci., Wrocław, Poland*)

5. **Photoexcitation Effects in Oxygen–Deficient $YBa_2Cu_3O_y$ Films.**
   H.Szymczak and R.Szymczak (*Inst.Phys., Polish Ac.Sci., Warsaw, Poland*)

6. **Angle Resolved Photoemission and the Superconducting Order Parameter in Bi-2212.**
   M.R. Norman (*Mater.Sci.Div., Argonne Natl.Lab., IL, USA*)

7. **Theory of Copper–Oxide Metals with Application to NMR.**
   C.M.Varma (*AT&T Bell Labs., Murray Hill, NJ, USA*)

8. **First Order Melting of the Flux–Line Lattice in High Temperature Superconductors.**
   P.L.Gammel (*AT&T Bell Labs., Murray Hill, NJ, USA*)

List of Posters

Posters presented at the Conference:

1. **The Influence of Anion Stoichiometry on the Electronic and Transport Properties of New Rare Earth – Copper Chalcogenides**
   I.Jacyna–Onyszkiewicz (*Inst.Phys., Univ., Poznań, Poland*), M.A.Obolensky (*Dept.Phys., Kharkov State Univ., Ukraine*), S.Robaszkiewicz (*Inst.Phys., Univ., Poznań, Poland*), V.Starodub (*Dept.Phys., Kharkov State Univ., Ukraine*) and M.Zimpel (*Inst.Phys., Univ., Poznań, Poland*)

2. **Detwinning of $YBa_2Cu_3O_{7-d}$ Single Crystals**
   W.Sadowski (*Dept.Appl.Phys.& Math., Techn.Univ. Gdańsk, Poland*) and E.Walker (*D.P.M.C. Univ. Geneva, Switzerland*)

3. **Anisotropy of the Electrical Resistivity and Hall Effect in Zn Doped $YBa_2Cu_3O_{7-d}$ Crystals**
   W.Sadowski (*Dept.Appl.Phys.& Math., Techn.Univ. Gdańsk, Poland*), M.Affronte (*Dipart.Fis., Univ.Modena, Italy*), B.Kusz, M.Gazda and O.Gzowski (*Dept.Appl.Phys.& Math., Techn.Univ. Gdańsk, Poland*)

4. **Influence of Oxygen Deficiency on Critical Currents in $YBa_2Cu_3O_{7-d}$ Melt-Textured Samples**
   A.Wiśniewski and M.Baran (*Inst.Phys., Polish Ac.Sci., Warsaw, Poland*)

5. **High Gas Pressure Liquid Phase Growth of Hg – High-T_c Layers**
   A.Morawski, T.Lada, A.Paszewin, R.Moliński, H.Marciniak (*Hi.Press.Res.Cent., Polish Ac.Sci., Warsaw*), R.Gatt, E.Olsson, T.Claeson (*Dept.Phys., Chalmers Univ.of Techn., Göteborg, Sweden*), H.Szymczak, M.Berkowski (*Inst.Phys., Polish Ac.Sci., Warsaw*), J.Karpiński, and K.Conder (*Sol.St.Phys.Lab., Swiss Fed.Inst.Techn., Zürich*)

6. **Synthesis, Melting, Crystallization of Hg–Family High-T_c Superconductors in a Gas Pressure System at Pressure of Ar up to 1.5 GPa**
   A.Morawski, T.Lada, A.Paszewin, R.Moliński (*Hi.Press.Res.Cent., Polish Ac.Sci., Warsaw*), J.Karpiński, and K.Conder (*Sol.St.Phys.Lab., Swiss Fed.Inst.Techn., Zürich*)

7. **DTA *in situ* Investigation of the Crystallization Temperature of the Hg-Family Phases Made under High Gas Pressure**
A.Morawski, T.Lada, A.Paszewin, R.Moliński (*Hi.Press.Res.Cent., Polish Ac.Sci., Warsaw*), J.Karpiński and K.Conder (*Sol.St.Phys.Lab., Swiss Fed.Inst.Techn., Zürich*)

8. **Fabrication and Properties of Superconducting Y–Ba–Cu–O Thick Films for Application in Microelectronics Hybrid Systems**
K.Przybylski, T.Brylewski, D.Kalarus (*Dept.Sol.St.Chem., Univ.Min.& Metall., Cracow, Poland*), J.Koprowski (*Dept.Electron., Univ.Min.& Metall., Cracow, Poland*), and M.Bućko (*Dept.Inorg.Chem., Univ.Min.& Metall., Cracow, Poland*)

9. **Temperature Dependence of Resistivity in Single Crystals $Nd_{2-x}Ce_xCuO_{4-d}$**
T.Klimczuk, W.Sadowski, M.Gazda and B.Kusz (*Dept.Appl.Phys.& Math., Techn.Univ. Gdańsk, Poland*)

10. **AC–Losses and Critical Current Densities in Bi-2223 and Tl-1223 Silver Sheathed Tapes**
M.Ciszek (*Inst.Low Temp.& Struct.Res., Polish Ac.Sci., Wrocław and IRC in Supercond., Univ.Cambridge, England, UK*), A.M.Campbell, B.A.Głowacki, and S.P.Ashworth (*IRC in Supercond., Univ.Cambridge, England, UK*)

11. **Electron Spin Resonance Study of Fe^{3+} Ions in $SrLaAlO_4$ and $SrLaGaO_4$ High-T_c Substrates**
R.Jabłoński, A.Gloubokov (*Inst.Electron.Mater.Tech., Warsaw, Poland*), and A.Pajączkowska (*Inst.Electron.Mater.Tech., Warsaw, Poland and Inst.Phys., Polish Ac.Sci., Warsaw*)

12. **Low Field Microwave Absorption in a YBCO Thin Films on Different Substrates**
R.Jabłoński, M.Palczewska (*Inst.Electron.Mater.Tech., Warsaw, Poland*), P.Przysłupski (*Inst.Phys., Polish Ac.Sci., Warsaw*), and A.Kłos (*Inst.Electron.Mater.Tech., Warsaw, Poland*)

13. **T_c vs. x Dependence for the System $Ho_{1-x}Pr_xBa_2Cu_3O_{7-d}$**
Z.Tomkowicz, A.Szytuła (*Inst.Phys., Jagiell.Univ., Cracow, Poland*), and A.Zygmunt (*Inst.Low Temp.& Struct.Res., Polish Ac.Sci., Wrocław*)

14. **Superconducting Properties of Ca,Zn,Ni Doped and Ion Irradiated $YBa_2Cu_4O_8$ Single Crystals**
K.Rogacki (*Inst.Low Temp.and Struct.Res., Polish Ac.Sci., Wrocław*), B.Dabrowski (*Dept.Phys., N.Illin.Univ., DeKalb, IL, USA*), J.Hettinger, and K.Grey (*Mat.Sci.Div., Argonne Natl.Lab., IL, USA*)

Index of Contributors